Springer Proceedings in Physics 299

Indexed by Scopus

The series Springer Proceedings in Physics, founded in 1984, is devoted to timely reports of state-of-the-art developments in physics and related sciences. Typically based on material presented at conferences, workshops and similar scientific meetings, volumes published in this series will constitute a comprehensive up to date source of reference on a field or subfield of relevance in contemporary physics. Proposals must include the following:

- Name, place and date of the scientific meeting
- A link to the committees (local organization, international advisors etc.)
- Scientific description of the meeting
- List of invited/plenary speakers
- An estimate of the planned proceedings book parameters (number of pages/articles, requested number of bulk copies, submission deadline).

Please contact:

For Americas and Europe: Dr. Zachary Evenson; zachary.evenson@springer.com
For Asia, Australia and New Zealand: Dr. Loyola DSilva; loyola.dsilva@springer.com

Jianqiao Liu · Yongjun Jiao
Editors

Proceedings of the 2023 Water Reactor Fuel Performance Meeting

WRFPM2023
July 17–21, Xi'an, China

Springer

Editors
Jianqiao Liu
Chinese Nuclear Society
Beijing, China

Yongjun Jiao
China National Nuclear Corporation
Beijing, China

ISSN 0930-8989 ISSN 1867-4941 (electronic)
Springer Proceedings in Physics
ISBN 978-981-99-7156-5 ISBN 978-981-99-7157-2 (eBook)
https://doi.org/10.1007/978-981-99-7157-2

This Springer imprint is published by the registered company Springer Nature Singapore Pte Ltd.
The registered company address is: 152 Beach Road, #21-01/04 Gateway East, Singapore 189721, Singapore

Paper in this product is recyclable.

Contents

Contributors

A. Ambard MMC Department, EDF R&D, Moret-sur-Loing, France

P. Aragón CIEMAT, Unit of Nuclear Safety Research, Madrid, Spain

A. Arkoma VTT Technical Research Centre of Finland, Espoo, Finland

M. C. Baietto INSA Lyon, Université de Lyon, Lyon, France

G. Bolsée Framatome SAS, Fuel, Lyon, France

F. Bourlier Framatome (Fuel Business Unit), Lyon, France

Liangzhi Cao Xi'an Jiaotong University, Shaanxi, China

F. Cappia Idaho National Laboratory, Idaho Falls, USA

Ou Changgui China National Nuclear Corporation, Lianyungang, China

Peng Chen Reactor Engineering and Safety Research Center, China Nuclear Power Technology Research Institute Co., Ltd, Shenzhen, China

J. Desquines IRSN, Cadarache, Saint Paul-Lez-Durance, France

B. Dif VTT Technical Research Centre of Finland, Espoo, Finland

Ma Fei Jiangsu Nuclear Power Corporation, Jiangsu, China

F. Feria CIEMAT, Unit of Nuclear Safety Research, Madrid, Spain

Liangqian Fu Nuclear Power Institute of China, Chengdu, China

Pengtao Fu China Nuclear Power Technology Research Institute Co, Ltd, Shenzhen, China

N. Garner Framatome Inc, Lynchburg, VA, USA

Gao Geng China National Nuclear Corporation, Lianyungang, China

G. Gentet Framatome SAS, Fuel, Lyon, France

A. Gérard IRSN, Les Angles, France

L. Gerken Framatome Inc, Lynchburg, VA, USA

Chen Guangjun Jiangsu Nuclear Power Corporation, Jiangsu, China

M. Guémas IRSN, Cadarache, France

S. Guilbert IRSN, Cadarache, Saint Paul-Lez-Durance, France

Xiaoyu Guo CNNC Jianzhong Nuclear Fuel Co., Ltd. (CJNF), Sichuan, China

Yan Guo China Nuclear Power Technology Research and Institute, Shenzhen, China

R. S. Hansen Idaho National Laboratory, Idaho Falls, USA

N. He Harbin Engineering University, Harbin, China

J. Heikinheimo VTT Technical Research Centre of Finland, Espoo, Finland

L. E. Herranz CIEMAT, Unit of Nuclear Safety Research, Madrid, Spain

B. Holden Framatome Inc, Lynchburg, VA, USA

Yang Hongye Jiangsu Nuclear Power Corporation, Jiangsu, China

Yu-Shan Huang Guangzhou Shinengine AM Technology Co.Ltd., Guangzhou, China

C. Naokazu Idota Tokyo Institute of Technology, Tokyo, Japan

I. Ioka Japan Atomic Energy Agency, Ibaraki, Japan

D. Jädernäs Studsvik Nuclear AB, Nyköping, Sweden

De-sheng Jin China Nuclear Power Technology Research Institute, Shenzhen, China

Shuai Jin Nuclear Power Institute of China, Chengdu, China

Xin Jin China Nuclear Power Technology Research Institute Co. Ltd, Shen-Zhen, China

D. W. Kamerman Idaho National Laboratory, Idaho Falls, USA

H. Kasahara Hitachi-GE Nuclear Energy, Ltd, Ibaraki, Japan

D. J. Kim Korea Atomic Energy Research Institute, Daejeon, Republic of Korea

D. S. Kim Korea Atomic Energy Research Institute, Daejeon, Republic of Korea

H. K. Kim Korea Atomic Energy Research Institute, Daejeon, Republic of Korea

R. Kliewer Framatome Inc, Lynchburg, VA, USA

A. M. Kpemou IRSN, Cadarache, Saint Paul-Lez-Durance, France

F. Kremer Institut de Radioprotection Et de Sûreté Nucléaire, Severe Accident Department, St Paul-Lez-Durance, France

V. Kuzin TVEL, JSC, Moscow, Russian Federation

K. Lafchiev Studsvik Nuclear AB, Nyköping, Sweden

H. S. Lee Korea Atomic Energy Research Institute, Daejeon, Republic of Korea

Xing Lei Department of Nuclear Engineering and Technology, Chongqing University, Chongqing, China;
Liangjiang Laboratory for New Energy (Nuclear Energy and Power), Chongqing, China

A. Wei Li State Key Laboratory for Strength and Vibration of Mechanical Structures, Xi'an Jiaotong University, Xi'an, China

Jinggang Li China Nuclear Power Technology Research Institute Co., Ltd. Reactor Engineering Software Research Institute, Shenzhen, China

Wenjie Li Science and Technology on Reactor System Design Technology Laboratory, Nuclear Power Institute of China, Chengdu, Sichuan, China

Zhijun Li China Nuclear Power Technology Research Institute Co, Ltd, Shenzhen, China

Ren Liang China Nuclear Power Technology Research Institute Co., Ltd., Shenzhen, China

Zhikang Lin China Nuclear Power Technology Research Institute Co., Ltd., Shenzhen, China

Mengying Liu Reactor Engineering and Safety Research Center, China Nuclear Power Technology Research Institute Co., Ltd, Shenzhen, China

Rong Liu School of Electric Power, South China University of Technology, Guangzhou, China

Shengyu Liu School of Electric Power, South China University of Technology, Guangzhou, China

Xiaohan Liu China Nuclear Power Technology Research Institute Co., Ltd. Reactor Engineering Software Research Institute, Shenzhen, China

Xingjun Liu College of Materials and Fujian Key Laboratory of Surface and Interface Engineering for High Performance Materials, Xiamen, Fujian, China

Zhouyu Liu Xi'an Jiaotong University, Shaanxi, China

Xianghui Lu China Nuclear Power Technology Research Institute Co., Ltd., Shenzhen, China

Yong Lu College of Materials and Fujian Key Laboratory of Surface and Interface Engineering for High Performance Materials, Xiamen, Fujian, China;
China Nuclear Power Technology Research Institute Co., Ltd. Reactor Engineering Software Research Institute, Shenzhen, China

V. E. Lukyanov Afrikantov OKBM, JSC Burnakovsky Proyezd, Nizhny Novgorod, Russia

Yayun Luo China Nuclear Power Technology Research Institute Co., Ltd. Reactor Engineering Software Research Institute, Shenzhen, China

Zehua Ma China Nuclear Power Technology Research Institute Co., Ltd., Shenzhen, China

J. Matsunaga Global Nuclear Fuel – Japan, Kanagawa, Japan

Xinying Miao China Nuclear Power Technology Research Institute Co., Ltd. Reactor Engineering Software Research Institute, Shenzhen, China

Gu Minqiang China National Nuclear Corporation, Lianyungang, China

Y. Miura Nippon Nuclear Fuel Development, Co., Ltd, Ibaraki, Japan

H. Miyata Hitachi-GE Nuclear Energy, Ltd, Ibaraki, Japan

B. Normand INSA Lyon, Université de Lyon, Lyon, France

J. O'Brian Framatome Inc, Lynchburg, VA, USA

M. Osaka Japan Atomic Energy Agency, Ibaraki, Japan

Yong Ouyang China Nuclear Power Technology Research Institute Co., Ltd., Shenzhen, China

B. Ki Chul Park Tokyo Institute of Technology, Tokyo, Japan

Zhen-xun Peng China Nuclear Power Technology Research Institute, Shenzhen, China

P. G. Petersen Idaho National Laboratory, Idaho Falls, USA

Zhang Qi China National Nuclear Corporation, Lianyungang, China

Guo-peng Qin CNNC Jianzhong Nuclear Fuel Co., Ltd., Yibin, China

A. Radostin TVEL, JSC, Moscow, Russian Federation

C. Sakaguchi Nippon Nuclear Fuel Development, Co., Ltd, Ibaraki, Japan

K. Sakamoto Nippon Nuclear Fuel Development, Co., Ltd, Ibaraki, Japan

C. Sartoris IRSN, Cadarache, France

Shihao Shao Xi'an Jiaotong University, Shaanxi, China

Guo Shaosheng China National Nuclear Corporation, Lianyungang, China

Guo Shuang CNNC Jianzhong Nuclear Fuel Co., Ltd. (CJNF), Beijing, China

Ki Seob Sim IAEA, Vienna, Austria

A. Slavickas Laboratory of Nuclear Safety, Lithuanian Energy Institute, Kaunas, Lithuania

J. Soulacroix EDF, DIPNN – DIRECTION TECHNIQUE, Lyon CEDEX 07, France

J. Strumpell Framatome Inc, Lynchburg, VA, USA

Dan Sun Science and Technology on Reactor System Design Technology Laboratory, Nuclear Power Institute of China, Chengdu, Sichuan, China

Barbara Szpunar University of Saskatchewan, Saskatoon, SK, Canada

T. Taurines IRSN, Cadarache, Saint Paul-Lez-Durance, France

A. Tidikas Laboratory of Nuclear Safety, Lithuanian Energy Institute, Kaunas, Lithuania

Xin Tong Guangzhou Shinengine AM Technology Co.Ltd., Guangzhou, China

D. Takehiko Tsukahara Tokyo Institute of Technology, Tokyo, Japan

M. Yu. Tuturkin Afrikantov OKBM JSC, Nizhny Novgorod, Russia

N. Vollmer Framatome GmbH, Erlangen, Germany

Nicolas Waeckel EDF (Retired), Lyon, France

Cuiping Wang College of Materials and Fujian Key Laboratory of Surface and Interface Engineering for High Performance Materials, Xiamen, Fujian, China

Kaiyuan Wang China Nuclear Power Technology Research Institute Co. Ltd, Shen-Zhen, China

Q. Wang Harbin Engineering University, Harbin, China

Weiwei Wang China Nuclear Power Technology Research Institute Co., Ltd., Shenzhen, China

Xiaoyan Wei China Nuclear Power Technology Research Institute, Shenzhen, China

Qing-long Wen Department of Nuclear Engineering and Technology, Chongqing University, Chongqing, China;
Key Laboratory of Low-Grade Energy Utilization Technologies and Systems, Chongqing University, Chongqing, China;
Liangjiang Laboratory for New Energy (Nuclear Energy and Power), Chongqing, China

Wu Wenqi Jiangsu Nuclear Power Corporation, Jiangsu, China

B. Xiaoli Wu Science and Technology on Reactor System Design Technology Laboratory, Nuclear Power Institute of China, Chengdu, China

Hai Wu China Nuclear Power Technology Research and Institute, Shenzhen, China

Hongchun Wu Xi'an Jiaotong University, Shaanxi, China

Zhang Xianggui China National Nuclear Corporation, Lianyungang, China;
Jiangsu Nuclear Power Corporation, Jiangsu, China

J. Xie Harbin Engineering University, Harbin, China

Qingyu Xie Reactor Engineering and Safety Research Center, China Nuclear Power Technology Research Institute Co., Ltd, Shenzhen, China

Duoting Xu China Nuclear Power Technology Research Institute, Shenzhen, China

Haode Xu Reactor Engineering and Safety Research Center, China Nuclear Power Technology Research Institute Co., Ltd, Shenzhen, China

Xiaobei Xu Xi'an Jiaotong University, Shaanxi, China

S. Yamashita Japan Atomic Energy Agency, Ibaraki, Japan

Ya-lun Yan China Nuclear Power Technology Research Institute, Shenzhen, China

J. H. Yang Korea Atomic Energy Research Institute, Daejeon, Republic of Korea

Shengzhi Yang China Nuclear Power Technology Research Institute, Shenzhen, China

Wenhua Yang Nuclear Power Institute of China, Chengdu, China

Yuhang Yang College of Materials and Fujian Key Laboratory of Surface and Interface Engineering for High Performance Materials, Xiamen, Fujian, China

H. Yokoyama Nippon Nuclear Fuel Development, Co., Ltd, Ibaraki, Japan

J. H. Yoon Korea Atomic Energy Research Institute, Daejeon, Republic of Korea

Xiaoyang Yuan School of Electric Power, South China University of Technology, Guangzhou, China

A. Yiwei Zhang Tokyo Institute of Technology, Tokyo, Japan

Guoliang Zhang China Nuclear Power Technology Research and Institute, Shenzhen, China

Jinzhao Zhang Tractebel (ENGIE), Brussels, Belgium

Li-ying Zhang CNNC Jianzhong Nuclear Fuel Co., Ltd., Yibin, China

Liang Zhang Nuclear Power Institute of China, Chengdu, China

Ming Zhang China Nuclear Power Technology Research Institute Co., Ltd. Reactor Engineering Software Research Institute, Shenzhen, China

T. Zhang Harbin Engineering University, Harbin, China

Yuanyue Zhang Nuclear Power Institute of China, Chengdu, China

Yuxiang Zhang China Nuclear Power Technology Research and Institute, Shenzhen, China

Wenbin Zhao Nuclear Power Institute of China, Chengdu, China

S. Zheng Framatome SaS, Lyon, France

Zhan Zhihe China National Nuclear Corporation, Lianyungang, China

Yanan Zhu China Nuclear Power Technology Research Institute Co., Ltd. Reactor Engineering Software Research Institute, Shenzhen, China

M. Zilly Framatome GmbH, Erlangen, Germany

Yufan Zong Xi'an Jiaotong University, Shaanxi, China

Numerical Investigation on the Effect of Fuel Pulvers on Axial Fuel Relocation

Zehua Ma(✉), Weiwei Wang, Ren Liang, Zhikang Lin, Yong Ouyang, and Xianghui Lu

China Nuclear Power Technology Research Institute Co., Ltd., Shenzhen, China
mazehua.xjtu@gmail.com, 342055908@qq.com, lr19891107@163.com, 37261279@qq.com, luxianghui@cgnpc.com.cn

Abstract. Burnup extension above the current regulatory burnup limit of 62 GWd/tU is an effective way economically increase cycle lengths of pressurized water reactors to 24 months. However, high burnup (>62 GWd/tU) fuel has been observed to severely fragment and even pulverize when subjected to temperature transient conditions such as loss-of-coolant accidents. After the cladding ballooning under loss-of-coolant conditions, the fuel fragments and fuel pulvers can axially relocate into the balloon region. The purpose of this work is to perform the three-dimension simulations of fuel relocation based on the validated simulation framework with coupled FEM-DEM approach to investigate the effect of fuel pulvers on axial fuel relocation. Parametric investigations are conducted on the relative increase of cladding volume and volume fraction of fuel pulvers. Based on the quantitative analysis, it is found that fuel pulvers can increase the filling ratio after fuel relocation, which means more fuel will accumulate in the ballooned region, probably resulting in higher local temperature during the loss-of-coolant transients. A series of fitting correlations of fuel stack reduction, mass fraction and filling ratio are proposed with respect to volume fraction of fuel pulvers. The proposed correlations can support the interpretation of experimental data, and can enhance the evaluation accuracy of fuel performance during loss-of-coolant accidents by the more realistic prediction of the critical parameters influenced by fuel relocation.

Keywords: Fuel relocation · Fuel pulver · High burnup · Loss-of-coolant accident

1 Introduction

The international nuclear industry is looking forward to increasing cycle lengths of pressurized water reactors (PWRs) from 12/18 months to 24 months [1]. The initiative is to strengthen the economic competitiveness of nuclear energy. To achieve this goal, the regulatory average burnup limit of 62 GWd/tU is supposed to extend to 75 GWd/tU [1]. However, high burnup fuel (>62 GWd/tU) may confront new challenges under design-basis accidents (DBAs), like loss of coolant accidents (LOCAs). According to the experimental observation, fuel fragmentation, relocation and dispersal (FFRD) could be

J. Liu and Y. Jiao (Eds.): WRFPM 2023, SPPHY 299, pp. 1–15, 2024.
https://doi.org/10.1007/978-981-99-7157-2_1

a special failure phenomenon for high burnup fuel under a LOCA [2]. Fuel fragmentation is a kind of serious fuel cracking caused by the relief of thermal stresses during operation. Additional fuel fragmentation would occur during a transient because of the thermal-mechanical response to the transient. Fine fragmentation is likely generated in the high burnup structure (HBS) at the fuel periphery [3]. During a LOCA, the overpressure and overheating of cladding will result in the clad ballooning at the peak axial power. Then fragmented fuel especially fine fragments at the top of fuel rod can move downwards and some of them will accumulate in the ballooned region of cladding. The appearance of fine fragmentation itself during a LOCA does not present a safety concern [4]. However, the fuel relocation may lead to a local increase in cladding temperature and oxidation level, raising the licensing concerns related to core coolability as described in 10 CFR 50.46 [5].

In order to support the licensing application of fuel burnup extension, many efforts have been undertaken in the field of fuel relocation modeling [6–8]. However, little attention was paid to the effect of fragment size distribution on fuel relocation. In our previous work [9], coupled finite element method (FEM) and discrete element method (DEM) was firstly employed in three-dimension (3D) fuel relocation simulation, in which the coarse fragments were built by Voronoi cells and the fine pulvers were modeled as isolated particles. The preliminary simulation results [9] revealed that current one-dimension (1D) fuel relocation model [7] failed to predict fuel relocation for high burnup fuel, since the fine pulvers could lead to a large difference for filling ratio. In addition, the DEM simulation based on Voronoi cells could get a more reliable result of fuel stack reduction and filling ratio. More currently, the aforementioned simulation framework was validated against FR-2 tests [10, 11], and used to perform the parameter study [11] that gained a series of fitting correlations to interpret the effect of mean diameter of coarse fragments on fuel relocation parameters (fuel stack reduction, mass fraction and filling ratio).

In this paper, the simulation framework [9] was used to investigate the effect of fuel pulvers on fuel relocation. Different volume fractions (20%, 40%, 60% and 80%) of fuel pulvers and different relative increase (40%, 60% and 80%) of cladding volume were considered in the parameter study. Fuel stack reduction, mass fraction and filling ratio of those cases were analyzed and the relative correlations were proposed with respect to volume fraction of fuel pulvers and relative increase of cladding volume. The paper was organized as follow: The model of pellet and cladding as well as related model setting were briefly introduced in Sect. 2. The simulation results of the parametric study considering different pulver volume fraction and different relative cladding volume increase were discussed in Sect. 3. And the final conclusions were given in Sect. 4.

2 Model Setup

2.1 Pellet Model

According to the characteristic of HBS, the pellet model can be separated into two regions: the inner cylindrical region was formed from coarse fragments, and the outer ring region was composed of fine fragments (pulvers). According to our experience [9–11], the coarse fragments can be characterized as Voronoi cells, and the pulvers can be characterized as separate spherical particles.

The 3D Voronoi cells were generated with Grasshopper [12], a visual programming language and environment within the Rhinoceros 3D computer-aided design (CAD) application. Voronoi cells can be generated by defining a boundary and the control points within it. The boundary is a 3D domain of any shape specified by the user. In this work, the boundary is a cylindrical pellet without chamfers and dishes. For simplicity, the control points (centroids of Voronoi cells) were created randomly by defining total number of points and random seed. Due to the random feature of modelling of Voronoi cells, the surface mean diameter (or Sauter mean diameter) was used as a metric to describe a group of Voronoi cells:

$$D_{32} = \frac{\sum \psi_i \left(d_i^{eq}\right)^3}{\sum \psi_i \left(d_i^{eq}\right)^2} \tag{1}$$

where ψ_i denotes the sphericity of the ith Voronoi cell in the group, d_i^{eq} denotes the equivalent diameter (or surface-volume diameter) of the ith Voronoi cell in the group. The sphericity was defined as:

$$\psi = \frac{\pi^{1/3}(6V)^{2/3}}{A} \tag{2}$$

where V is the volume of Voronoi cell and A is the surface area of Voronoi cell. The equivalent diameter was defined as:

$$d^{eq} = \frac{6V}{A} \tag{3}$$

The equivalent diameter and the sphericity of Voronoi cells can be easily obtained and monitored in Grasshopper. Then the mean diameter of Voronoi cells can be tuned by adjusting the number of control points. And different groups of Voronoi cells can be generated by adjusting the random seed. According to the statistical analysis of measured data in various experimental programs [11], for the average rod burnup $<$60 GWd/tU, mean diameter of fuel fragments is roughly in the range from 2.0 to 3.0 mm. As to the average rod burnup $\geq$60 GWd/tU, mean diameter of fuel fragments (coarse + fine) will be in the range from 1.5 to 2.0 mm. In this work, a constant mean diameter of 2.0 mm was used for Voronoi cells. From our experience [9], the particle size can be set to 0.5 mm which was typical for fuel pulvers and reasonable to save the computational time.

As listed in Table 1, nominal fuel rod data of the Halden IFA-650.4 test were employed to build the pellet model. The 48 pellets were labelled #1 - #48 from bottom to top of the rodlet. It was assumed that all pellets in the model had the same burnup, and their Voronoi cell regions had the same fragmentation pattern. Four kind of volume fraction (20%, 40%, 60% and 80%) of fuel pulvers were considered in the parametric study. In order to consider the effect of random modelling of Voronoi cells, five fragmentation patterns were selected for the Voronoi cells. Therefore, there are 20 kind of pellets in the parametric study. Figure 1 showed the five fragmentation patterns of Voronoi cells with mean diameter of 2.0 mm for 20% pulver volume fraction cases.

Table 2 listed the dimensional data of pellet model. According to the dimensional data and the pre-specified Voronoi cell mean diameter (2.0 mm) and the particle size

Table 1. Nominal fuel rod data of the Halden IFA-650.4 test. Data compiled from Ref. [13].

Parameter	Value
Cladding outer diameter, mm	10.75
Cladding wall thickness, mm	0.725
Pellet diameter, mm	9.13
Rodlet active length, mm	480
Assumed pellet height, mm	10
Pellet density, g/cm^3	10.421

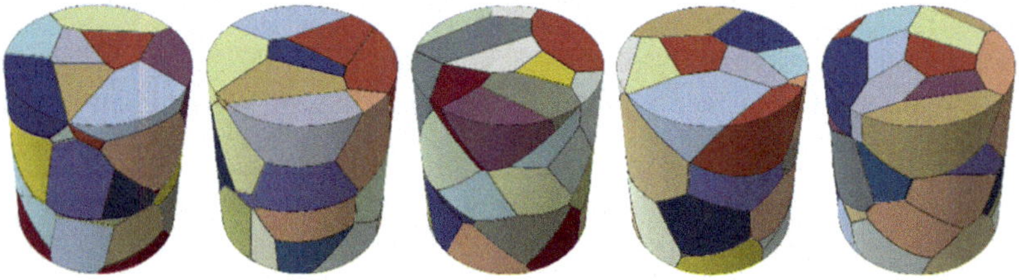

(a) (b) (c) (d) (e)

Fig. 1. Five fragmentation patterns of Voronoi cells for 20% pulver volume fraction cases

(0.5 mm), the pellet models of the different cases can be generated (shown in Fig. 2). Due to limitation of modelling, the void fraction of pulver region was about 46%-54%, which was larger than the porosity of a real HBS ($\leq$22%) [14]. Therefore, more particles should be added in the model to ensure the mass conservation. From the experimental observation [3], much severer fragmentation was found around the ballooned region due to the higher local power and larger cladding circumferential strain. Therefore, in this work, the supplementary particles were placed in the ballooned region, and the midplane of supplementary particles was near (as close as possible) to the burst center. For example, for the cases with pulver volume fraction of 80%, supplementary particles were placed between pellet #11 and #12 (shown in Fig. 3).

Table 2. Dimensional data of pellet model

Volume fraction of fuel pulvers, %	20	40	60	80
Thickness of pulver region, mm	0.482	1.029	1.678	2.523
Radius of Voronoi cell region, mm	4.083	3.536	2.887	2.042
The number of particles per pellet	1080	2000	2760	4000
Void fraction of pulver region, %	46.0	50.0	54.0	50.0
Total number of supplementary particles	39429	86609	141203	172881

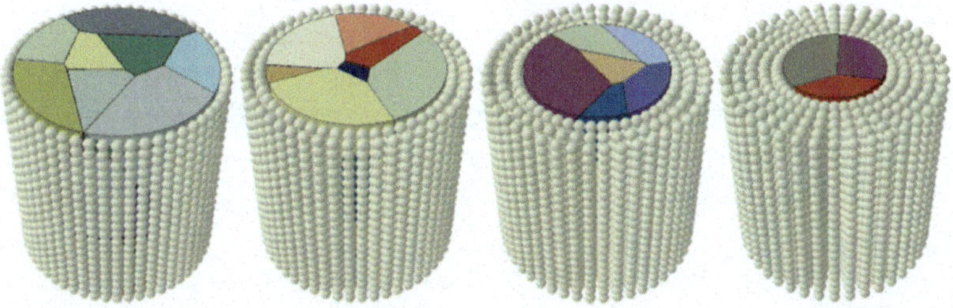

Fig. 2. Pellet models of the cases with different pulver volume fractions: (a) 20%, (b) 40%, (c) 60%, (d) 80%.

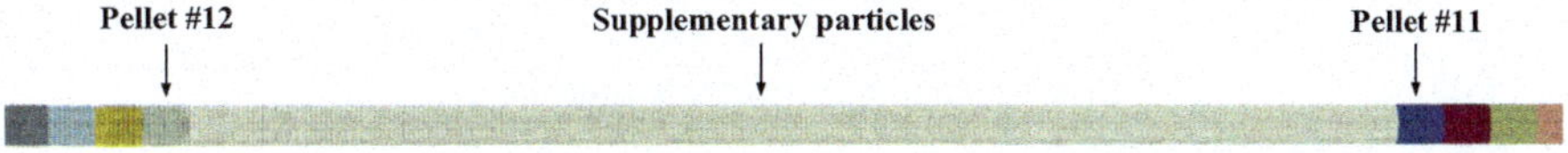

Fig. 3. Location of supplementary particles in the cases with pulver volume fraction of 80%

2.2 Cladding Model

Three kind of relative increase (40%, 60% and 80%) of cladding volume were considered in this work. Considering the 20 kind of pellets, there were 60 cases built in the parametric study. Table 3 listed the case matrix for the parametric study. Case number "P20C40-1" represented the case with 20% volume fraction of fuel pulvers, 40% relative increase of cladding volume and the first fragmentation pattern of Voronoi cells.

Based on our previous work [15], the calculated cladding ballooning at different time can be utilized as the cladding geometry for the 60 cases. Figure 4 depicted the radius and circumferential strain profile of cladding with volume increase of 40%, 60% and 80%. Based on Fig. 4, three azimuthal symmetrical claddings can be built.

Table 3. Case matrix for the parametric study

Volume fraction of fuel pulvers, %	Relative increase of cladding volume, %		
	40	60	80
20	P20C40–1—P20C40–5	P20C60–1—P20C60–5	P20C60–1—P20C60–5
40	P40C40–1—P40C40–5	P40C60–1—P40C60–5	P40C60–1—P40C60–5
60	P60C40–1—P60C40–5	P60C60–1—P60C60–5	P60C60–1—P60C60–5
80	P80C40–1—P80C40–5	P80C60–1—P80C60–5	P80C60–1—P80C60–5

2.3 Model Setting

In the ABAQUS implementation, the Voronoi cells used discrete rigid shell elements (R3D3) by trilateral free meshing with the grid size of 0.5 mm, while the cladding

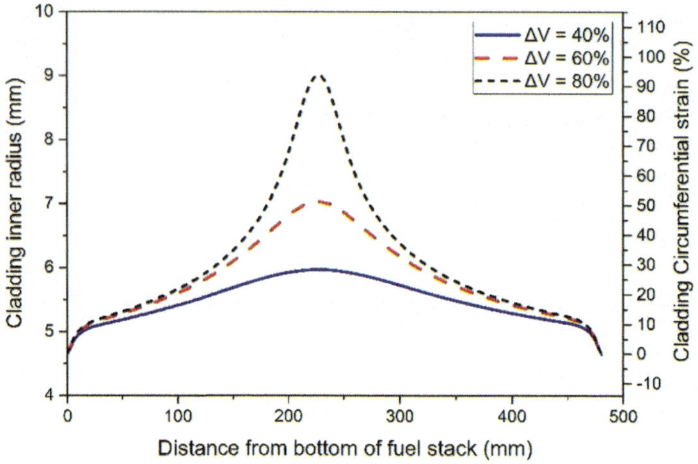

Fig. 4. Radius and circumferential strain profile of cladding with relative volume increase of 40%, 60% and 80%

used discrete rigid shell elements (R3D4) by quadrilateral-dominated free meshing with the grid size of 1.0 mm. Since the parameters like stress, strain and temperature are out of concern in the relocation simulation, rigid shell elements (R3D3/R3D4) can be used to replace solid elements (like C3D8) for the fragments and the cladding, which can dramatically reduce the element number and effectively save the computational cost. The particles were built by PD3D elements in ABAQUS. For each Voronoi cell, a reference point was created in its centroid to assign the basic material properties like mass and rotary inertia. For inter-particle contact, the tangential friction coefficient was assumed to be 0.3 and the linear pressure-overclosure algorithm was adopted in normal direction with the contact stiffness of 1000 N/mm. For other contact conditions, the tangential friction coefficient was assumed to be 0.3 and the "hard" contact algorithm was adopted in normal direction. In the simulation, the cladding was fixed and the gravity of 9.81 m/s^2 was applied to Voronoi cells and particles as the driving force for fuel relocation. The time period and time increment of simulation was 3.0 s and 5×10^{-6} s respectively, and the coordinates and displacement of each Voronoi cell/particle were given every 0.1 s. Other detailed settings of the model can be found in our previous work [9].

3 Results and Discussion

3.1 Fuel Stack Reduction

Missing fuel length was defined as the difference of fuel stack (active region) height before and after the transient. It should be mentioned that the fuel stack height in the model did not include the length of supplementary particles. Figure 5 demonstrated the fuel stack status in Case P80C40-1 before relocation and after relocation. The missing fuel length in Case P80C40-1 was 71.41 mm. Fuel stack reduction was defined as the percentage of the missing fuel length to the original fuel stack height. Then the fuel stack reduction for Case P80C40-1 was about 14.88%.

Fig. 5. Fuel stack status in Case P80C40–1: (a) before relocation, stack height = 480 mm, (b) after relocation, stack height = 408.59 mm. Only upper part of the fuel stack was shown, and the cladding was hidden for better visualization.

Figure 6 depicted the fuel stack reduction versus relative increase of cladding volume for the 60 cases in the parametric study. For a certain pulver volume fraction, the fuel stack seemed to linearly increase with the relative increase of cladding volume. For a certain relative increase of cladding volume, the higher the pulver volume fraction was, the larger the fuel stack reduction would be. That was because the pulvers were able to fill the inter-fragment gap, resulting in severer axial fuel relocation and larger fuel stack reduction. Considering the cases with different fragmentation patterns, it seemed that the deviation of fuel stack reduction increased as the pulver volume fraction decreased. The reasons were obvious: (1) The randomness of the Voronoi cells modeling implied the great dispersion of inter-fragment gap after relocation, (2) for the cases with low pulver volume fraction (like 20%), there were not enough pulvers to fill the inter-fragment gaps.

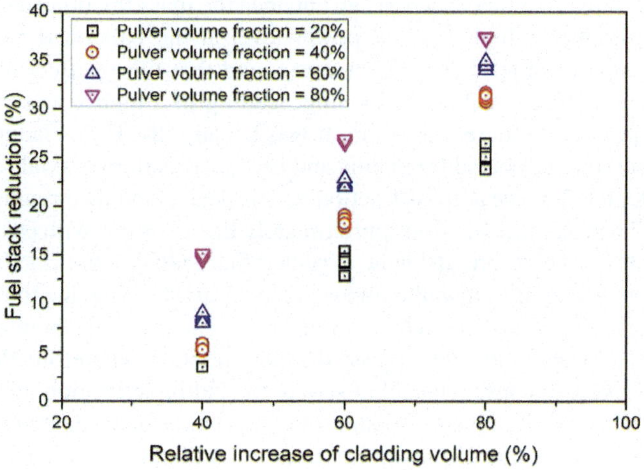

Fig. 6. Fuel stack reduction versus relative increase of cladding volume for all cases

In this work, linear fitting was employed to obtain the correlation of fuel stack reduction with respect to relative increase of cladding volume. Table 4 listed fitting coefficients for the correlations for different volume fraction of fuel pulvers. Adjusted R-squared indicated how well data points fit a curve or line. An adjusted R-squared value of 1 indicated all data points can be perfectly explained by the fitting curve or line. It was clear that for a large volume fraction ($\geq$40%) of fuel pulvers, the fitting correlation had a relatively high accuracy. It should be mentioned that the fitting coefficients were verified to be suitable for the fuel containing the coarse fragments with mean diameter of 2.0 mm. The circumstances of other coarse fragment mean diameter combined with different pulver volume fraction were out of scope of this paper and necessary to be investigated in the future work.

Table 4. Fitting coefficients for the correlation of fuel stack reduction ΔH (unit: %) with respect to relative increase of cladding volume ΔV (unit: %), $20 \leq \Delta V \leq 80$.

Volume fraction of fuel pulvers, %	Linear fitting: $\Delta H = A_1 \Delta V + B_1$		Adjusted R^2
	A_1	B_1	
20	0.5028	−15.2512	0.9851
40	0.6431	−20.1932	0.9987
60	0.6507	−17.1560	0.9969
80	0.5615	−7.3464	0.9984

3.2 Fuel Mass Fraction

Mass fraction is defined as the ratio of the local fuel mass after relocation to that before relocation. A mass fraction >1 indicates that fuel has accumulated in this region and a mass fraction <1 corresponds to a region partially (or completely) void of fuel. Figure 7 showed the mass fraction profiles for the cases. For a certain relative increase of cladding volume, higher pulver volume fraction can enhance the mass fraction near the ballooned region, which implies a higher decay power level at the ballooned region after fuel relocation. Comparing Fig. 7 (i) – (l), it was found that filling effect of pulvers were more evident in a large cladding ballooning, leading to a higher mass fraction at the location with peak circumferential strain. It was because the large cladding ballooning allowed well mixture of coarse fragments and fine pulvers during axial relocation.

Figure 8 depicted the peak mass fraction versus peak cladding circumferential strain for all cases. The peak mass fraction approximately linearly grew with peak cladding circumferential strain. Therefore, it can be expected that reducing the cladding ballooning was an effective way to diminish the impact of local overheating due to fuel relocation. For a certain cladding ballooing, a higher pulver volume fraction can improve the compactness of the relocated fuel and increase the mass fraction. Therefore, the high burnup fuel was more likely to cause a local hot spot at the ballooned region during LOCAs.

Linear fitting was employed to obtain the correlation between peak mass fraction and peak cladding circumferential strain. Table 5 listed the fitting coefficients for the correlation of peak mass fraction with respect to peak cladding circumferential strain. According to the adjusted R-squared, the fitting correlation had a relatively high accuracy.

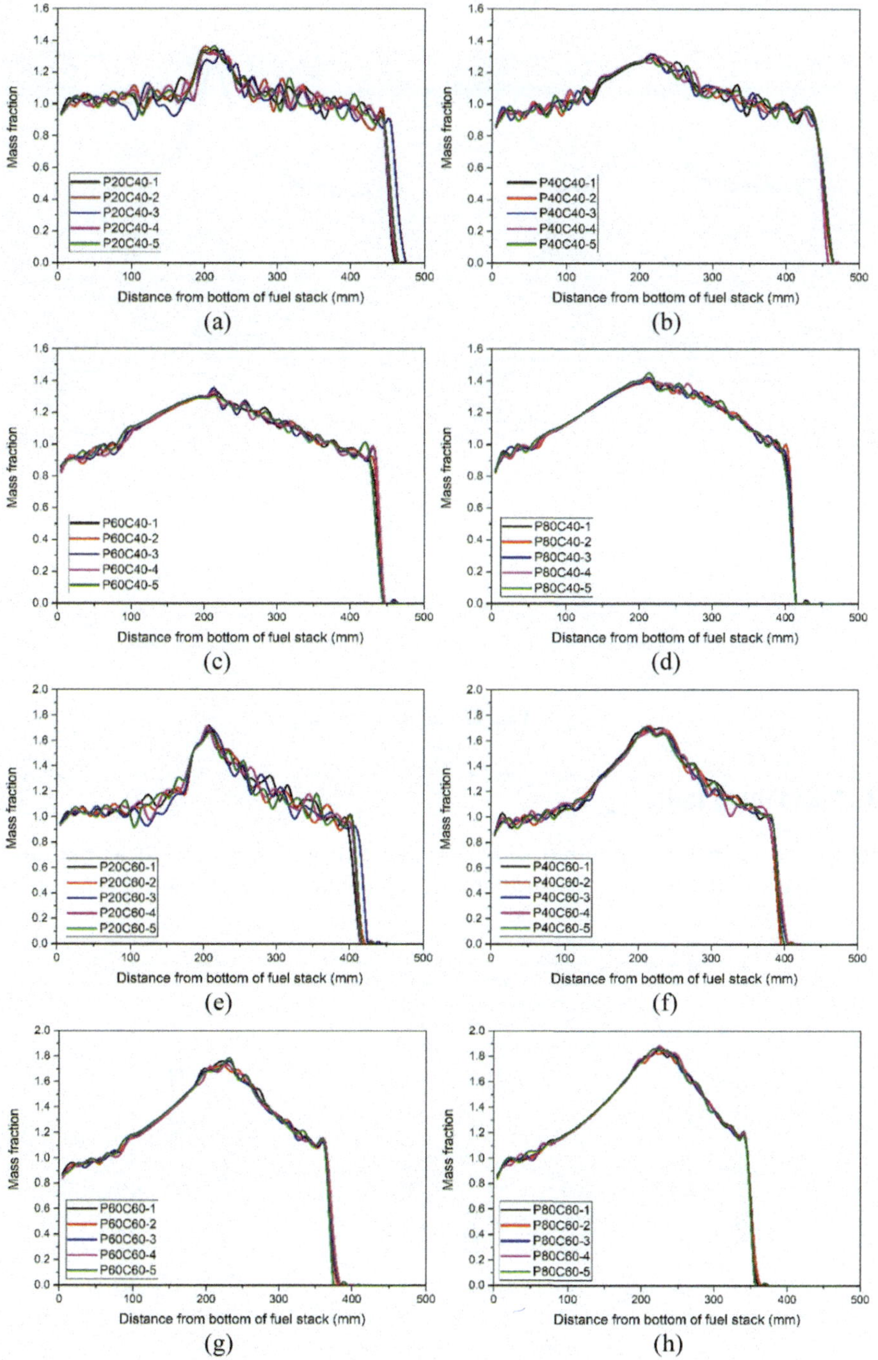

Fig. 7. Mass fraction profiles for all cases: (a) P20C40, (b) P40C40, (c) P60C40, (d) P80C40, (e) P20C60, (f) P40C60, (g) P60C60, (h) P80C60, (i) P20C80, (j) P40C80, (k) P60C80, (l) P80C80

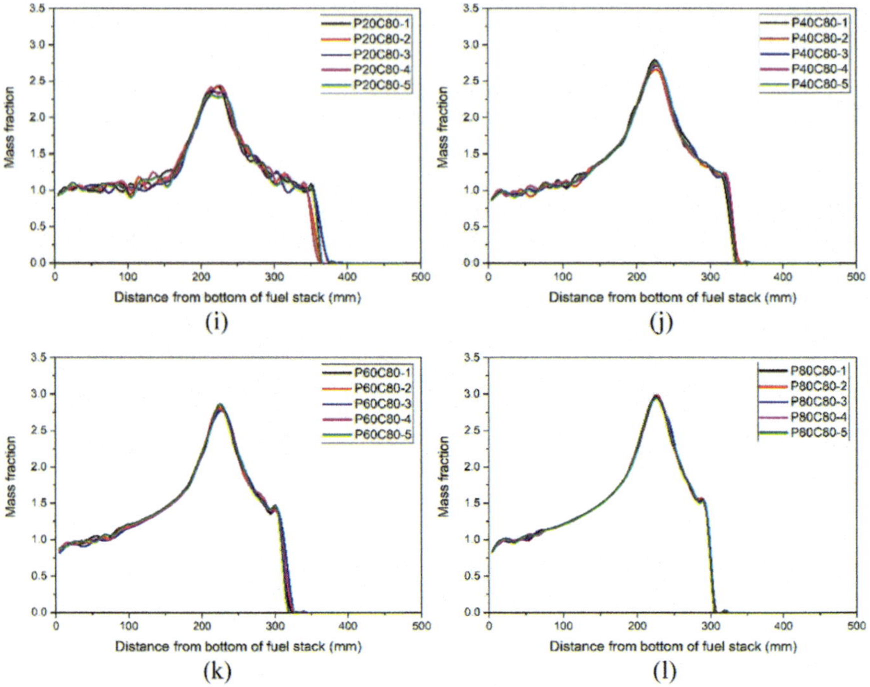

Fig. 7. (*continued*)

3.3 Fuel Filling Ratio

Filling ratio, sometimes referred to as packing fraction, was defined as the ratio of the volume of fuel to the total available local volume. Filling ratio represented the

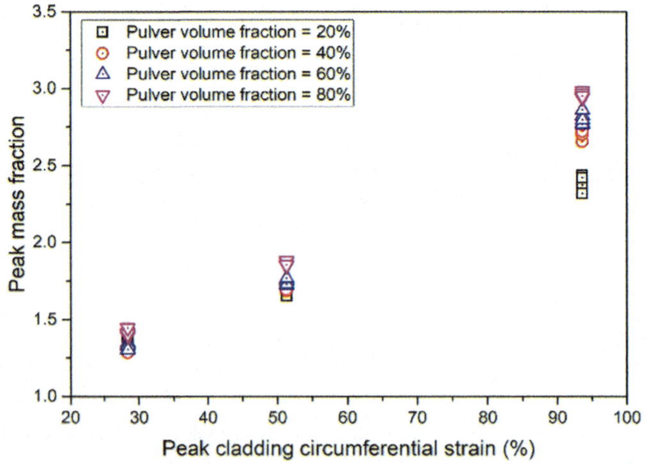

Fig. 8. Peak mass fraction and peak cladding circumferential strain for all cases

Table 5. Fitting coefficients for the correlation of peak mass fraction φ_{peak} with respect to peak cladding circumferential strain ε_{peak} (unit: %), $28.4 \leq \varepsilon_{peak} \leq 93.6$.

Volume fraction of fuel pulvers, %	Linear fitting: $\varphi_{peak} = A_2\varepsilon_{peak} + B_2$		Adjusted R^2
	A_2	B_2	
20	0.01619	0.8749	0.9986
40	0.02215	0.6319	0.9872
60	0.02295	0.6350	0.9885
80	0.02391	0.6986	0.9910

compactness of fuel fragments. A filling ratio close to 1 indicated that the fragments were tightly packed with minimal void spaces between them, while a filling ratio close to 0 corresponded to insufficient fuel to fill the local cladding volume. Figure 9 demonstrated the filling ratio profiles for all cases. In our previous work with mere coarse fragments [11], the filling ratio decreased with the cladding circumferential strain at the fully packed segments. By comparison, as shown in Fig. 9, the fine pulvers was able to increase the filling ratio especially at the ballooned region. Particularly, for a fuel with a large pulver volume fraction (≥40%), it showed a rather flat trend of filling ratio along the fuel length, indicating the good mobility of pulvers to fill the inter-fragment gaps.

Given that the coarse fragments and fine pulvers were fully-relocated and well-mixed within the lower half of cladding, the data points at the location <240 mm were taken to investigate the relationship between filling ratio and cladding circumferential strain. Figure 10 depicts the filling ratio versus the cladding circumferential strain for all cases. Compared with previous work [11], the filling ratio profile in Fig. 10 no longer exhibited an exponential decay shape.

As to cases with 20% pulver volume fraction (as shown in Fig. 10 (a)), the filling ratio was in the range of 0.55 to 0.8. For the slight ballooned region (like <30% for the cases of 60% relative cladding volume increase), the filling ratio decreased with cladding circumferential strain, whose trend was still similar to the situations without pulvers [11]. That was because the pellet within this region basically remained the original compactness due to the restraint of cladding. And a slightly increasing strain in this region can not provide enough space to accommodate the pulvers from the upper part of fuel rod. However, for the medium ballooned region (like 30%–46% for the cases of 60% relative cladding volume increase), the filling ratio increased with cladding circumferential strain, since the coarse fragments and fine pulvers can gradually be fully mixed in this region. But, for the significant ballooned regions (like >46% for the cases of 60% relative cladding volume increase), the filling ratio decreased again with cladding circumferential strain, because the coarse fragments generated more and more inter-fragment gaps after relocation in this region, but there were not enough pulvers can fill those inter-fragment gaps.

As to cases with higher volume fraction (≥40%) of pulvers, the filling ratio was roughly a constant along the cladding length. That was because there were enough

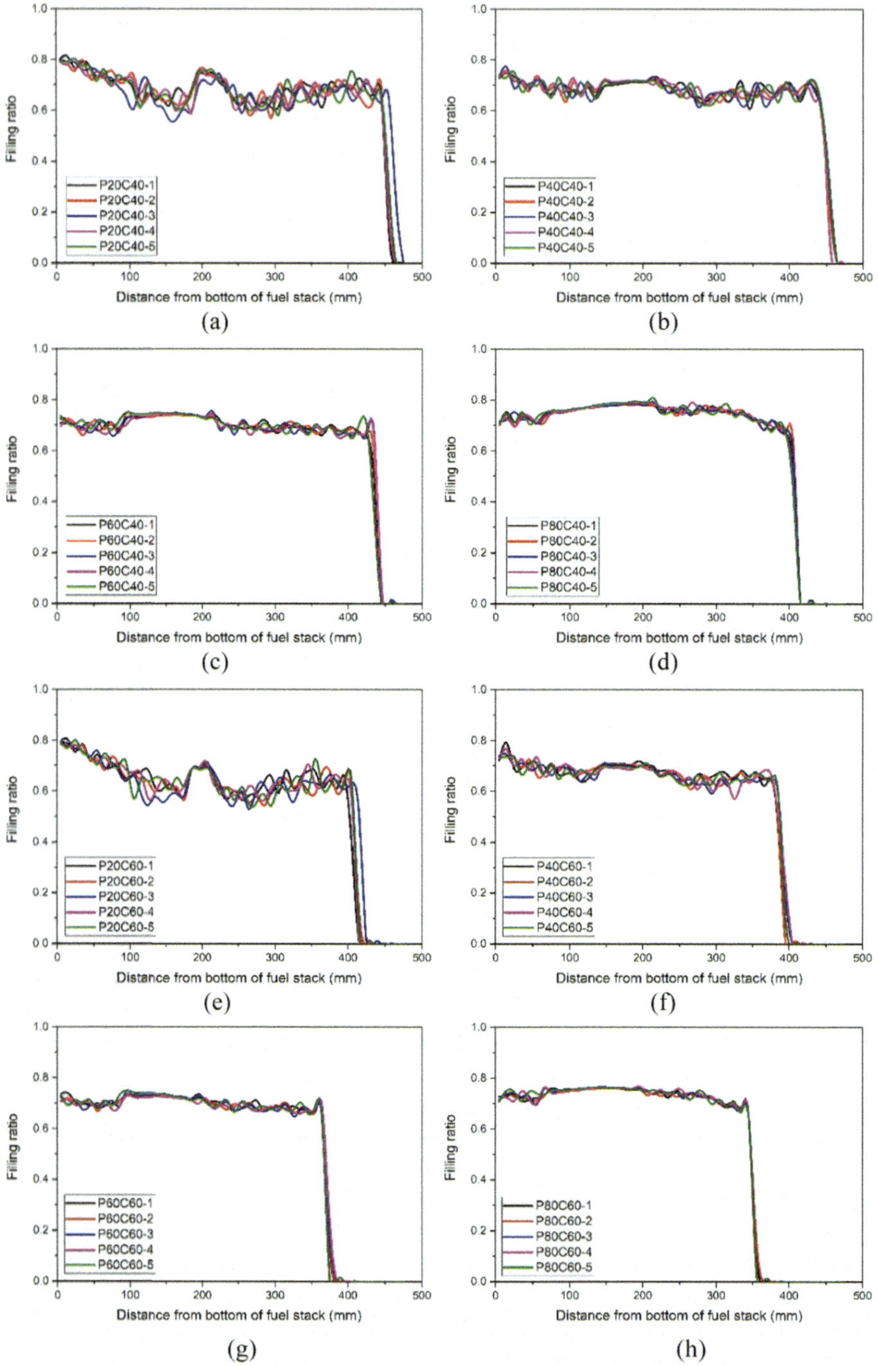

Fig. 9. Filling ratio profiles for all cases: (a) P20C40, (b) P40C40, (c) P60C40, (d) P80C40, (e) P20C60, (f) P40C60, (g) P60C60, (h) P80C60, (i) P20C80, (j) P40C80, (k) P60C80, (l) P80C80

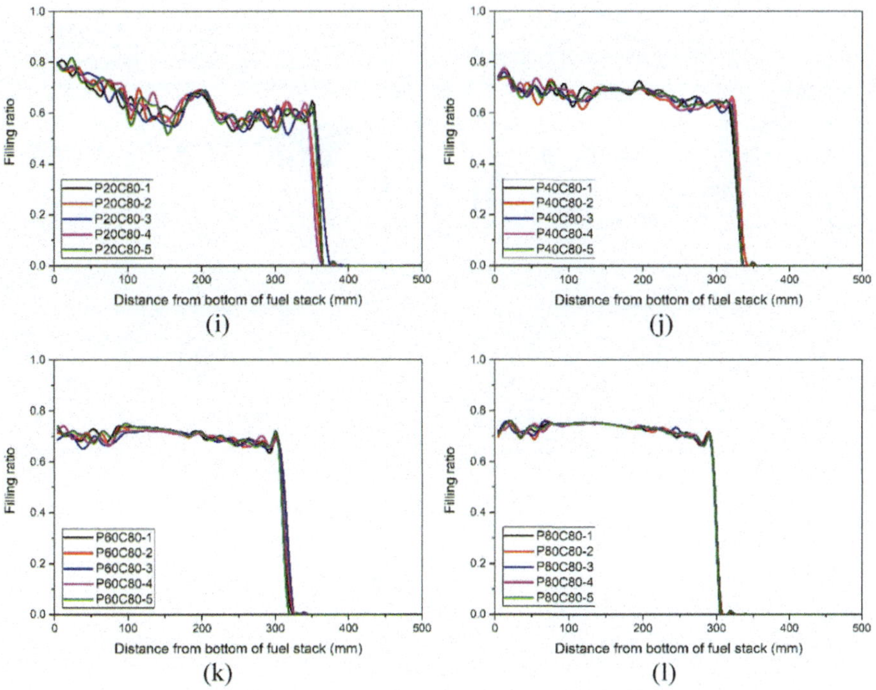

Fig. 9. (*continued*)

pulvers to fill most of the inter-fragment gaps. And the equivalent (constant) filling ratio was determined by the size of pulvers. The equivalent filling ratio for the cases with 40%, 60% and 80% puvler volume fraction can be simply calculated by averaging all the data in Fig. 10 (a), (b) and (c), respectively. It can be calculated that the equivalent filling ratio was 0.696, 0.714 and 0.748 for the cases with 40%, 60% and 80% puvler volume fraction, respectively. For an arbitrary puvler volume fraction between 40% and 80%, a correlation of filling ratio was proposed:

$$\eta = 2 \times 10^{-5}\left(x_p\right)^2 - 0.0011x_p + 0.708 \tag{3}$$

where η denoted filling ratio, and x_p denoted puvler volume fraction in %, ($x_p \geq 40\%$).

4 Conclusions

Three-dimension simulation for fuel relocation based on coupled FEM-DEM approach was performed to investigate the effect of fuel pulver on axial fuel relocation. Simulation results of fuel stack reduction, mass faction and filling ratio were given and discussed. A series of correlations of fuel stack reduction, mass faction and filling ratio with respect to volume fraction of fuel pulvers were proposed. Some conclusions could be drawn according to the simulation results:

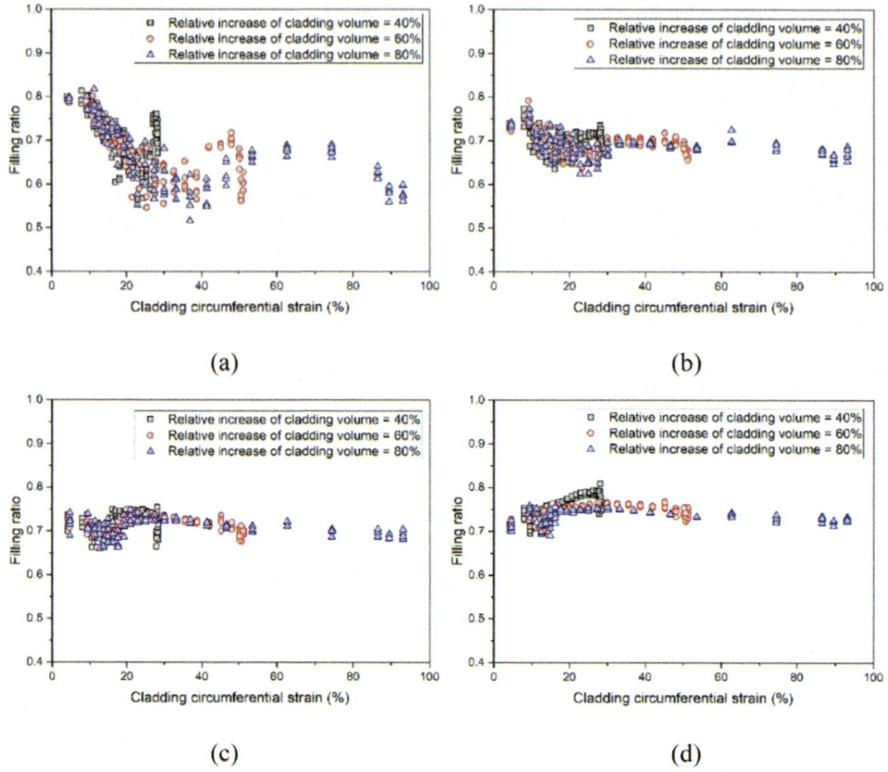

Fig. 10. Filling ratio versus cladding circumferential strain for the cases with pulver volume fraction of: (a) 20%, (b) 40%, (c) 60%, (d) 80%.

(a) Fuel stack reduction was a linear function of relative increase of cladding volume. And a higher volume fraction of fuel pulvers could improve fuel stack reduction.
(b) Peak mass fraction was a linear function of peak cladding circumferential strain. And a higher volume fraction of fuel pulvers could enhance peak mass fraction, indicating that high burnup fuel had a higher potential to cause local hot spot during fuel relocation.
(c) Filling ratio exhibited as a constant for the fuel puvler volume fraction $\geq$40%.

In the future work, more parametric study will be conducted to investigate axial fuel relocation behaviors with other coarse fragment mean diameter ($\neq$ 2.0 mm) combined with different pulver volume fraction.

Acknowledgement. This work is supported by the Project funded by China Postdoctoral Science Foundation (Grant no. 2022M712983).

References

1. Zhang, H., Blakely, C., Yu, J., Stewart, R., Asgari, M.: Fuel Rod Burst Potential Evaluation under LOCA Conditions for an Existing Plant with Extended Burnup Exceeding the Current Limit by 20%. Technical Report: INL/EXT-19–55888, Idaho National Laboratory (2019).
2. Raynaud, P.: Fuel Fragmentation, Relocation, and Dispersal during the Loss-Of-Coolant Accident. Technical Report: NUREG-2121, US NRC (2012).
3. Sonnenburg. H., et. al.: Report on Fuel Fragmentation, Relocation and Dispersal. Technical Report: NEA/CSNI/R(2016)16, OECD (2016).
4. Bales, M., Chung, A., Corson, J., Kyriazidis, L.: Interpretation of Research on Fuel Fragmentation, Relocation, and Dispersal at High Burnup. Technical Report: RIL 2021–13, US NRC (2021).
5. US Nuclear Regulatory Commission: Acceptance criteria for emergency core cooling systems for light water nuclear power reactors. Code of Federal Regulations: 10 CFR 50.46, Washington, D.C. (2017).
6. Govers, K., Verwerft, M.: Simulation of ballooning & relocation in the Halden LOCA tests with FRAPTRAN. In: Proceedings of the Enlarged Halden Programme Group Meeting, Enlarged Halden Programme Group Meeting, Roros, Norway (2014).
7. Jernkvist, L.O., Massih, A.R.: Models for Axial Relocation of Fragmented and Pulverized Fuel Pellets in Distending Fuel Rods and its Effect on Fuel Rod Heat Load. Technical Report: 2015:37, Quantum Technologies AB (2015).
8. Govers, K., Verwerft, M.: Discrete element method study of fuel relocation and dispersal during loss-of-coolant accidents. Journal of Nuclear Materials 478, 322-332 (2016).
9. Ma, Z.H, Shirvan, K., Wu, Y.W., Su, G.H.: A three-dimensional axial fuel relocation framework with discrete element method to support burnup extension. Journal of Nuclear Materials 541, 152408 (2020).
10. Ma, Z.H., Liang, R., Lin, Z.K., Ouyang, Y., Lu, X.H., Su, G.H.: Numerical Simulation of Fuel Relocation at High Burnup with Discrete Element Method. In: Proceedings of the 2022 29th International Conference on Nuclear Engineering, Shenzhen, China (2022).
11. Ma, Z.H., Liang, R., Lin, Z.K., Ouyang, Y., Lu, X.H., Wu, Y.W., Su, G.H.: Three-dimensional simulation of fuel relocation during loss-of-coolant accidents based on the coupled FEM-DEM approach. Journal of Nuclear Materials 570, 153978 (2022).
12. Molina-Siles, P.: Parametric Environment: The Handbook of Grasshopper Nodes & Exercises. ISBN 978–84- 9048–499–9, Universitat Politecnica de Val encia (2016).
13. Jernkvist, L.O.: Computational Assessment of LOCA Simulation Tests on High Burnup Fuel Rods in Halden and Studsvik. Technical Report: 2017:12, Quantum Technologies AB (2017).
14. Romano, A., Horvath, M.I., Restani R.: Evolution of porosity in the high-burnup fuel structure. Journal of Nuclear Materials 361, 62–68 (2007).
15. Ma, Z.H., Shirvan, K., Li, W., Wu, Y.W.: Modeling axial relocation of fragmented fuel during loss of coolant conditions by using ABAQUS. In: Proceedings of the 2020 28th Conference on Nuclear Engineering, Anaheim, California, USA (2020).

On the Creep Collapse of the Cladding Considering the Irradiation Growth Effect

Ming Zhang, Yayun Luo, Yong Lu, Yanan Zhu, Xiaohan Liu, Jinggang Li, and Xinying Miao(✉)

China Nuclear Power Technology Research Institute Co., Ltd. Reactor Engineering Software Research Institute, Shenzhen, China
mxinying@163.com

Abstract. Creep collapse of the cladding is one of the most important performance to be verified at the fuel rod's design stage to guarantee safe operation. The irradiation growth effect should be considered to improve the conservatism. In this study, finite element software ABAQUS was utilized to model the creep and irradiation growth of the cladding. The creep collapse criterion was established. The results show that irradiation growth can promote creep collapse of the cladding tube. Thus, the irradiation growth should be considered in the design of fuel rods in order to increase conservatism and safety.

Keywords: Cladding tube · Creep collapse · Irradiation growth

1 Introduction

The zirconium alloy cladding tubes suffer high temperature, high pressure, and high neutron irradiation. Such service conditions can cause circumferential creep deformation of the cladding tubes inwardly, and ultimately leading to the collapse of the cladding tubes. In the design phase of nuclear reactor fuel rods, creep collapse performance analysis of fuel rod cladding is an important part of fuel rod design verification. Generally, specialized software is used for the creep collapse analysis of the cladding tube. Common specialized software includes foreign BUCKLE [1], COLAPX [2], COVE [3], CEPAN [4], and CGN's FROCO [5] and BINE [6]. However, the above software adopts the assumption of plane strain and cannot consider the irradiation growth of fuel rods in three-dimensional conditions, nor the impact of irradiation growth on creep collapse. Therefore, the engineering application of the above software is relatively limited.

In order to accurately simulate the behavior of the cladding tube, this study will use the finite element software ABAQUS and its subroutine CREEP to establish a three-dimensional creep and radiation growth of the cladding model, The influence of the irradiation growth of the cladding tube on the creep collapse behavior of the cladding tube will be analyzed.

J. Liu and Y. Jiao (Eds.): WRFPM 2023, SPPHY 299, pp. 16–19, 2024.
https://doi.org/10.1007/978-981-99-7157-2_2

2 Methods

2.1 Constitutive Model of the Cladding

The cladding is made of zirconium alloy. Plastic behavior is not considered in the creep collapse simulation in general. thus linear elastic model is utilized in this study. The elastic modulus is 78000 MPa. The Possion's ratio 0.345 and the yield strength is 160 MPa.

The total creep is the combination of thermal and irradiation creep.The creep rate:

$$\dot{\varepsilon}^{th} = A\frac{E}{T}\left(\sinh\frac{\alpha_i\sigma_{eff}}{E}\right)^n \exp\left(\frac{-Q}{RT}\right)$$
$$\dot{\varepsilon}^{irr} = C_0\phi^{C_1}\sigma_{eff}^{c_2}f(T) \tag{1}$$

The irradiation growth is a function of the neutron flux with the follow rate form:

$$\dot{\varepsilon}_{grow} = 0.00075 \cdot \phi \cdot 10^{-21} \tag{2}$$

2.2 Finite Element Model

The diameter of the cladding is 9.5 mm, with the thickness 0.57 mm and axial height 50 mm. The initial ovality of the cladding is 0.07 mm. The mesh number of the cladding is 40,100 and 4 along the axial, circumferential and radial direction, correspondingly.

2.3 Loading History

The pressure of the magnitude 13 MPa and set constant during the life. The temperature and flux history are shown in Fig. 1.

2.4 Creep Collapse Failure Criterion

The cladding will failure due to the following criterion for simplicity:

(1) Strength criterion: when the stress exceeds 250 MPa.
(2) Ovality rate criterion: when the ovality rate exceeds 0.001mm/h.

3 Results and Discussion

The ovality increases with time because of the loading and creep. The results are shown in Fig. 2: When the irradiation growth is not considered, the cladding collapse failure due to criterion (2) at time 4148.8 h. The maximuim stress is 189.5 MPa. When considering the irradiation growth, the collapse time is 3833.00 h with maximum stress 195.1 MPa. The irradiation growth can accelerate the collapse of the cladding, which leading the failure 278.65 h ahead of time (6.72%). At time = 3833.0 h,the ovality is 0.341 mm without growth and 0.442 mm with growth.The growth can increase the 29.6% of the ovality, which is important on the creep collapse.

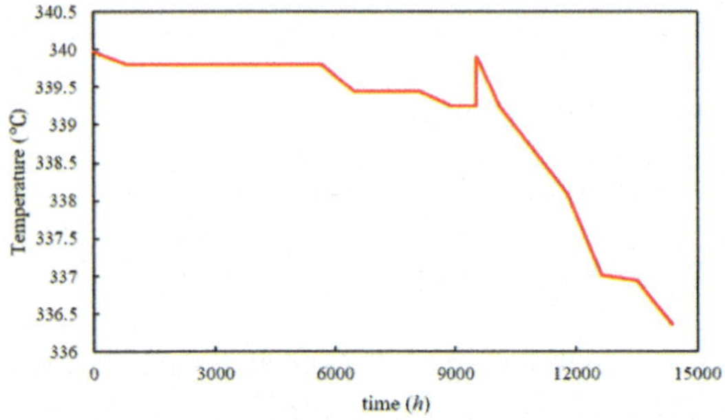

(a) Temperature history

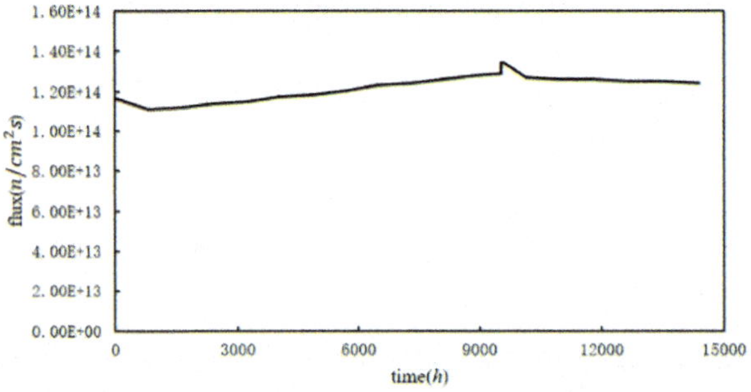

(b) Flux history

Fig. 1. Temperature history and flux history of the cladding tube

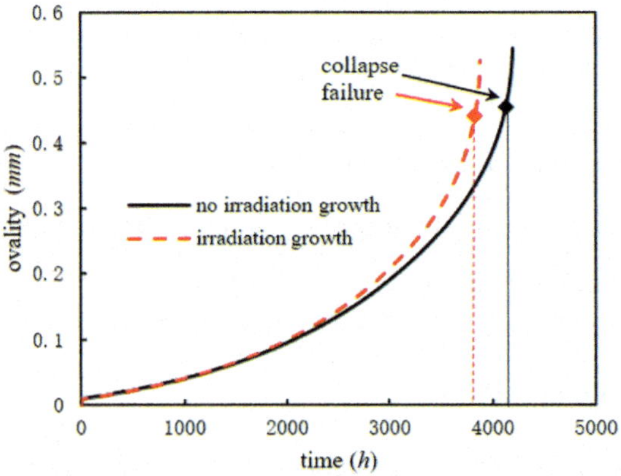

Fig. 2. The influence of the irradiation growth to the ovality of the cladding

Irradiation growth is an anisotropic volume growth, the introduction of a growth model not only generates creep strain, but also generates a certain amount of growth strain that needs to be reassembled to achieve a new equilibrium state inside the cladding tube structure. This process intensifies the collapse of the cladding tube, thus accelerating the collapse failure of the cladding tube without significantly changing the stress.

4 Conclusion

In this study, a finite element model of a cladding tube with radiation growth and creep is established with the ABAQUS and its subroutine CREEP. The conclusion is: Irradiation growth has a promoting effect on creep collapse, which can advance the critical time of collapse failure by 6%. Therefore, the irradiation growth model should be integrated into the creep collapse analysis software of the cladding tube in order to increase the conservatism.

References

1. PANKASKIE P J. BUCKLE: an analytical computer code for calculating creep buckling of an initially oval tube[J]. Rivista Italiana Delle Sostanze Grasse, 1974.
2. MERCKX K R. Calculational procedure for determining creep collapse of LWR fuel rods[J]. Nuclear Engineering & Design, 1974, 31(1): 95–101.
3. MOHR C L. COVE-1: a finite difference creep collapse code for oval fuel pin cladding material[R]. Richland: Battelle Pacific Northwest Labs, 1975.
4. FRANKLIN D G, LUCAS G E, BEMENT A L. Creep of zirconium alloys in nuclear reactors[M]. West Conshohocken: ASTM International, 1983.
5. JIN Xin, LIU Xiaohan, LU Yong, et al. Introduction of fuel rod cladding creep collapse code FROCO [C]. Beijing: China Conference on Theoretical and Applied Mechanics, 2017.
6. ZHANG Ming, LI Jinggang, LIU Xiaohan, et al. Three-dimensional mechanical modeling of the cladding and its deformation analysis software development[J]. Science technology and engineering, 2022, 22(29): 100–105.

Practical Development of Accident Tolerant Fecral-Ods Fuel Claddings for BWRs in Japan

K. Sakamoto[1(✉)], C. Sakaguchi[1], Y. Miura[1], H. Yokoyama[1], J. Matsunaga[2], H. Kasahara[3], H. Miyata[3], I. Ioka[4], S. Yamashita[4], and M. Osaka[4]

[1] Nippon Nuclear Fuel Development, Co., Ltd, Ibaraki, Japan
Kan.Sakamoto@nfd.co.jp
[2] Global Nuclear Fuel – Japan, Kanagawa, Japan
[3] Hitachi-GE Nuclear Energy, Ltd, Ibaraki, Japan
[4] Japan Atomic Energy Agency, Ibaraki, Japan

Abstract. An oxide-dispersion-strengthened FeCrAl (FeCrAl-ODS) has been continuously developed in Japan as a promising candidate alloy for the accident tolerant fuel cladding of BWRs (boiling water reactors). This paper will introduce the progress in practical development of accident tolerant FeCrAl-ODS fuel claddings for BWRs in the program fully or partially supported and organized by the Ministry of Economy, Trade and Industry (METI) of Japan.

The experimental studies have been conducted to obtain and accumulate key material properties of FeCrAl-ODS fuel claddings to support the evaluations in the analytical studies. For the evaluation at normal operation condition, fatigue test of unirradiated fuel cladding and tensile test of irradiated sheet specimen were conducted. In the fatigue test, a tensile-compressive bending strain was loaded on the C-shaped specimens by cyclic movement of a push-pull rod. Test temperature was 623 K, frequency was 1 Hz, and strain amplitude were 0.27, 0.34 and 0.55%. The results of fatigue tests demonstrated that cycles to failure of the FeCrAl-ODS cladding were higher than that of the O'Donnell and Langer fatigue curve of Zr-based alloy. The tensile test was conducted in a hot cell using the SS-J2 type specimens at ambient temperature, 573 K and 623 K at a strain rate of 10^{-3} s^{-1}. The specimens were irradiated up to 7.8 and 13 dpa at 573 K in the High Flux Isotope Reactor at ORNL. The irradiation hardening and ductility loss obtained at 7.8 and 13 dpa were comparable to those at 3.9 dpa.

Keywords: ATF · Fuel cladding · FeCrAl · ODS

1 Introduction

The heat and hydrogen generation due to the exothermic reaction of Zircaloy cladding tubes with water and steam was one of the primary factors exacerbating the course of severe accidents [1]. Therefore, accident tolerant fuel (ATF) cladding tubes capable of suppressing the oxidation with water and steam and the associated heat and hydrogen generation have been extensively developed [2, 3]. In Japan, FeCrAl-ODS has been continuously developed as a promising candidate alloy for the accident tolerant fuel cladding

J. Liu and Y. Jiao (Eds.): WRFPM 2023, SPPHY 299, pp. 20–28, 2024.
https://doi.org/10.1007/978-981-99-7157-2_3

of BWRs [4]. Although much effort has been devoted to its practical development, some material properties and evaluations are still lacking.

In Japanese fiscal year 2022, fatigue property of unirradiated FeCrAl-ODS claddings and mechanical properties of irradiated FeCrAl-ODS sheets were evaluated. The fatigue tests were conducted at 3 different strain amplitude up to 0.55% at 623 K. Evolution of micro-hardness, strength and ductility with neutron irradiation were evaluated up to 13.0 dpa.

2 Experiments

2.1 Fatigue Test

A tensile-compressive bending strain was loaded on the C-shaped specimens by cyclic movement of a push-pull rod. The test method of the fatigue test has been described in [5]. Overview of experimental setup is illustrated in Fig. 1. The C-shaped specimen was fixed at two circumferential locations (90 and 270 degrees in Fig. 1) and loaded with the tensile-compressive bending strain by applying reciprocal movement of push-pull rods. The maximum tensile/compressive strains were generated to the opposite side of the opening location of the C-type specimens (180 degrees in Fig. 1). The test temperature was 623 K, frequency was 1 Hz, and strain amplitude were 0.27, 0.34 and 0.55%.

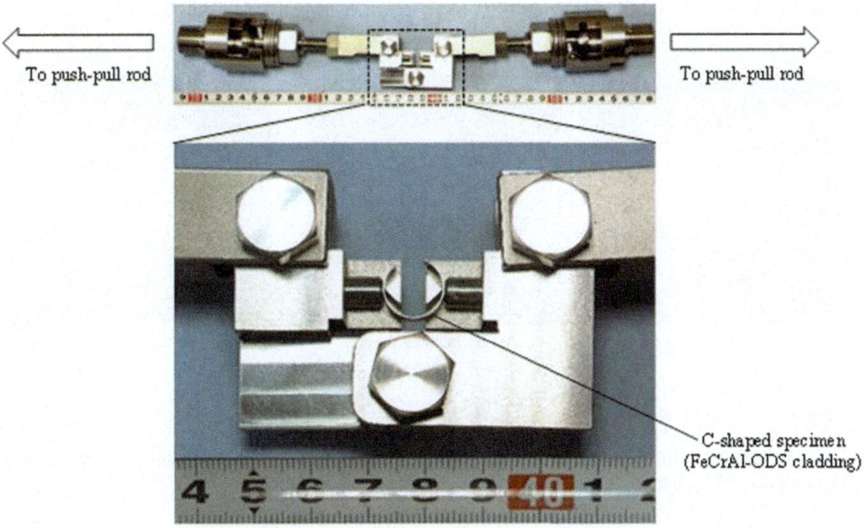

Fig. 1. Overview of fixture applied to fatigue test of FeCrAl-ODS cladding.

2.2 Micro-Hardness and Tensile Tests of Irradiated Specimens

The flat-sheet (16 mm × 4 mm × 0.5 mm) and SS-J2 type (16 mm × 4 mm × 0.5 mm, gage length: 5 mm) specimens fabricated from the FeCrAl-ODS sheet materials were irradiated up to 7.8 and 13 dpa at 573 K in the High Flux Isotope Reactor (HFIR) at the Oak Ridge National Laboratory (ORNL). The axial center of some specimens was beaded with an electron beam prior to irradiation test. All specimens were transported to the hot laboratory at Nippon Nuclear Fuel Development (NFD) and subjected to the dimensional measurement to evaluate swelling induced by neutron irradiation prior to the micro-hardness and tensile tests. The micro-hardness test was conducted on the surface of cross-section of the flat-sheet specimens in a hot cell at ambient temperature. The tensile tests were conducted in a hot cell using the SS-J2 type specimens at ambient temperature, 573 K and 623 K at a strain rate of 10^{-3} s^{-1}.

3 Results and Discussions

3.1 Fatigue Test

The deformation of the C-type specimen was recorded to evaluate a local strain during the fatigue test. As an example, the deformations of the C-type specimen tested at a strain amplitude of 0.34% are shown in Fig. 2 at maximum tensile, zero and maximum compressive bending strains. Image analysis of the recorded images confirmed that the strain amplitude was correctly applied as intended in all tests. The relationship between strain amplitude and failure cycle of FeCrAl-ODS cladding is shown in Fig. 3, which also includes that of Zr-based alloys [5]. As shown in Fig. 3, it is demonstrated that the fatigue property of FeCrAl-ODS cladding is comparable to Zr-based alloys.

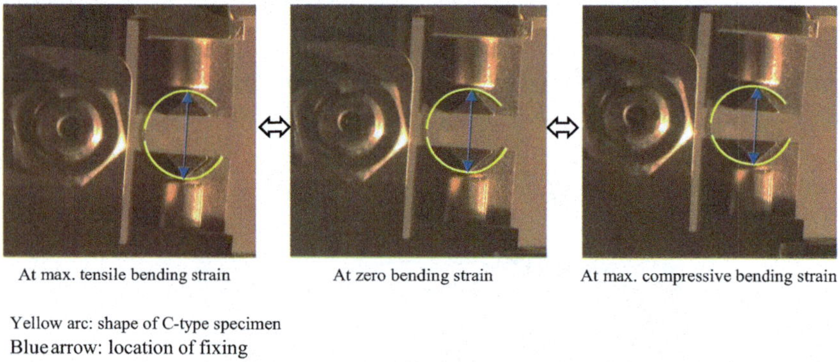

Fig. 2. Local deformation of C-type specimen in fatigue test at strain amplitude of 0.34% and test temperature of 623 K.

3.2 Micro-Hardness and Tensile Tests of Irradiated Specimens

The dependence of swelling on irradiation damage is summarized in Fig. 4. The mean and standard deviation of swelling at the longest length of the specimens (16 mm) were evaluated for 23 specimens at each irradiation damage. Data from specimens with 2.6 dpa irradiation damage were not included due to high uncertainty. No breakaway of

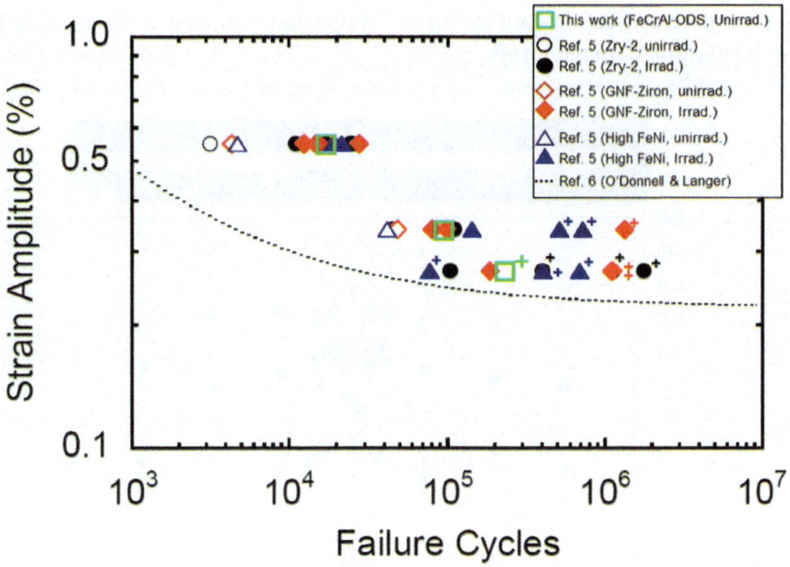

Fig. 3. The relationship between strain amplitude and failure cycle of FeCrAl-ODS cladding [5, 6] (+: no failure).

swelling was observed in the measured range and the swelling of FeCrAl-ODS sheet at 13.0 dpa was almost half of irradiation growth of Zircaloy-2 at the same irradiation damage [7], although the uncertainty of the evaluated value was high.

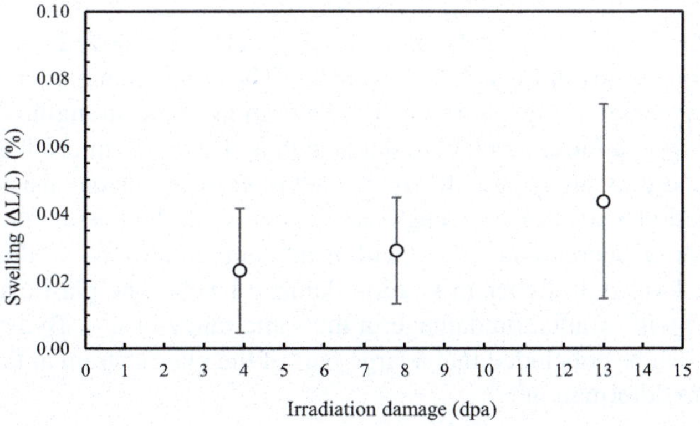

Fig. 4. Dependence of irradiation growth of FeCrAl-ODS sheet materials on irradiation damage.

The evolution of micro-hardness with irradiation damage of specimens with and without electron beam bead is shown in Fig. 5 and Fig. 6, respectively. The relationship between the micro-hardness and the irradiation damage is shown in Fig. 7 for the specimens without electron beam bead, where the mean values and standard deviations of micro-hardness are calculated from all measured points (21 points). The micro-hardness increases with the irradiation damage up to 7.8 dpa and is almost at the saturation value at higher irradiation damage levels.

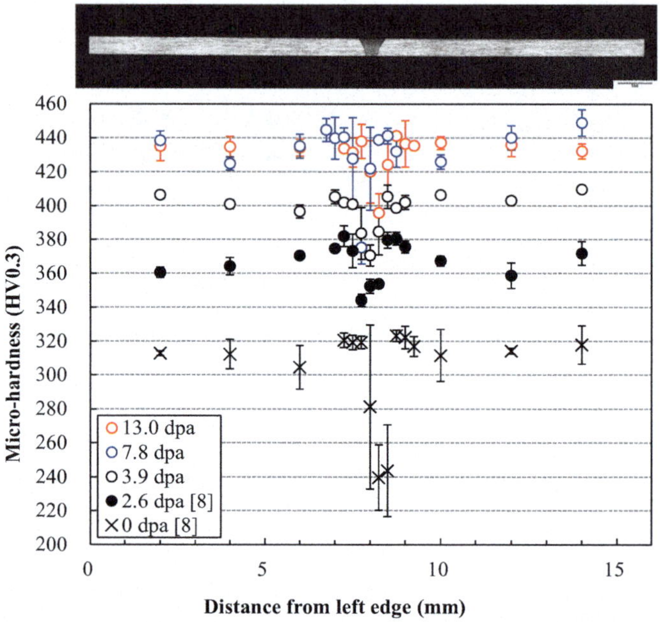

Fig. 5. Evolution of micro-hardness with irradiation damage of FeCrAl-ODS sheet specimens with electron beam bead [8].

The results of tensile tests at 7.8 and 13.0 dpa were summarized with those at lower irradiation damages [4] in Figs. 8, 9, 10 and 11. The irradiation hardening saturated at 2.6–3.9 dpa as shown in Figs. 8 and 9. The saturation of the strengths (UTS and yield stress) started from a lower irradiation damage than that of the micro-hardness. Further experiments and evaluations are clearly needed to explain this inconsistency, since the trend of irradiation hardening of strength are generally similar to that of micro-hardness. The uniform strain decreased to $<1\%$ with irradiation damage to 3.9 dpa and is almost at the saturated value at higher irradiation damage levels. The plastic strain at failure also decreased with irradiation damage in the same range of dpa. By comparing these two strains, it can be concluded that a large part of the plastic strain at failure is a result of non-uniform deformation.

4 Conclusions

The experimental studies have been conducted to obtain and accumulate key material properties of FeCrAl-ODS fuel cladding to support the evaluations in the analytical studies. In JFY2022, fatigue tests on unirradiated fuel claddings and tensile tests on neutron irradiated sheet specimens were conducted for evaluation under normal operating conditions. The results of the fatigue tests showed that the cycles to failure of the

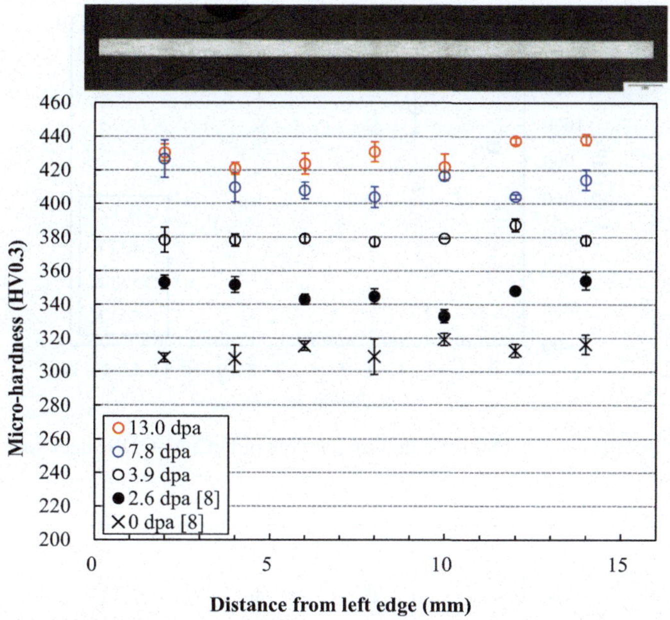

Fig. 6. Evolution of micro-hardness with irradiation damage of FeCrAl-ODS sheet specimens [8].

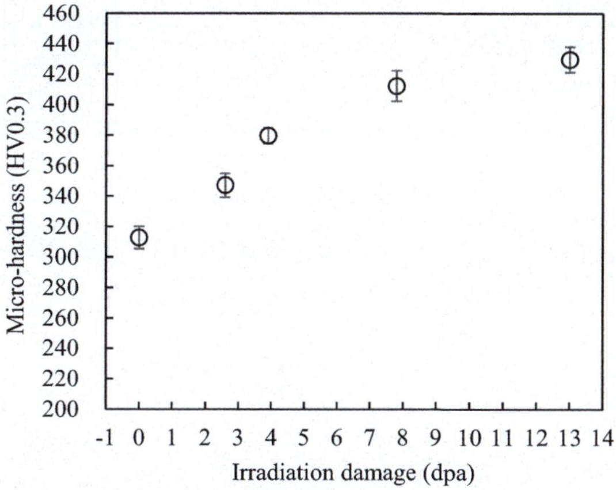

Fig. 7. Relationship between micro-hardness and irradiation damage for specimens without electron beam bead [8].

FeCrAl-ODS cladding were higher than the O'Donnell and Langer fatigue curve of the Zr-based alloy. In the tensile tests, the irradiation hardening and ductility reduction were almost saturated around the irradiation damage of 3.9 dpa.

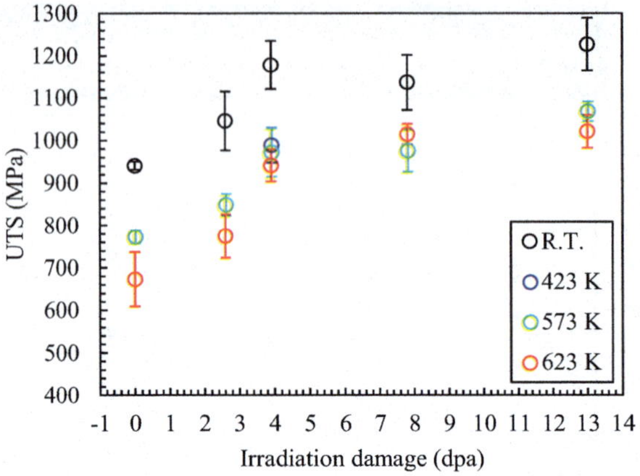

Fig. 8. Dependence of UTS on irradiation damage in FeCrAl-ODS sheet specimens [4].

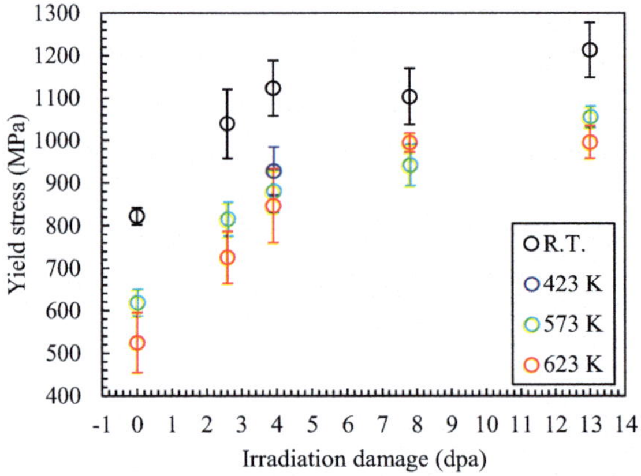

Fig. 9. Dependence of yield stress on irradiation damage in FeCrAl-ODS sheet specimens [4].

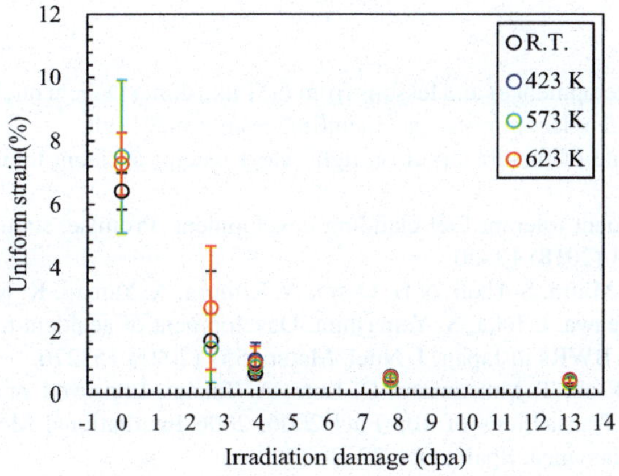

Fig. 10. Dependence of uniform strain on irradiation damage in FeCrAl-ODS sheet specimens [4].

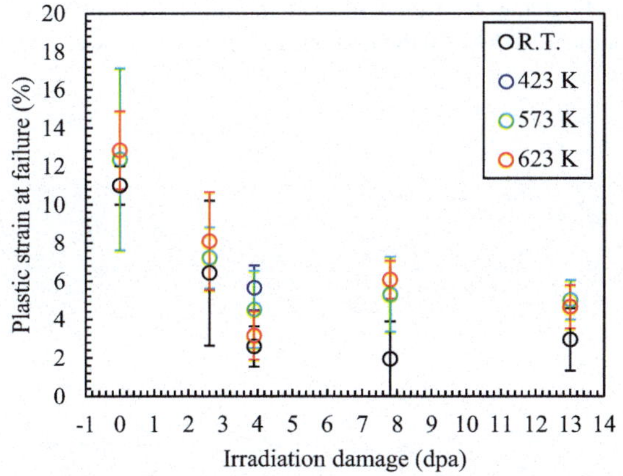

Fig. 11. Dependence of plastic strain at failure on irradiation damage in FeCrAl-ODS sheet specimens [4].

Acknowledgments. Part of this study is the result of "Development of Technical Basis for Introducing Advanced Fuels Contributing to Safety Improvement of Current Light Water Reactors" program carried out by JAEA under the project on technical development for improving nuclear safety, supported by Ministry of Economy, Trade and Industry (METI) of Japan.

References

1. TEPCO, The development of and lessons from the Fukushima Daiichi nuclear accident, https://www.tepco.co.jp/en/decommision/accident/images/outline01.pdf.
2. OECD/NEA, State-of-the-art report on light water reactor accident-tolerant fuels, NEA No. 7317 (2018).
3. K. Terrani, Accident tolerant fuel cladding development: Promise, status, and challenges, J. Nucl. Mater., 501 (2018) 13–30.
4. K. Sakamoto, Y. Miura, S. Ukai, N.H. Oono, A. Kimura, A. Yamaji, K. Kusagaya, S. Takano, T. Kondo, T. Ikegawa, I. Ioka, S. Yamashita, Development of accident tolerant FeCrAl-ODS fuel cladding for BWRs in Japan, J. Nucl. Mater., 557 (2021) 153276.
5. S. Ishimoto, Y. Etoh, T. Matsumoto, D. Lutz, A. Takagi, Improved Zr alloys for high burnup BWR fuel, Transactions of Top Fuel 2006, 2006 International Meeting on LWR Fuel Performance, Salamanca, Spain, Oct. 22–26 (2006).
6. W. J. O'Donnell and B. F. Langer, Fatigue Design Basis for Zircaloy Components, Nuclear Science and Engineering, 20, 1–12 (1964).
7. R.B. Adamson, C.E. Coleman, M. Griffiths, Irradiation creep and growth of zirconium alloys: A critical review, J. Nucl. Mater., 521 (2019) 167–244.
8. J. Gao, P. Song, Y.-J. Huang, K. Yabuuchi, A. Kimura, K. Sakamoto, S. Yamashita, Effects of neutron irradiation on 12Cr6Al-ODS steel with electron-beam weld line, J. Nucl. Mater, 524 (2019) 1–8.

Experimental Study on Pool Boiling Heat Transfer Characteristics of SiC Cladding Under Atmospheric Pressure

Xing Lei[1,3], Qing-long Wen[1,2,3](✉), Zhen-xun Peng[4], De-sheng Jin[4], and Ya-lun Yan[4]

[1] Department of Nuclear Engineering and Technology, Chongqing University, Chongqing, China
qlwen@cqu.edu.cn, 1204057181@qq.com

[2] Key Laboratory of Low-Grade Energy Utilization Technologies and Systems, Chongqing University, Chongqing, China

[3] Liangjiang Laboratory for New Energy (Nuclear Energy and Power), Chongqing, China

[4] China Nuclear Power Technology Research Institute, Shenzhen, China

Abstract. Since the Fukushima accident in Japan, Accident-Tolerant Fuel (ATF) was investigated to mitigate the zircaloy-water oxidation reaction at a high temperature. Silicon carbide (SiC) cladding is one of the most promising ATF claddings as a replacement for zirconium-based cladding in the light water reactor. To verify the enhanced safety margin of the SiC cladding, a pool boiling experiment was conducted in deionized water at atmospheric pressure. The measured maximum heat flux of the SiC was 600.2 kW/m^2. Based on the bubble behavior at low heat flux (23.69 kW/m^2) recorded by a high-resolution camera, a heat transfer model of SiC cladding was developed. The measured and predicted heat flux values are in good agreement within ±20%.

Keywords: SiC cladding · Pool boiling · Bubble behavior · Model of heat transfer

Nomenclature

Roman letters

$\vec{a}$	Normal vector of solid phase
$\vec{b}$	Normal vector of gas phase
c	Constant
m	Constant
n	Constant
C_d	Constant
D_b	Departure diameter (mm)
u_a	Data acquisition card accuracy
u_b	Vapor phase velocity (m/s)
u_f	Liquid phase velocity (m/s)
u_i	Instrument accuracy

J. Liu and Y. Jiao (Eds.): WRFPM 2023, SPPHY 299, pp. 29–42, 2024.
https://doi.org/10.1007/978-981-99-7157-2_4

u_s Signal amplifier accuracy
$c_{p,f}$ Liquid specific heat (J/kg·K)
Bo Boiling number
Re_b Bubble Reynolds number
$Re_{b'}$ New bubble Reynolds number
Pr_f Prandtl number
$Nu_{b'}$ Bubble Nusselt number
I Current (A)
V Voltage (V)
T Temperature of SiC inner wall (°C)
T_w Temperature of SiC outer wall (°C)
T_s Saturation temperature (=99.5 °C)
A Area of SiC cladding outer wall (m^2)
P Power (W)

Greek letters

σ Surface tension (N/m)
σ_x Direct measurement uncertainty
σ_y Indirect measurement uncertainty
ρ_f Density of the vapor phase (kg·m^{-3})
ρ_g Density of the liquid phase (kg·m^{-3})
λ_f Liquid heat conductivity coefficient (W/m·K)
β Contact angle (°)
h_{fg} Latent heat of vaporization (KJ/kg)

1 Introduction

The Fukushima accident in Japan demonstrates that a large amount of hydrogen released by the zircaloy-water oxidation reaction at a high temperature is potentially dangerous. To replace zirconium-based cladding in light water reactors, the program of accident-tolerant fuel cladding was proposed. As a strong candidate for ATF cladding materials, SiC cladding has various advantages, including extremely high thermal conductivity, high radiation resistance, low neutron absorption cross-section, high melting point(~1800 °C), and low hydrogen production under accidents [1, 2].

Historically, SiC has been used in fusion reactors and generation IV (Gen-IV) nuclear reactors such as GFR, LFR, and VHTR due to its excellent performance at high temperatures [1]. In the study of Carpenter [3], the SiC duplex was first proposed as cladding material. In this study, the SiC cladding and traditional zircaloy alloy were compared to prove that SiC has excellent creep and radiation resistance. Snead et al. [4], provide a summary of the properties of non-irradiated and irradiated SiC. Carpenter [5] evaluated the irradiation behavior of SiC cladding in his Ph.D. thesis and found that SiC has an acceptable low irradiation-enhanced corrosion rate and predictable expansion behavior. Wiesche [6] studied the influence of the properties of the heating wall of the SiC plate heater on the heat transfer performance which shows that it has rarely an effect on the natural convection and bubble dynamic. Kim et al. [7] discussed the overall trend of SiC and its future applicability in light water reactor fuels. Lee et al. [8] evaluated the oxidation rate of SiC cladding in the coolant loss accident (LOCA) scenario. The experimental results show that the oxidation rate of SiC is (about 3 orders of magnitude) slower than that of zircaloy-4. Sub Lee Song [9] studied the CHF enhancement of SiC nanofluids, which further proved the good heat transfer of SiC materials. The effect of SiC coating on zircaloy-based cladding was studied at a low pressure by Dong Hoon Kam et al. [11]. The results show the contact angle of the SiC-coated surface decreases, indicating that the SiC coating increases the hydrophilicity of the surface, thereby improving the heat transfer performance and increasing the CHF value. According to this study, the SiC-coated surface shows enhanced CHF with thickened SiC coatings showing better results. A pool boiling experiment of horizontal SiC cladding was carried out by Gwang Hyeok Seo al, and the CHF value of SiC cladding was approximately 63% higher than that of Zr-4 cladding [12].

Pool boiling is an effective way to investigate the mechanism of heat transfer. In the past few decades, atmospheric pool boiling heat transfer experiments on the surface of various materials have been carried out extensively, and predictive correlations under different working fluids have been developed. After the Fukushima nuclear accident, the pool boiling heat transfer phenomenon on the surface of SiC cladding as one of the most promising accident-tolerant fuel (ATF) claddings has attracted wide attention [13, 14].

However, experimental data is insufficient and the mechanism of heat transfer is still not well understood. Therefore, a pool boiling experiment was conducted in deionized water at atmospheric pressure. Bubble behaviors on heat transfer surface were carefully observed with high speed camera. Based on the experimental study of pool boiling heat transfer on the surface of the SiC cladding, a heat transfer model of SiC cladding was developed.

2 Experimental Apparatus and Procedure

2.1 Boiling Apparatus

The boiling apparatus is schematically shown in Fig. 1, consisting of a pool boiling vessel, test section, high-speed camera system, and data acquisition system. The pool boiling vessel was placed on a stainless-steel platform with two visible windows to observe the bubble behavior on the SiC cladding surface. The DC power supply is used to heat the test section to simulate the fission heat in the reactor. A pre-heater was immersed in the pool boiling vessel to pre-heat to the saturation temperature of water at atmospheric pressure and compensate for unavoidable heat losses. The temperature, current and voltage signals of all measured points were recorded by computer through the data acquisition system for process and analysis.

Figure 2 shows the structure diagram of the test section. The test section mainly includes SiC cladding, copper bar, alumina ceramic, thermal conductive filler, Ni-Cr(80/20) rod and so on. The designed parameters of the test section are shown in Table 1.

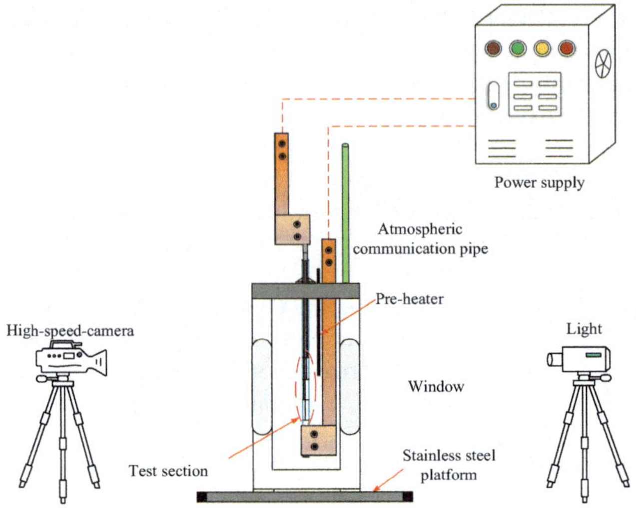

Fig. 1. Schematic of pool boiling apparatus

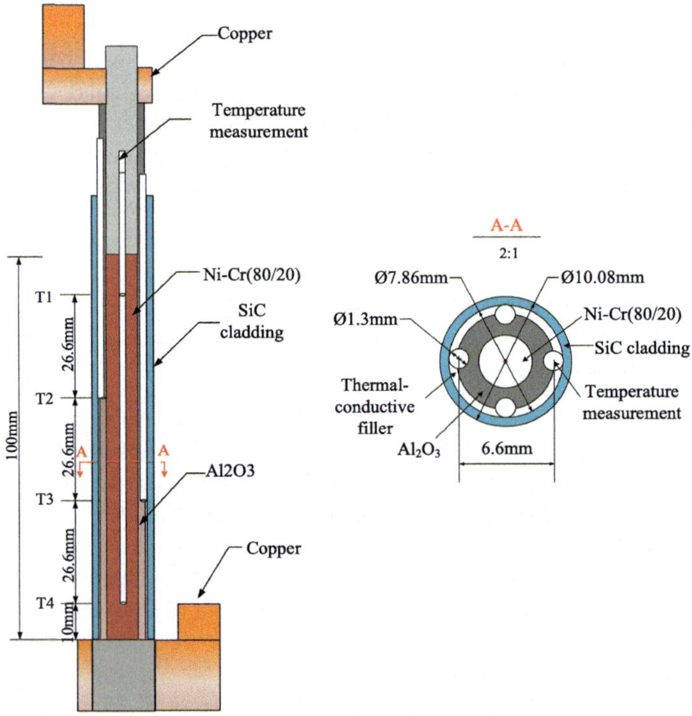

Fig. 2. Schematic of test element

Table 1. Main parameters of test section (mm)

	Total length	O.D	I.D	Thickness	Heated length
SiC	352	10.08	7.86	1.11	100

2.2 Experimental Procedure

Before taking the measurement, the deionized water in the boiling vessel was heated to the saturation temperature of atmospheric pressure by the pre-heater. It is worth noting that when heated to saturation temperature, it must also be heated for some time until no bubbles were generated to exclude non-condensable gases dissolved in deionized water. Voltage, current, and temperature signals will be continuously recorded when the steady state was reached. The heat flux on the cladding surface increases with the increase of the voltage applied to Ni-Cr (80/20) rod until the maximum heat flux was reached. When maximum heat flux was reached, the system will automatically reduce the power by 15%, while the high-speed camera will capture the bubble behavior on the SiC cladding surface.

2.3 Data Acquisition

The heat flux was calculated by the voltage and current applied to the center heating rod (as shown in Fig. 2, the red part is the heating section):

$$Q=\frac{IU}{1000 \cdot A} \tag{1}$$

where, I is the current, A, U is the voltage, V; and A is the outer surface area of the test section, m^2.

Previous studies have shown, the bubble contact angle was used to measure the surface wettability of different materials, which directly determined the heat transfer of the cladding. The liquid is spread on the surface, and the contact angle is called at the junction of gas, solid and liquid phases. From the solid-liquid interface to the liquid, the vapor-liquid image was programmed to obtain the vapor-liquid boundary, and the identified vapor-liquid boundary result is shown in Fig. 3.

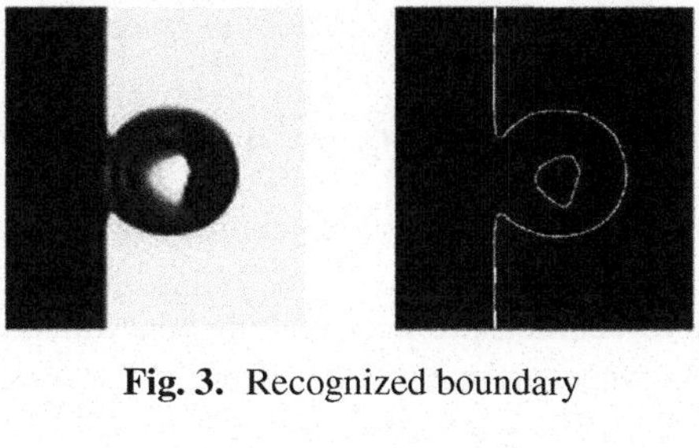

Fig. 3. Recognized boundary

It can be seen from Fig. 3 that the vapor-liquid phase edge is a set of pixels with different gray levels. The connection between the centers of two adjacent pixels is regarded as a tangent, and the normal vector of a point on the curve can be obtained. The normal vector along the solid phase and the normal vector along the gas phase is obtained by MATLAB programming (as shown in Fig. 4 the red arrow line in the figure is the normal vector of the point). In this paper, the average value of five adjacent normal vectors of a liquid phase and the average value of five adjacent normal vectors of a solid phase is selected as the measurement at the intersection of the solid liquid interface.

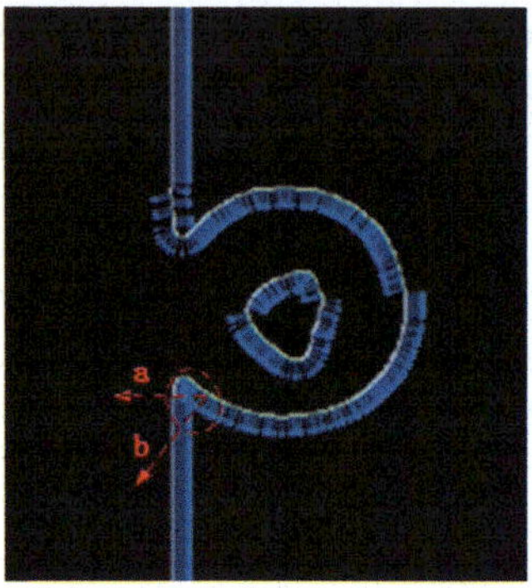

Fig. 4. Schematic of contact angle measurement

In the formula, $\overrightarrow{a}$ is the normal vector along the solid phase; $\overrightarrow{b}$ is the normal vector along the gas phase; $\cos\langle\overrightarrow{a}, \overrightarrow{b}\rangle$ is the cosine value of the angle between vector $\overrightarrow{a}$ and vector $\overrightarrow{b}$. According to the trigonometric function relationship:

$$\overrightarrow{a} \cdot \overrightarrow{b} = \left|\overrightarrow{a}\right| \left|\overrightarrow{b}\right| \cos\langle\overrightarrow{a}, \overrightarrow{b}\rangle \tag{2}$$

The contact angle can be obtained by solving Eq. (2).

2.4 Uncertainty Analysis

The basic parameters measured directly in this study include fluid temperature, inner wall temperature of the test section, current, voltage, etc., as shown in Table 2. The uncertainty of the direct measurement parameters is given by:

$$\sigma_x = \sqrt{u_i^2 + u_s^2 + u_a^2} \tag{3}$$

where u_i is the instrument precision, u_s is the accuracy of the signal amplifier and u_a is the data acquisition card accuracy.

The parameters of indirect measurement include heat transfer area, power, heat flux. The uncertainty of indirect measurement parameters can be calculated by the uncertainty of direct measurement parameters according to the error transfer formula:

$$y = f(x_1, x_2 \cdots x_n)$$

$$\sigma_y = \frac{\Delta y}{y} = \frac{1}{y}\sqrt{\sum_{i=1}^{n}\left(\frac{\partial f(x_1, x_2, \cdots, x_n)}{\partial x_i}\cdot \Delta x_i\right)^2} \tag{4}$$

where $x_1, x_2 \cdots x_n$ are independent direct measurement parameters.

Table 2. Uncertainties of the measurement parameters

Parameter	I	V	T
Uncertainties	0.27%	0.14%	0.12%
Parameter	A	P	Heat flux
Uncertainties	0.1%	0.18%	0.21%

3 Observations and Results

3.1 Boiling Heat Transfer Curve

The pool boiling heat transfer curves of the SiC cladding surface are shown in Fig. 5. According to the observed bubble images on the SiC cladding surface, the whole pool boiling heat transfer process can be divided into the following four sections: natural convection section, isolated bubble nucleate boiling section, slug bubble nucleate boiling section, and film bubble nucleate boiling section.

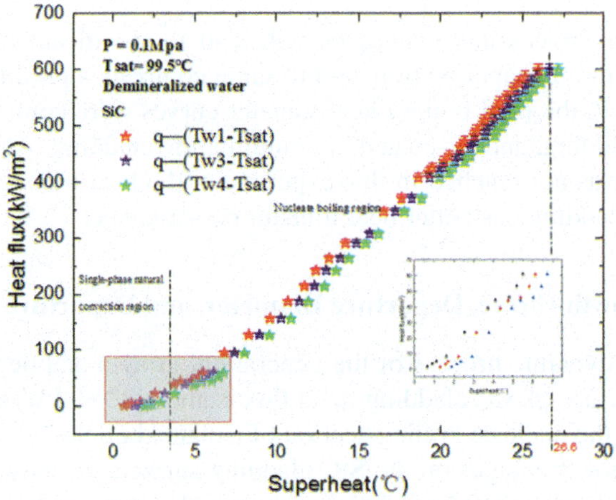

Fig. 5. Pool boiling heat transfer curve of SiC cladding

In the first picture in Fig. 6, the surface superheat of the SiC cladding was too low to cannot reach the minimum wall superheat required for surface bubble nucleation, the bubble has not yet been generated. At this time, natural convection was the dominant factor of heat transfer on the cladding surface. With the increase of surface heat flux, the temperature of onset of nucleate boiling (ONB) was reached, and the cladding surface began to generate bubbles. With the further increase of the heat flux on the cladding surface, the bubble growth rate increased, and the bubble departure frequency increased. At this time, the previous detached bubble is not far away from the bubble detachment point, and the next bubble has already been generated and detached. These bubbles collide and aggregate with each other during the upward movement, forming a larger bubble to move upward. When the heat flux of the cladding surface was greater than 200 kW/m^2, a large number of bubbles on the cladding surface accumulated together to form a long steam film, and the steam film quickly moves upward along the cladding surface. It can be seen that the slope of the pool boiling curve increases steadily in the film bubble nucleate boiling section.

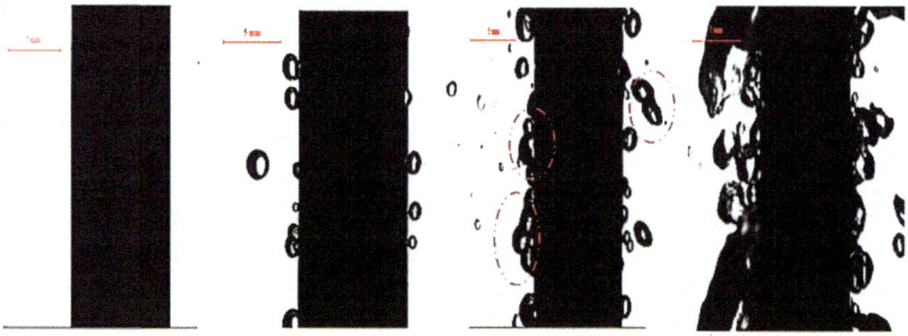

Fig. 6. The bubble behavior changes with increasing heat flux

Since different processing techniques will lead to significant differences in heat transfer, the data in this paper were from the same company's production of cladding. As shown in Fig. 7, the pool boiling heat transfer curves were compared between SiC cladding, Zr-4 cladding, and Cr-coated zirconium alloy cladding. Although the critical heat flux (CHF) was not reached in this experiment, the heat transfer before CHF has proved that SiC cladding has better heat transfer performance.

3.2 Bubble Growth Curve, Departure Diameter, and Departure Frequency

In this study, the dynamic process of the generation, growth and departure of isolated bubbles on the surface of SiC cladding (heat flux is about 23.69 kW/m^2) was studied in the early stage of the nucleate boiling section of isolated bubbles.

The bubble growth images on the SiC cladding surface are shown in Fig. 8. In the early stage of bubble growth (0.5 ms–5.0 ms), the bubble grows on the cladding surface in a relatively regular spherical shape. At this time, the forward contact angle of the bubble is almost equal to the backward contact angle. With the further growth of the bubble, due to the upward buoyancy, the bubble gradually shows an upward trend, the backward contact angle gradually increases, the forward contact angle gradually decreases, and

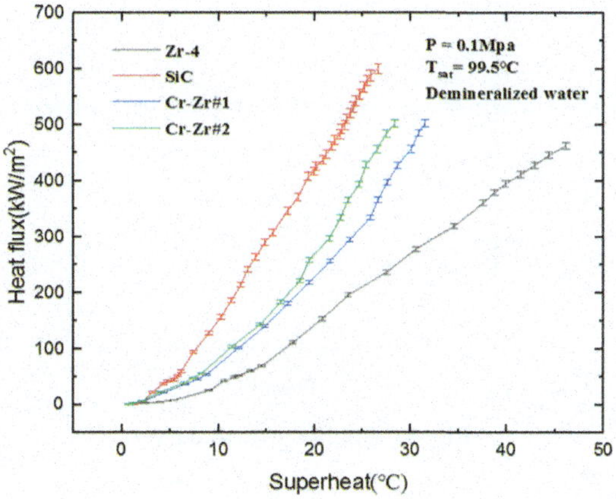

Fig. 7. Boiling heat transfer curves on different claddings

the bubble's center of mass moves up. When the bubble grows to a certain diameter, the bubble begins to depart the nucleation point.

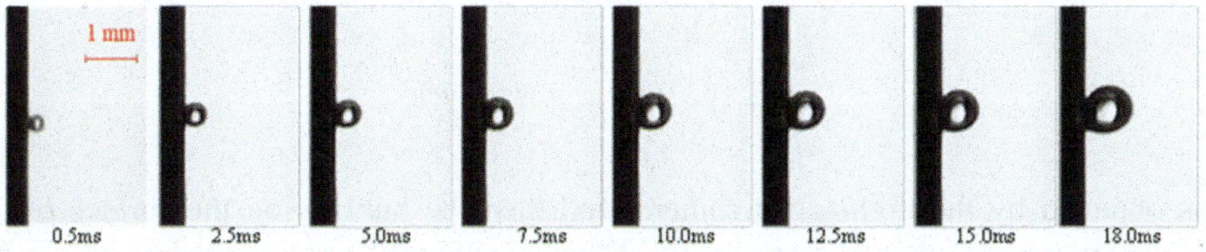

Fig. 8. Behavior of bubble growth

The bubble growth curve on the surface of SiC cladding is shown in Fig. 9. During the whole bubble growth process before the bubble departed from the nucleation point, the bubble growth shows a complex nonlinear growth trend. In the early stage of bubble growth (0.5 ms ~ 5.0 ms), the bubble growth rate is faster, and as the bubble grows, the bubble growth rate gradually slows down.

The bubble is subjected to various forces during the growth process, such as buoyancy, surface tension, etc., therefore, the shape of the bubble is irregular when it is departed from the nucleation point, and it is difficult to determine its true diameter. Previous investigators generally deal with the bubble detachment diameter with the equivalent diameter method. Bubble departure frequency is another key bubble dynamics parameter in boiling heat transfer analysis. It is defined as the reciprocal of the time period of two bubbles in the nucleate boiling process. In this paper, the bubble detachment frequency data within 0.5 s on the surface of each sample are averaged (Fig. 10).

As the surface heat flux increases, the surface superheat of the cladding surface reached the minimum wall superheat required for surface bubble nucleation, and the cladding surface began to generate bubbles, as shown in Fig. 11. At this time, due to the low heat flux in this section, the bubbles are firmly attached to the cladding surface due to the large surface tension. The bubble image attached to the surface of the cladding

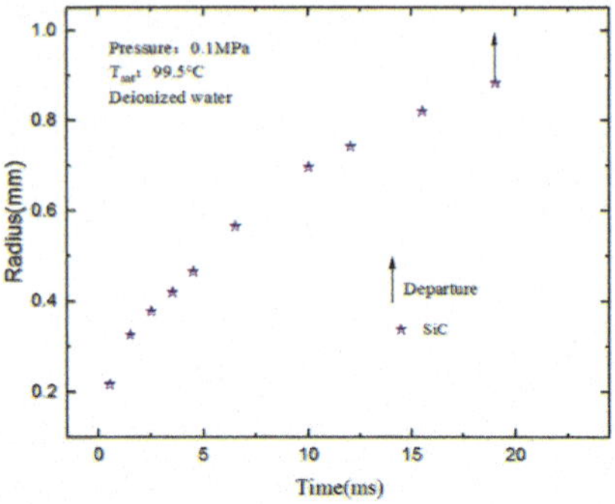

Fig. 9. Bubble growth curve

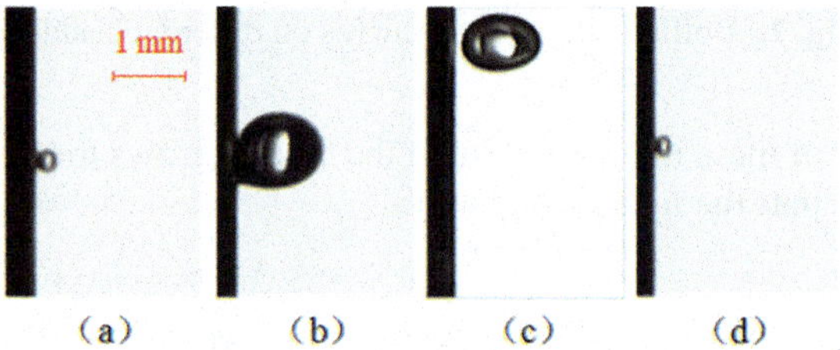

Fig. 10. Bubble growth cycle

was obtained by the high-speed camera, and then the bubbles on the surface of the SiC cladding were counted respectively. The contact angle was obtained by MATLAB program (Table 3).

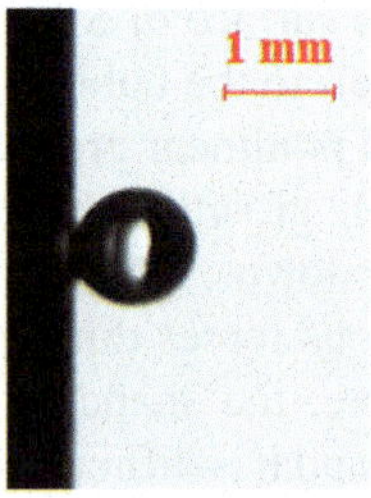

Fig. 11. Attached bubble

Table 3. Bubble dynamic parameters

	Diameter	Frequency	ONB	Contact angle
SiC	1.771 mm	25.74 Hz	2.2°C	52.13°

When the bubble is detached from the heated surface, it is subjected to buoyancy, surface tension, liquid inertia force, viscous force, and the force of surrounding fluid disturbance on the bubble due to the bubble growth squeezing the surrounding fluid. The size of the bubble, when it is departed from the heating surface, can be obtained by establishing a dynamic force balance equation for a single bubble based on bubble dynamics. According to this analysis, Fritz [15] obtained the bubble departure diameter by

$$D_b = C_d\,\beta\left[\frac{\sigma}{g(\rho_f-\rho_g)}\right]^{1/2} \tag{5}$$

where C_d is the constant obtained from the experimental fitting, β is the contact angle, g is the acceleration of gravity, and ρ_f is the density of the vapor phase and ρ_g is the density of the liquid phase respectively.

According to the contact angle and departure diameter measured by this study, the relationship between the departure diameter of the bubble from the heated surface can be obtained by nonlinear fitting of Eq. 5.

$$D_b = 0.01296\beta\left[\frac{\sigma}{g(\rho_f-\rho_g)}\right]^{1/2} \tag{6}$$

3.3 Heat Transfer Model

In this experiment, the higher the bubble departure frequency, the better the heat transfer. Therefore, the boiling number (*Bo*) can be used to describe this phenomenon.

$$Bo = \frac{u_b}{u_f} \tag{7}$$

where u_b is the vapor phase velocity and u_f is the liquid phase velocity.

In pool boiling, bubble flow is an important parameter to measure heat transfer. Therefore, the bubble Reynolds number Re_b is developed.

$$Re_b = \frac{\rho_g u_b D_b}{\mu_f} \tag{8}$$

where u_b is the bubble velocity and D_b is the bubble departure diameter. It reflects the disturbance of bubble motion to the surrounding liquid.

Re_b and Bo can be developed a new dimensionless number. The bubble growth caused by wall evaporation and its influence on the surrounding flow field are unified in this dimensionless number, which is defined as the Reynolds number of nucleate boiling bubble:

$$Re_{b'} = Re_b Bo = \frac{q''}{\mu_f h_{fg}} \times 0.01296\beta\left[\frac{\sigma}{g(\rho_f-\rho_g)}\right]^{1/2} \tag{9}$$

In the Forster vapor-liquid exchange mechanism [16], the flow of the liquid boundary layer also affects the nucleate boiling heat transfer, that is, the liquid momentum and the liquid thermal diffusivity of the boundary layer. The dimensionless number associated with heat transfer is defined as the ratio between the momentum and energy diffusion of the fluid:

$$\mathrm{Pr}_f = \frac{c_{p,f}\mu_f}{\lambda_f} \tag{10}$$

In saturated pool boiling, the bubble temperature can be regarded as the saturation temperature.

$$Nu_{b'} = \frac{c_{p,f}}{h_{fg}}(T_w - T_s) \tag{11}$$

The above dimensionless criterion numbers describing pool boiling are arranged in a similar form to the convective heat transfer in the tube:

$$Nu_{b'} = f(Re_{b'}, \mathrm{Pr}_f) = c(\mathrm{Pr}_f)^m (Re_{b'})^n \tag{12}$$

where c, m, n are constants.

Combined with the experimental data, the correlation coefficient was obtained, and the following heat transfer model was developed.

$$\begin{aligned}\frac{c_{p,f}}{h_{fg}}(T_w - T_s) &= c\left(0.01296\beta\frac{q''}{\mu_f h_{fg}}\left[\frac{\sigma}{g(\rho_f - \rho_g)}\right]^{1/2}\right)^n \left(\frac{c_{p,f}\mu_f}{\lambda_f}\right)^m \\ &= 0.121(0.01296\beta)^{0.71}\left(\frac{q''}{\mu_f h_{fg}}\left[\frac{\sigma}{g(\rho_f - \rho_g)}\right]^{1/2}\right)^{0.71}\left(\frac{c_{p,f}\mu_f}{\lambda_f}\right)^{-2.103}\end{aligned} \tag{13}$$

At high heat flux, the predicted value of the model is in good agreement with the experimental value. The reason for the large error at low heat flux is that the model prediction results are poor when the heat flux on the cladding surface is low. This may be due to the natural convection dominating the early stage of nucleate boiling, and the micro-layer evaporation theory [17] is introduced into the model so that the heat flux predicted by the heat transfer relationship is higher than the experimental heat flux (Fig. 12).

4 Summary and Conclusions

SiC cladding is considered to be one of the most promising new cladding materials. The boiling pool heat transfer characteristics of this new material surface have attracted global attention, which will help to evaluate whether it has better heat transfer. In this paper, the bubble behavior on the surface of SiC cladding under the condition of saturated pool boiling at atmospheric pressure was studied. The following conclusions are obtained: The quantitative relationship between the departure diameter and the contact angle was established by studying the surface bubble behavior. Based on the experimental data, a heat transfer model of SiC cladding was developed. The measured and predicted heat flux values are in good agreement within $\pm 20\%$.

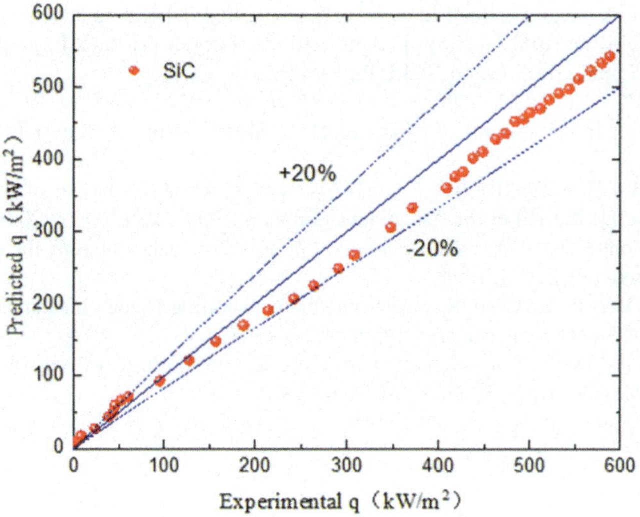

Fig. 12. Evaluation of heat transfer model

References

1. L. Hallstadius, S. Johnson, E. Lahoda, Cladding for high performance fuel, Prog. Nuclear. Energy 57 (2012) 71–76.
2. C.R.F. Azevedo, Selection of fuel cladding material for nuclear fission reactors, Eng. Fail. Anal. 18 (2011) 1943–1962.
3. Carpenter, D.M., 2006. Assessment of Innovative Fuel Designs for High Performance Light Water Reactors (MS thesis), MIT.
4. L.L. Snead, T. Nozawa, Y. Katoh, T.-S. Byun, S. Kondo, D.A. Petti, Handbook of SiC properties for fuel performance modeling, J. Nuclear. Mater. 371 (2007) 329–377.
5. D.M. Carpenter, An assessment of silicon carbide as a cladding material for light water reactors, Ph.D. thesis, Nuclear Science and Engineering, Massachusetts Institute of Technology, Cambridge, MA, USA, 2010.
6. S. aus der Wiesche, U. Bardas, S. Uhkötter, Boiling heat transfer on large diamond and SiC heaters: the influence of thermal wall properties, Int. J. Heat Mass Transf. 54 (2011) 1886–1895.
7. Kim, W.J., Kim, D.J., Park, J.Y., 2012. Material Issues for the Application of SiC Composites to LWR Fuel Cladding. Korean Nuclear Society Spring Meeting, Jeju, Korea.
8. Lee, T., Lee, J.H., Jeong, Y.H., 2013. Flow boiling critical heat flux characteristics of magnetic nanofluid at atmospheric pressure and low mass flux conditions. Int. J.Heat Mass Transf. 56 (1-2), 101–106.
9. Sub Lee Song, Ju Hyung Lee, Soon Heung Chang, CHF enhancement of SiC nanofluid in pool boiling experiment, Experimental Thermal and Fluid Science, Volume 52,2014,Pages 12-18.
10. TERRANI K A. Accident tolerant fuel cladding development: promise, status, and challenges[J]. Journal of Nuclear Materials, 2018, 501: 13–30.
11. Dong Hoon Kam, Jong Hyuk Lee, Taeseung Lee, Yong Hoon Jeong et al. Critical heat flux for SiC- and Cr-coated plates under atmospheric condition. Annals of Nuclear Energy 76 (2015) 335–342

12. Gwang Hyeok Seo, Gyoodong Jeun, Sung Joong Kim, Pool boiling heat transfer characteristics of zircaloy and SiC claddings in deionized water at low pressure. Experimental Thermal and Fluid Science, Volume 64, 2015, Pages 42–53.
13. TERRANIK A, YANG Y, KIM Y J, et al. Hydrothermal corrosion of SiC in LWR coolant environments in the absence of irradiation[J]. Journal of Nuclear Materials,2015, 465:488–498.
14. HIRAYAMA H, KAWAKUBO T, GOTO A, et al. Corrosion behavior of silicon carbide in 290°C water[J].Journal of the American Ceramic Society, 1989, 72(11):2049–2053.
15. Mikic B, Rohsenow W. A new correlation of pool-boiling data including the effect of heating surface characteristics[J]. 1969.
16. Engelberg-Forster K,-Greif R.-Heat transfer to a boiling-liquid- mechanism-and correlations[J]. Journal of Heat Transfer, 1959, 81(1):-43–52.
17. Cooper M, Lloyd A. The microlayer in nucleate pool boiling[J]. International Journal of Heat-and Mass Transfer, 1969,-12(8): 895–913.

Preliminary Development of a Simulation Capability for Zircaloy Clad Ballooning in LOCA

A. Wei Li[1(✉)] and B. Xiaoli Wu[2]

[1] State Key Laboratory for Strength and Vibration of Mechanical Structures, Xi'an Jiaotong University, Xi'an 710049, China
leeczway@163.com

[2] Science and Technology on Reactor System Design Technology Laboratory, Nuclear Power Institute of China, Chengdu 610213, China

Abstract. Zircaloy clad is widely known to be vulnerable to ballooning and burst in a loss-of-coolant accident (LOCA) typical of high-temperature steam environment, due to internal and external pressure difference and degraded strength. Ballooning can be detrimental in light of nuclear reactor safety. Due to the large deformation and complex constitute models in the ballooning process, detailed numerical simulation of Zircaloy clad ballooning has long been a challenge. This paper presented a preliminary Finite-Element-Method (FEM) implementation for analyzing Zircaloy clad ballooning. The developed capability can be used to predict progressive clad ballooning up to failure determined by empirical burst limit. Included physic models are currently phase transformation and high-temperature creep. Especially, anisotropic creep was considered for the α phase Zircaloy. The stress update algorithm is implicit in time integration, which ensures relatively large time increment during the simulation. The PUZRY separate effect tests were used to validate the FEM capability. The predicted time to failure (burst) is acceptable when compared with the experimental data. Besides, it was found that the effect of anisotropic coefficients is significant on the prediction of ballooning.

Keywords: Anisotropic creep · LOCA · Ballooning

1 Introduction

It is broadly recognized that Zr clad in LOCAs is prone to ballooning and burst, which blocks flow channel within the reactor core and especially for high-burnup fuel (rod-average burnup $>$ 62 MWd/kgU), causes relocation and dispersal of fuel fragments. High-temperature fuel particles accumulate in the ballooned region, which heats the thinned clad by direct contact, causing the risk of accelerating steam oxidation. This phenomenon is part of what is known as fuel fragmentation, relocation and dispersal

J. Liu and Y. Jiao (Eds.): WRFPM 2023, SPPHY 299, pp. 43–55, 2024.
https://doi.org/10.1007/978-981-99-7157-2_5

(FFRD) of high-burnup UO_2 fuel pellet. While FFRD is mainly relevant to the high-burnup characteristic of UO_2 pellet, the correspondence between the axial location of Zr clad ballooning and relocation of high-burnup UO_2 pellet fragments is evidently not negligible, since clad ballooning leads to the loss of radial constraint for the fuel fragments to relocate radially and axially (e.g., relocation occurs if the clad hoop strain is below 8% [1]). As for clad, the simulated LOCA experiment conducted by Ozawa etc. [2] shows that high-burnup condition has little influence on clad mechanical behavior due to the recovery of irradiated microstructure, compared with low-burnup condition. Considering the complex phenomena of the FFRD associated nuclear fuel safety, it is indeed a practical need to develop a reliable and high-fidelity simulation capability for Zircaloy clad ballooning study in the context of nuclear power plant life extension.

It is accepted that excessive creep deformation is the dominant mechanism for Zr clad ballooning. Not only would the thermal creep rate depend on temperature and stress, but also on Zr phase transformation between α and β phases which is related to the clad heating or cooling rate. In the α-phase, Zr clad exhibits anisotropic creep deformation as compared to the isotropic creep behavior in the β-phase. Several well-known LWR fuel rod transient analysis codes like FRAPTRAN [3] and BISON [4] have been extended or taken advantage of for Zr clad LOCA behavior simulation. The multidimensional finite element method adopted in BISON makes it suited for high-fidelity analysis of Zr clad LOCA behavior. Kim et al. [5] also adopted finite element method to study the ballooning behavior of Zircaloy clad in their axisymmetric code MERCURY, which reads pre-LOCA irradiation results from FRAPCON. Goldberg et al. [6] developed a finite element code DIONISIO to simulate Zircaloy clad ballooning. It is capable of both axisymmetric and 3D geometries. On the other hand, much less work has been reported about the implementations of anisotropic creep model in nuclear fuel performance codes.

This paper is intended to develop a simulation capability for the simulation of Zr clad ballooning behavior during LOCAs. First, an algorithm to update stress and consistent tangential stiffness was derived, taking into account of anisotropic high-temperature creep of the α-Zr clad based on the Hill criterion. Then the algorithm was implemented in the FEM software ABAQUS via its user-defined material interface (UMAT). In addition, the α-β phase transformation is included as is required by the creep model. The simulation capability is benchmarked against the literature data of PUZRY bursting tests. The role of anisotropic coefficients in ballooning calculation is also discussed.

2 Material Properties and Behavior Models

The phase transformation model, which describes the fraction of β-phase (body-centered cubic crystal structure) Zr under constant heating or cooling rate, is taken from [7]. The β-phase fraction is required as the creep rate across the α-phase, α-β mixed phase and β-phase regions varies significantly. The empirical burst limits of Zr-4 clad, i.e., the burst strain limit of FRAPTRAN [8] and the burst stress limit of Rosinger [4] are included in this study as the stop criteria for the calculations. The basic thermo-mechanical properties including thermal conductivity, specific heat, Young's modulus, Poisson ratio and transversely isotropic thermal expansion coefficients are all taken from the MATPRO manual [9].

There is significant difference in the magnitude of thermal creep rate between normal operation (<700 K) and LOCA (>900 K). A linear interpolation is applied to the region of 700–900 K, following the routine in BISON [4]. For Zr-4, Table 1 lists three widely-used high-temperature creep models, which are referred to as the Rosinger model [10], the Donaldson model [11] and the Kaddour model [12] respectively in this paper. It should be emphasized that if the creep rate constants in a creep model are originally based on uniaxial loading test data, they need to be properly adjusted in terms of the anisotropic constants before a simulation get started.

Table 1. High-temperature creep models in public literature

Creep model of Zr-4	Expression
Rosinger model	$\dot{\varepsilon}_e^c = A_R \cdot \sigma_e^{n_R} \cdot \exp\left(-\frac{Q_R}{RT}\right)$
Donaldson model	$\dot{\varepsilon}_e^c = A_D \cdot \frac{G}{T}\left(\frac{\sigma_e}{G}\right)^{n_D} \cdot \exp\left(-\frac{Q_D}{RT}\right)$
Kaddour model	$\dot{\varepsilon}_e^c = \frac{A_K}{T} \cdot \sigma_e^{n_K} \cdot \exp\left(-\frac{Q_K}{RT}\right)$

The high-temperature creep model is coupled with the phase transformation model via the creep rate parameters (A, n, Q) which are interpolated between 0 and 1 according to the value of β-phase fraction. In this paper, the conventional phase-mixing approach which makes linear interpolation of $\ln(A)$, n and Q between 100%α and (50%α + 50%β), and between (50%α + 50%β) and 100%β is adopted. Masshin [13] also developed another approach, i.e., the two-phase Ashby-Verrall method in place of the conventional phase-mixing method, by which a fairly good agreement is reached between the experimental data by Kaddour [12] and predictions for the mixed-phase region when using the Kaddour creep model for pure phases. However, the Rosinger creep model is utilized in this paper due to the simplicity of the phase-mixing approach in the mixed phase region.

The simulation would be terminated if either the maximum hoop stress or maximum hoop strain exceeds the burst limit, e.g., FRAPTRAN overstrain burst model [8] and the overstress burst limit model by Erbacher et al. [14].

3 Implicit Integration Algorithm of Creep

While the elastic deformation of Zr clad is taken to be isotropic, the high-temperature creep deformation of α-phase Zr (typically below 850 °C) is anisotropic which can be described with the Hill criterion [15]. The numerical integration algorithm has been inspired by the work of Autay et al. [16] for plastic deformation simulation. Displayed equations are centered and set on a separate line. The yield function is defined as

$$\varphi = \left(\boldsymbol{\sigma}^T \boldsymbol{P} \boldsymbol{\sigma}\right)^{\frac{1}{2}} \tag{1}$$

where $\boldsymbol{\sigma}$ is the Cauchy stress tensor (in a six-component vector form for numerical convenience), $\boldsymbol{P}$ is a symmetric matrix which for 3D general loading case is given by

$$\boldsymbol{P} = \frac{2}{3}\begin{bmatrix} H+G & -H & -G & 0 & 0 & 0 \\ -H & H+F & -F & 0 & 0 & 0 \\ -G & -F & F+G & 0 & 0 & 0 \\ 0 & 0 & 0 & 2N & 0 & 0 \\ 0 & 0 & 0 & 0 & 2M & 0 \\ 0 & 0 & 0 & 0 & 0 & 2L \end{bmatrix} \tag{2}$$

Hill's effective stress can be related to the yield function as follows

$$\sigma_e = \sqrt{1.5}\varphi \tag{3}$$

It can be checked that when the anisotropic constants are set as $F = G = H = 0.5$ and $L = M = N = 1.5$, Eq. (3) is just the definition of the von Mises stress. The concept of creep potential function $W(\sigma_{eq})$ [17] is employed such that

$$\dot{\varepsilon}_e^c = \frac{\partial W(\sigma_e)}{\partial \sigma_e} \tag{4}$$

where $\dot{\varepsilon}_e^c$ is the effective creep strain rate given in Table 1. Creep strain component is obtained according to the flow rule

$$\dot{\boldsymbol{\varepsilon}}^c = \frac{\partial W(\sigma_e)}{\partial \boldsymbol{\sigma}} = \sqrt{1.5}\dot{\varepsilon}_e^c \frac{\boldsymbol{P}\boldsymbol{\sigma}}{\varphi} = \sqrt{1.5}\dot{\varepsilon}_e^c \boldsymbol{n} \tag{5}$$

where $\dot{\boldsymbol{\varepsilon}}^c$ is the creep strain rate tensor and $\boldsymbol{n}$ is the flow direction vector. The numerical integration is composed of two steps, i.e., local iteration and consistent tangent stiffness.

For the local iteration, the equality to be met is as follows

$$f\left(\Delta\varepsilon_e^c\right) = \dot{\varepsilon}_e^c \Delta t - \Delta\varepsilon_e^c \tag{6}$$

where $\Delta\varepsilon_e^c$ is the effective creep strain increment, Δt is the time step size and $f(\Delta\varepsilon_e^c)$ represents the residual. Newton-Raphson iteration is employed to find the root of $f(\Delta\varepsilon_e^c) = 0$, which requires the derivative to be known, i.e.,

$$f'\left(\Delta\varepsilon_e^c\right) = \left(\sqrt{1.5}\frac{\partial \dot{\varepsilon}_e^c}{\partial \sigma_e}\Delta t\right)\frac{\partial \varphi}{\partial \Delta\varepsilon_e^c} - 1 = \left(\sqrt{1.5}\frac{\partial \dot{\varepsilon}_e^c}{\partial \sigma_e}\Delta t\right)\left(\boldsymbol{n}\frac{\partial \boldsymbol{\sigma}}{\partial \Delta\varepsilon_e^c}\right) - 1 \tag{7}$$

Therefore, the key is to further expand Eq. (7). To this end, the stress tensor at the end of time step is written as

$$\boldsymbol{\sigma} = \boldsymbol{\sigma}_t + \boldsymbol{D}^e\left(\Delta\boldsymbol{\varepsilon} - \Delta\boldsymbol{\varepsilon}^c\right) = \boldsymbol{\sigma}^{tr} - \sqrt{1.5}\Delta\varepsilon_e^c\boldsymbol{D}^e\boldsymbol{n} \tag{8}$$

so that

$$\boldsymbol{\sigma} = \boldsymbol{I}_c^{-1}\boldsymbol{\sigma}^{tr} \tag{9}$$

where

$$\boldsymbol{I}_c = \boldsymbol{I} + \sqrt{1.5}u\boldsymbol{D}^e\boldsymbol{P} \tag{10}$$

$$u = \frac{\Delta\varepsilon_e^c}{\varphi} \tag{11}$$

$$\boldsymbol{\sigma}^{tr} = \boldsymbol{\sigma}_t + \boldsymbol{D}^e\Delta\boldsymbol{\varepsilon} \tag{12}$$

with $\boldsymbol{I}$ representing the identity matrix, $\boldsymbol{D}^e$ the elasticity matrix, $\Delta\boldsymbol{\varepsilon}$ the total mechanical strain tensor, $\Delta\boldsymbol{\varepsilon}^c$ the creep strain tensor, $\boldsymbol{\sigma}^{tr}$ the trial stress tensor, and $\boldsymbol{\sigma}_t$ the stress tensor at the beginning of time step. Therefore, the partial derivative of the stress tensor to the effective creep strain increment can be expanded as

$$\frac{\partial\boldsymbol{\sigma}}{\partial\Delta\varepsilon_e^c} = -\boldsymbol{I}_c^{-1}\frac{\partial\boldsymbol{I}_c}{\partial\Delta\varepsilon_e^c}\boldsymbol{I}_c^{-1}\boldsymbol{\sigma}^{tr} = -\sqrt{1.5}\frac{\boldsymbol{I}_c^{-1}\boldsymbol{D}^e\boldsymbol{P}\boldsymbol{I}_c^{-1}\boldsymbol{\sigma}^{tr}}{\varphi}\left(1 - u\frac{\partial\varphi}{\partial\Delta\varepsilon_e^c}\right) \tag{13}$$

which in combination with Eq. (1) gives

$$\frac{\partial\varphi}{\partial\Delta\varepsilon_e^c} = \frac{\sqrt{1.5}\boldsymbol{n}\boldsymbol{I}_c^{-1}\boldsymbol{D}^e\boldsymbol{n}}{\sqrt{1.5}u\boldsymbol{n}\boldsymbol{I}_c^{-1}\boldsymbol{D}^e\boldsymbol{n} - 1} \tag{14}$$

It can also be proved that Eq. (13) returns to -3μ which accords with the von Mises case, when the anisotropic constants are given by $F = G = H = 0.5$ and $L = M = N = 1.5$, where μ is the shear modulus. The stress update is also completed according to Eq. (9), if convergence of the above local iterations (Eq. (6)) is reached after a given number of iterations.

For the consistent tangent stiffness or material Jacobian matrix, as is required by implicit time integration to allow larger time-step size, it is defined as

$$\boldsymbol{J} = \frac{\partial\Delta\boldsymbol{\sigma}}{\partial\Delta\boldsymbol{\varepsilon}} \tag{15}$$

Starting from the differential incremental form of Eq. (8), i.e.,

$$d(\Delta\boldsymbol{\sigma}) = d\boldsymbol{\sigma} = \boldsymbol{D}^e\left[d(\Delta\boldsymbol{\varepsilon}) - d\left(\Delta\boldsymbol{\varepsilon}^c\right)\right] \tag{16}$$

plus Eq. (5)

$$d\left(\Delta\boldsymbol{\varepsilon}^c\right) = \sqrt{1.5}\boldsymbol{P}\left[ud\boldsymbol{\sigma} + \boldsymbol{\sigma}u_{\Delta\varepsilon_e^c}d\left(\Delta\varepsilon_e^c\right)\right] \tag{17}$$

with

$$u_{\Delta\varepsilon_e^c} = \frac{\partial}{\partial\left(\Delta\varepsilon_e^c\right)}\left(\frac{\Delta\varepsilon_e^c}{\varphi}\right) = \frac{\varphi - \Delta\varepsilon_e^c\varphi_{\Delta\varepsilon_e^c}}{\varphi^2} \tag{18}$$

leads to another form of Eq. (16), i.e.,

$$d\boldsymbol{\sigma} = \boldsymbol{D}^* d(\Delta\boldsymbol{\varepsilon}) - \boldsymbol{D}^*\boldsymbol{P}\boldsymbol{x}d\left(\Delta\varepsilon_e^c\right) \tag{19}$$

where

$$\boldsymbol{D}^* = \boldsymbol{I}_c^{-1}\boldsymbol{D} \tag{20}$$

$$\boldsymbol{x_c} = \sqrt{1.5}u_{\Delta\varepsilon_e^c}\boldsymbol{\sigma} \tag{21}$$

Differentiation of Eq. (4) leads to the following relationship

$$d\left(\Delta\varepsilon_e^c\right) = \sqrt{1.5}\frac{\partial\dot{\varepsilon}_e^c}{\partial\sigma_e}\frac{\boldsymbol{\sigma}^T\boldsymbol{P}d\boldsymbol{\sigma}}{\varphi} = \sqrt{1.5}\frac{\partial\dot{\varepsilon}_e^c}{\partial\sigma_e}\boldsymbol{n}^T d\boldsymbol{\sigma} \tag{22}$$

which can be further simplified to

$$d\left(\Delta\varepsilon_e^c\right) = \frac{1}{\frac{\sqrt{\frac{2}{3}}}{\frac{\partial\dot{\varepsilon}_e^c}{\partial\sigma_e}} + \boldsymbol{n}^T\boldsymbol{D}^*\boldsymbol{P}\boldsymbol{x_c}}\boldsymbol{n}^T\boldsymbol{D}^*d(\Delta\boldsymbol{\varepsilon}) \tag{23}$$

after Eq. (19) is plugged into Eq. (22).

Therefore, the consistent tangent stiffness matrix is obtained as

$$\boldsymbol{J} = \boldsymbol{D}^*\left(\boldsymbol{I} - \frac{\boldsymbol{P}\boldsymbol{x_c}\boldsymbol{n}^T\boldsymbol{D}^*}{q_c}\right) \tag{24}$$

with

$$q_c = \frac{\sqrt{\frac{2}{3}}}{\frac{\partial\dot{\varepsilon}_e^c}{\partial\sigma_e}} + \boldsymbol{n}^T\boldsymbol{D}^*\boldsymbol{P}\boldsymbol{x_c} \tag{25}$$

Finally, Eqs. (1)–(25) are numerically implemented through the UMAT interface of ABAQUS.

4 Simulations and Discussion

The PUZRY test series [18] were conducted for the purpose of studying the ballooning and burst of Zr-4 clad tube subject to inner pressure transients at elevated temperatures. The data of the temperature, deformation (tangential and axial strain) and pressure at burst were published. The 50 mm long specimens were fresh (unirradiated and unoxidized) Zr-4 tubes, with inner and outer diameters of 9.3 and 10.75 mm respectively. The schematic drawing of the tube specimen is reported in Fig. 1. The range of experimental constant temperature is 698–1201 °C that covers α, mixed and β phase regions, and the constant pressurization rate is between about $7\cdot10^{-4}$ and $2.6\cdot10^{-2}$ MPa/s.

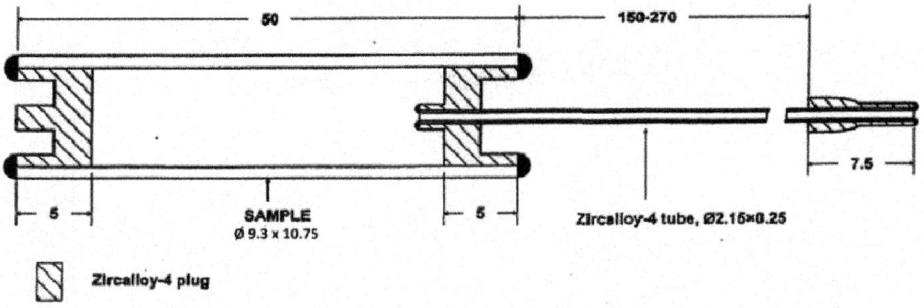

Fig. 1. Drawing of the PUZRY Zr-4 clad tube ballooning and burst tests.

16 of the PUZRY tests were simulated in this paper. Table 2 summarizes experimental conditions for the PUZRY cases simulated here. Axisymmetric geometry is taken advantage of in order to reduce the computational time. The simulation setup is given in Fig. 2. The plug is treated as rigid body. Argon gas pressure is applied to the boundary as indicated by an arrow. For the pure β phase, the anisotropic constants were set as $F = G = H = 0.5$ and $L = M = N = 1.5$. As for the pure α phase, the anisotropic constants were taken from [19], i.e., $F = 0.95$, $G = 0.304$, $H = 0.24$ and $L = M = N = 1.5$ for this work.

Table 2. PUZRY test cases selected for comparison

Test case number	Temperature (°C)	Pressure ramp rate (MPa/s)	Phase region
From 1 to 4, 8–12	1001.0–1201.3	$6.2 \cdot 10^{-4}$-$7.63 \cdot 10^{-3}$	100%β
17, 18, 20	850.1, 900.2, 849.7	$1.162 \cdot 10^{-2}$, $1.151 \cdot 10^{-2}$, $2.25 \cdot 10^{-2}$	80%α + 20%β
14, 16, 26, 30	702.2, 750.3, 698.4, 800.4	$1.19 \cdot 10^{-2}$, $1.224 \cdot 10^{-2}$, $1.193 \cdot 10^{-2}$, $2.63 \cdot 10^{-2}$	100%α

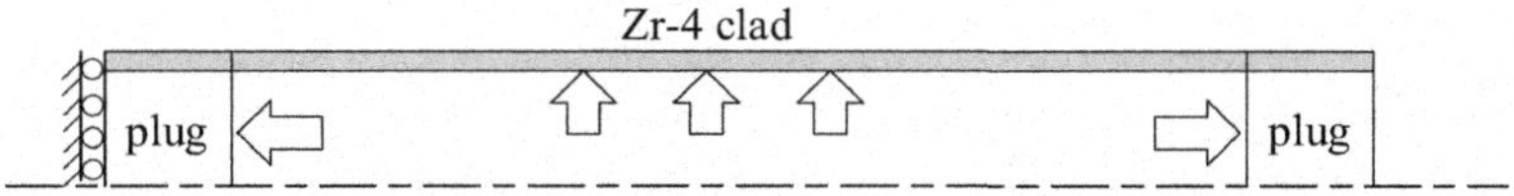

Fig. 2. Simulation setup.

Figure 3(a) plots the hoop strain at burst of all 31 cases of the PUZRY against the burst strain model. The comparison is not satisfactory in the mixed and β phase. Figure 3(b) shows that the Rosinger best-estimate overstress model conforms to test data in the mixed phase region, and the Rosinger lower bound overstress model predicts closer results in the β phase region. It is noted that the burst stress limit seems to be dependent on the

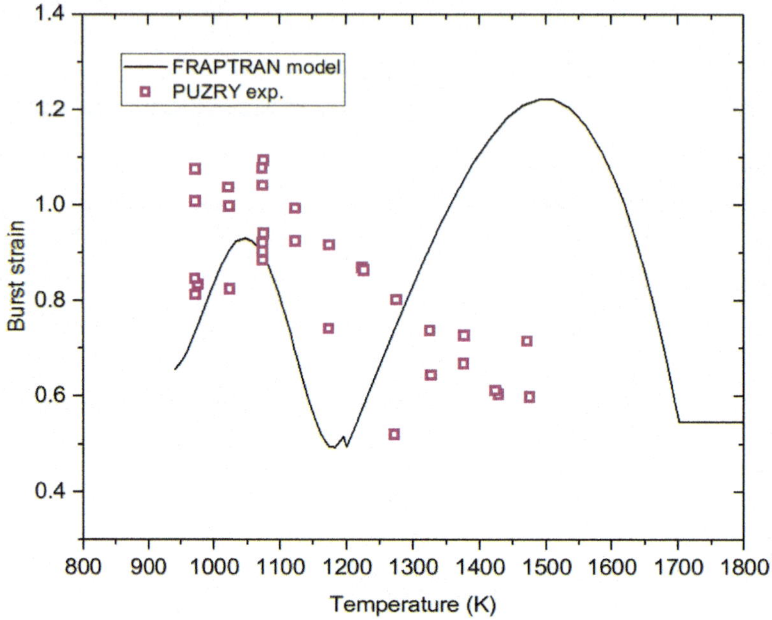

(a)

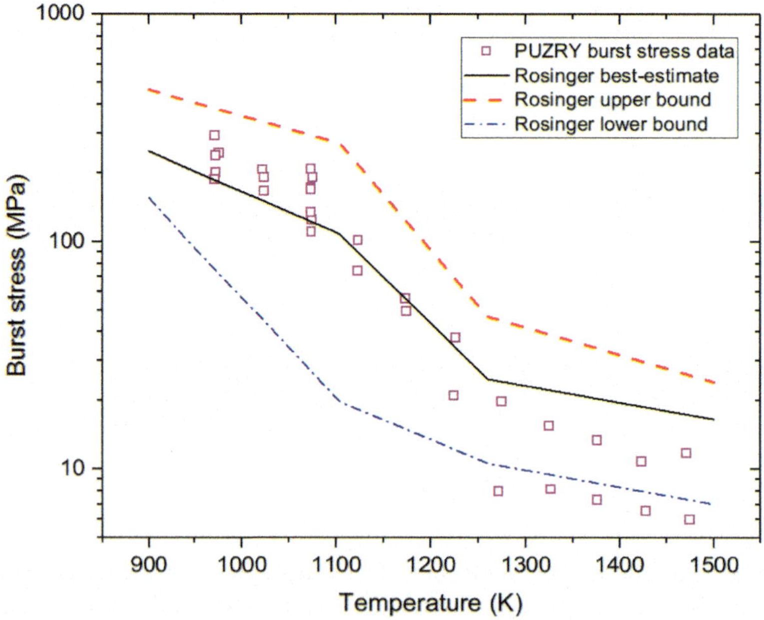

(b)

Fig. 3. Comparison of the PUZRY test data with the FRAPTRAN burst strain model (a) and Rosinger burst stress models (b).

pressurization rate, where higher pressurization rate leads to higher burst stress limit. Here the test data of burst hoop stress is calculated according to [14]. Thus we used the combination of the FRAPTRAN burst strain model, the Rosinger best-estimate burst stress model and the Rosinger lower bound model in the α, mixed and β phase regions respectively.

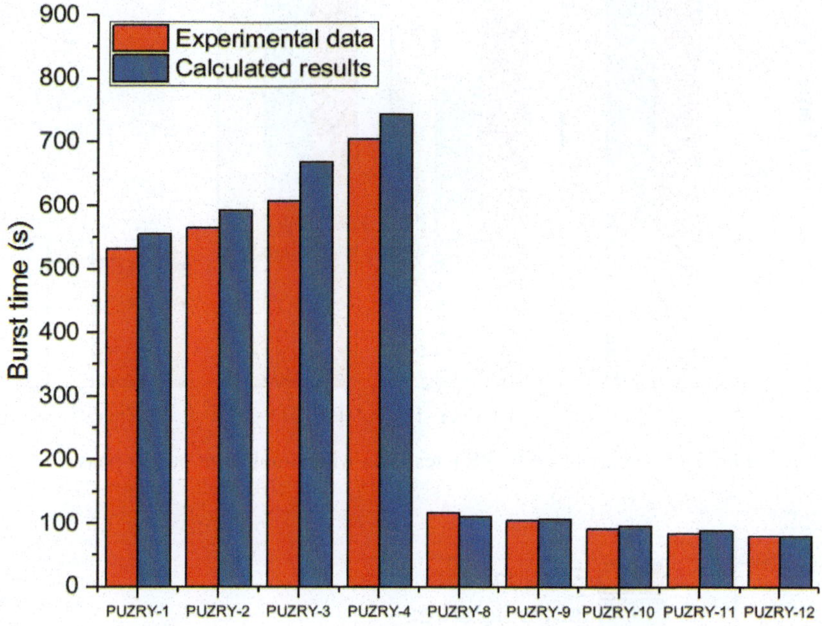

Fig. 4. Comparison of PUZRY test data with simulation: pure β phase

The results of the simulation and test result comparison are summarized in Figs. 4, 5 and 6. All the simulate cases were terminated due to overstrain. Although the burst limit model is not consistent with the PUZRY data of the pure β phase, the calculated burst time was fairly close to the experimental burst time, which is reasonable considering the high rate of deformation just before burst occurred. It is also demonstrated that the Rosinger creep model is sufficiently accurate in the pure β phase. For the pure α phase, the anisotropic creep algorithm shows better agreement with the experimental data than the isotropic creep algorithm does. The isotropic creep algorithm overpredicts the creep rate leading to shorter time to burst. As noted above, the creep rate model is typically based on uniaxial stress test data. The resultant creep model can be used as it is for isotropic case, with the uniaxial stress being replaced by effective stress. For anisotropic case, however, the anisotropic constants enter into the creep model via the definition of the effective stress and thus modify the creep rate. In the mixed phase region, from Fig. 6 it seems that isotropic creep model is more appropriate. Another source of uncertainty of the creep rate calculation for the mixed phase is possibly the conventional phase mixing homogenization approach.

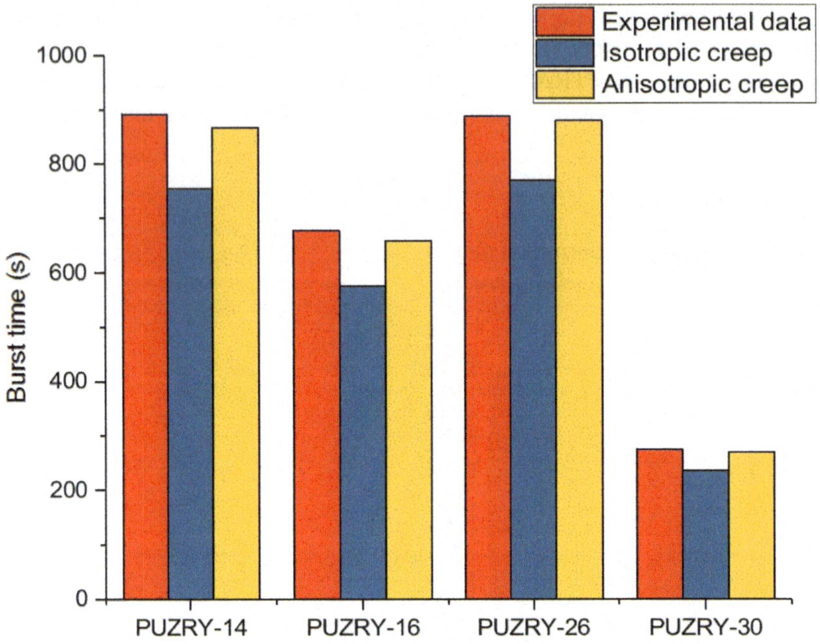

Fig. 5. Comparison of PUZRY test data with simulation: pure α phase

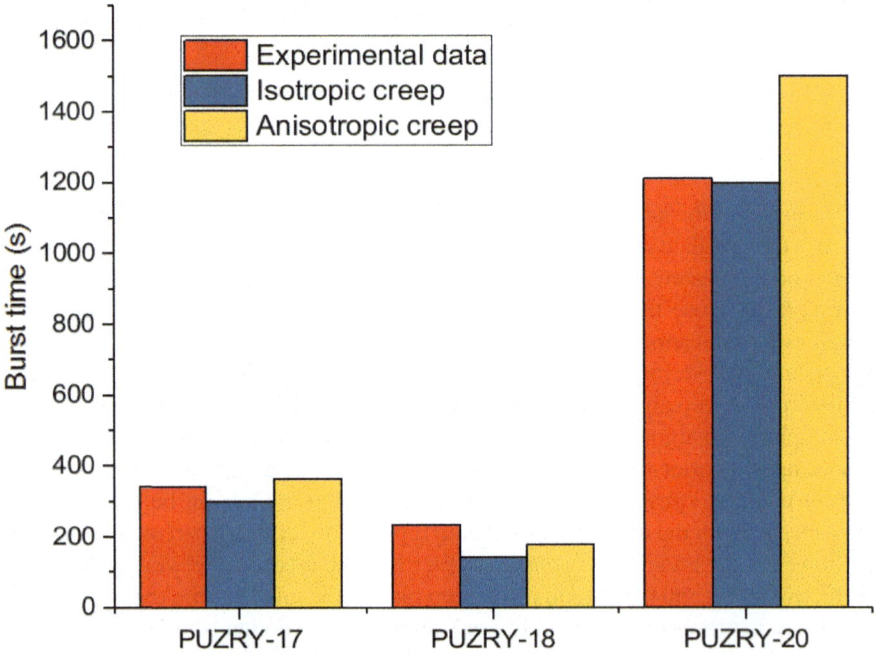

Fig. 6. Comparison of PUZRY test data with simulation: $\alpha + \beta$ mixed phase

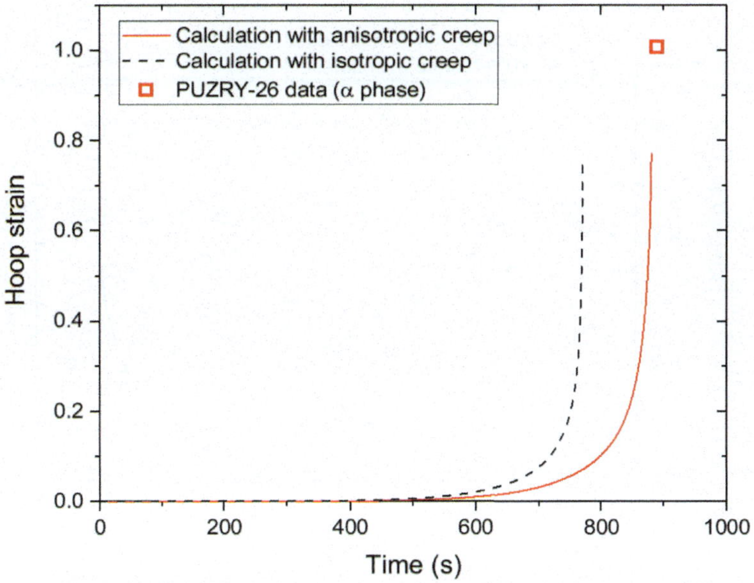

Fig. 7. The maximum hoop strain vs time for the PUZRY-26 test case

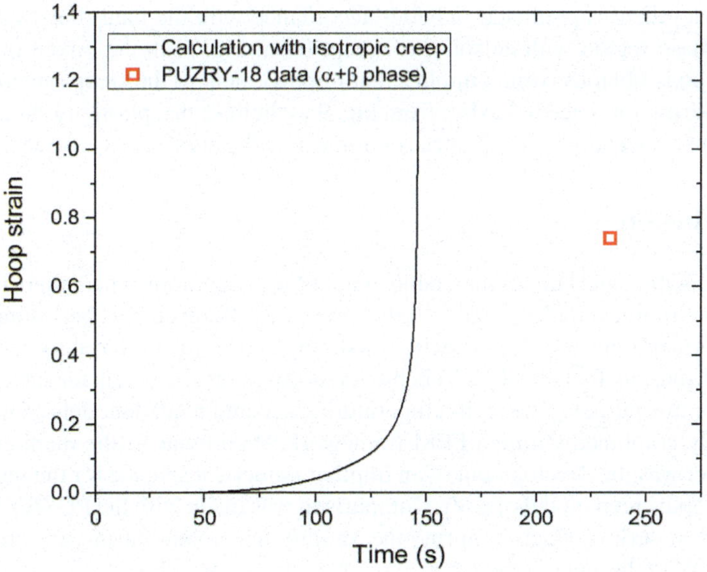

Fig. 8. The maximum hoop strain vs time for the PUZRY-18 test case

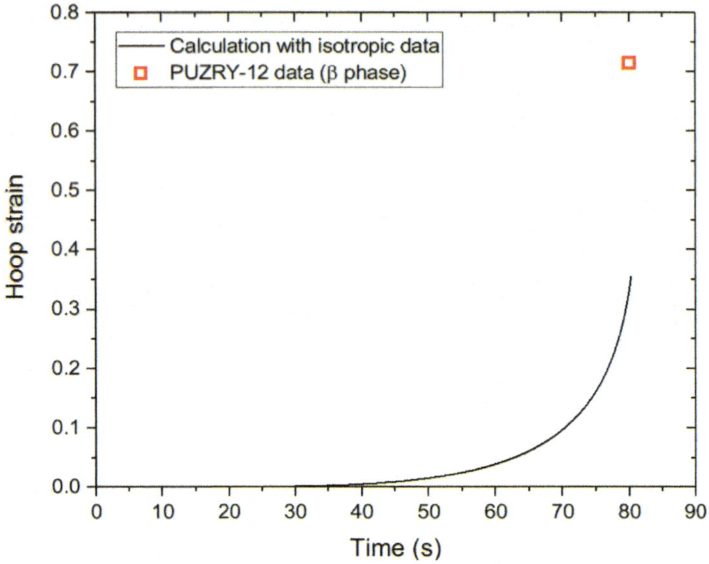

Fig. 9. The maximum hoop strain vs time for the PUZRY-12 test case

Figures 7, 8 and 9 illustrate the time history of the maximum hoop strain. The hoop strain development is predicted in better accordance with the final (burst) strain of the test in α phase region with anisotropic creep. From Fig. 8, in the mixed phase region the deviation is obvious, which implies that improvement to the creep rate model in the mixed phase region is needed. Also, from Fig. 9 we believe that plasticity should be taken into account, especially for the β phase region where the burst stress is often rather small.

5 Conclusion

Zr clad ballooning and burst is inevitable under LOCA condition where there is high temperature environment and internal-external pressure difference. It is becoming more and more important to develop high-fidelity modeling and simulation tool for the investigation of high-burnup fuel clad LOCA behavior as this is related to reactor safety reconsideration. For this purpose, the high-temperature creep and related models of Zr clad were successfully combined within a FEM framework. Preliminarily, the main contribution of this paper was the development of an implicit numerical scheme for the integration of the anisotropic creep model of α-Zr. Comparison was made with the PUZRY ballooning and burst test series, which confirms the validity and robust of the anisotropic creep algorithm. With the anisotropic creep algorithm, the predicted progressive ballooning of the pure phases agrees fairly well with test data. However, more mechanistic creep model instead of interpolation between phases is required for the mixed phase. Besides, the anisotropic constants are actually dependent on other parameters such as temperature. Additionally, plasticity should also be included in the permanent mechanical strain to make it more realistic especially for transient heating or pressurization condition. These effects will be considered in subsequent work.

References

1. Sonnenburg, H., et al. "Report on fuel fragmentation, relocation, and dispersal." Nuclear Energy Agency Committee on Safety of Nuclear Installations (2016).
2. Ozawa, Masaaki, et al. "Behavior of irradiated Zircaloy-4 fuel cladding under simulated LOCA conditions." ASTM Special Technical Publication 1354 (2000): 279–302.
3. Govers, K., and M. Verwerft. "Simulation of ballooning and relocation in the Halden LOCA tests with FRAPTRAN." EHPG meeting. 2014.
4. Hales, J. D., et al. BISON theory manual the equations behind nuclear fuel analysis. No. INL/EXT-13–29930. Idaho National Lab.(INL), Idaho Falls, ID (United States), 2016.
5. Kim, Hyochan, et al. "Development of MERCURY for simulation of multidimensional fuel behavior for LOCA condition." Nuclear Engineering and Design 369 (2020): 110853.
6. Lemes, Martín, Alicia Denis, and Alejandro Soba. "Simulation of nuclear fuel behavior in accident conditions with the DIONISIO code." Journal of Nuclear Engineering and Radiation Science 5.2 (2019).
7. Massih, Ali R. "Transformation kinetics of zirconium alloys under non-isothermal conditions." Journal of nuclear materials 384.3 (2009): 330–335.
8. Geelhood, K. J., et al. "FRAPTRAN 1.4: a computer code for the transient analysis of oxide fuel rods." US Nuclear Regulatory Commission, Office of Nuclear Regulatory Research, NUREG/CR-7023 1 (2011).
9. Allison, C. M., et al. "SCDAP/RELAP5/MOD3. 1 code manual, volume IV: MATPRO–A library of materials properties for light-water-reactor accident analysis." DT Hagrman, NUREG/CR-6150, EGG-2720 4 (1993): 4–234.
10. Rosinger, H. E. "A model to predict the failure of Zircaloy-4 fuel sheathing during postulated LOCA conditions." Journal of Nuclear Materials 120.1 (1984): 41–54.
11. Donaldson, A. T., and T. Healey. Creep deformation of Westinghouse zircaloy-4 fuel cladding tubes in the alpha plus beta phase temperature range. No. CEGB-TPRD/B--0564/N85. Central Electricity Generating Board, 1984.
12. Kaddour, Djillali, et al. "Experimental determination of creep properties of zirconium alloys together with phase transformation." Scripta materialia 51.6 (2004): 515–519.
13. Massih, A. R., and Lars O. Jernkvist. "Solid state phase transformation kinetics in Zr-base alloys." Scientific Reports 11.1 (2021): 1–16.
14. Erbacher, F. J., et al. Burst criterion of Zircaloy fuel claddings in a loss-of-coolant accident. ASTM International, 1982.
15. Hill, Rodney. "A theory of the yielding and plastic flow of anisotropic metals." Proceedings of the Royal Society of London. Series A. Mathematical and Physical Sciences 193.1033 (1948): 281–297.
16. Autay, R., et al. "Numerical implementation of coupled anisotropic plasticity-ductile damage in sheet metal forming process." Journal of Mechanics 34.4 (2018): 417–430.
17. Dunne, Fionn, and Nik Petrinic. Introduction to computational plasticity. OUP Oxford, 2005.
18. Perez-Fero, E., et al. "IFPE/AEKI-EDB-E110, Experimental Database of E110 Claddings under Accident Conditions." (2008).
19. Hunt, C. E. L. "Anisotopic Theory and the Measurement and Use of the Anisotropic Factors for Zircaloy-4 Fuel Sheaths." (1975).

Updates to the IAEA Guide on Fuel Reliability and Performance

Jinzhao Zhang[1(✉)], Nicolas Waeckel[2], and Ki Seob Sim[3]

[1] Tractebel (ENGIE), Brussels, Belgium
jinzhao.zhang@tractebel.engie.com
[2] EDF (Retired), Lyon, France
nicolas@waeckel.fr
[3] IAEA, Vienna, Austria
k.s.sim@iaea.org

Abstract. The IAEA Nuclear Energy Series (NES) Guide on the quality and reliability aspects in nuclear power reactor fuel engineering (NF-G-2.1) was published in 2015. The guide was intended to provide an overview of the interconnected issues of design, manufacturing and operation in fuel quality and reliability assurance. After the Fukushima Daiichi accident in 2011, the IAEA Safety Standards on the design of the reactor core has been fully revised as a new specific safety guide (SSG-52) which has been published in 2019. Therefore, the IAEA Guide NF-G-2.1 has been updated to reflect state of the art information on fuel reliability and performance issues, to maintain consistency with the recommendations stated in the new IAEA Safety Standards on the design of the reactor core (SSG-52) and to apply to newly developed advanced technology or accident tolerant fuels (ATFs) for light water reactors (LWRs) or fuels for LWR-based small modular reactors (SMRs). This revision provides comprehensive recommendations on fuel design changes, fuel manufacturing, qualification, in-reactor operation, and on-site services, with the objective to assist the IAEA member states to achieve excellence in fuel reliability and performance. This paper presents the major updates to the IAEA Guide NF-G-2.1 Revision 1.

Keywords: Fuel performance · Fuel reliability · IAEA Guide · Margin management · Quality management

1 Introduction

The IAEA Nuclear Energy Series (NES) Guide on the quality and reliability aspects in nuclear power reactor fuel engineering (NF-G-2.1) was published in 2015 [1]. Since then, the nuclear fuel designers, manufacturers and utilities have made significant improvements to the fuel design and materials, manufacturing techniques, and fuel operation, to ensure fuel reliability and performance under more challenging and flexible operational conditions, and with longer fuel cycles and higher burnups.

After the Fukushima Daiichi accident in 2011, the IAEA Safety Standards on the design of the reactor core has been fully revised as a new specific safety guide (SSG-52)

J. Liu and Y. Jiao (Eds.): WRFPM 2023, SPPHY 299, pp. 56–67, 2024.
https://doi.org/10.1007/978-981-99-7157-2_6

which has been published in 2019 [2]. At the same period, accident tolerant or advanced technology fuels (ATFs) have been developed which are being qualified and licensed for implementation in the existing nuclear power plants [3]. More recently, small modular reactors (SMR) are being developed [4], and the adaptability of the current fuels and the ATFs to LWR-based SMRs is being considered.

Therefore, the IAEA Guide NF-G-2.1 has been updated to be consistent with the SSG-52 recommendations and to reflect these progresses. This paper presents the major updates to the IAEA Guide NF-G-2.1 Revision 1, covering the following aspects:

- Review of in-reactor performance issues impacting fuel reliability and proposed mitigation measures (Sect. 2);
- Guidance on the fuel design changes, quantification and management of fuel design and operating margins, fuel design and safety evaluation, fuel qualification and licensing (Sect. 3);
- Good practices to improve fuel reliability during operation (Sect. 4).

2 In-Reactor Fuel Performance Issues and Mitigation Measures

The nuclear fuel present in the reactor core is regularly renewed but, opposite to the early days of the nuclear era, it can now stay in the reactor up to 6 years, experiencing aggressive irradiation conditions. As a result, it is crucial to know as accurately as possible how irradiation impacts the overall fuel rod behavior under operational states (i.e., normal operation and anticipated operational occurrences or AOOs), as well as the long-term storage conditions.

The main fuel performance issues and the corresponding mitigation measures are summarized in NF-G-2.1 Revision 1. The major fuel degradation or failures in water cooled reactors include [5]:

- *Manufacturing defects* may lead to unreliable in-reactor fuel behaviors, e.g., internal hydriding of cladding due to undesired moisture in the fuel pellets; pollution of the end plugs welds leading to through the wall cracks and leaks; grid springs stress corrosion cracking leading to debris fretting and leakage; fuel cladding shavings resulting from fuel rod insertion in the fuel assembly skeleton leading to fretting and leakage, etc..
- Excessive *corrosion/hydriding* of the Zr alloy cladding can lead to its embrittlement and to a potential risk of failure during in-reactor transients, either due to excessive power or to lack of cooling, or during storage/shipment conditions.
- Fuel assembly or fuel rod induced vibration causing *grid to rod fretting* (GTRF) that leads to cladding wear and to a potential risk of local loss of integrity of the fuel rod during normal operation.
- *Stress corrosion cracking* (SCC) can affect various structural components and can lead to their failure, eventually hamper the fuel assembly to comply with its functional requirements.
- Foreign *debris fretting* trapped in the grid cells, and endogenous *debris fretting* due to material coming from the fuel assembly itself.
- In-reactor *dimensional changes* of Zr alloy components.
- *Mixed cores* including various fuel designs features require specific care to avoid damageable incompatibilities between fuel assemblies particularly in the case of un-channeled fuel designs.

- *Excessive fuel rod internal pressure*, due to the fission gas releases during irradiation (and fuel rod free volume reduction), can lead to a pellet-clad gap re-opening but experience has shown it is not an in-reactor issue within design and operating envelopes, but can become a concern in SFP or in dry cask storage conditions if the cladding is highly corroded and the hydrides radially oriented.
- *Pellet-Cladding Interaction* assisted by *stress corrosion cracking* (PCI-SCC) can generate local cladding failures in case of excessive power transients. There are 3 parameters that need to occur simultaneously to induce PCI-SCC: high enough cladding tensile stresses (induced by the power ramp), proper concentration of corrosive fission products and sufficient time for the first 2 parameters to initiate and propagate a through the wall crack [6]. Unless the cladding is highly corroded/hydrided with radially oriented hydrides and subjected to a high enough stress field, *Pellet-Cladding Mechanical Interaction* (PCMI) results rarely in cladding failure.

Once the fuel is leaking, it is important from a safety and operational/economical point of view that the primary failures do not degenerate into secondary failures such as:

- *Axial propagation of PCI-SCC cracks* (this phenomenon is unlikely in PWRs but can happen in BWRs),
- *Cladding secondary hydriding* as a consequence of a primary defect: a highly corroded/hydrided zone appears a few meters above or below the primary leak and can result in a guillotine type of fuel rod failure and fuel pellets fragments loss,
- *Fuel-coolant interaction* between water sensitive fuel pellets and coolant: among the proposed ATF concepts, uranium nitride for instance can react with water.

To ensure that the fuel behave adequately during normal and off-normal conditions, the following good practices should be followed:

- To develop appropriate tools (i.e., relevant experimental protocols and validated calculation codes) to verify the various fuel designs proposed by the vendors and to select the best ones,
- To review carefully international and domestic in-reactor experience of each type of fuel design,
- To implement any fuel design change, or any change in the plant system (i.e., power uprates, modifications in coolant flow and temperatures, or water chemistry), under a well-conceived surveillance program. If the changes are significant the relevance of irradiating lead test rods (LTRs) or lead test assemblies (LTAs) need to be considered.
- To consider the use of different types of fuel designs to balance the risk to experience a detrimental "common failure mode effect" versus the risk to experience a deleterious unexpected incompatibility between fuel designs.
- To use appropriate poolside examination techniques to measure cladding corrosion evolution, fuel assembly and fuel bundle dimensional changes, fuel assembly distortion distribution within the core, or use specific cleaning tools to re-move the crud layers.
- To perform post-irradiation examinations in the hot cells to better understand the fuel degradation or failure mechanisms [5].

3 Fuel Design Changes to Improve Reliability and Performance

3.1 Fuel Design Changes, Verification and Validation

Fuel design changes are initially assessed and categorized as major or minor changes. *Minor changes* are those that remain within the conditions of the qualified fuel design. Changes beyond this envelop are *major changes* that require further investigation to characterize potential impact of the changes.

Fuel design changes to improve reliability and performance are verified via design verification process that includes checking against operational experiences, design analyses using computer codes, out-of-reactor tests to assess thermal hydraulic characteristics, mechanical response and material properties, on-site inspections, experimental irradiation tests at material test reactors, and independent design review by experts who have not participated in the fuel design and testing.

In design analyses, although the impact or each minor change on reactor core performance may be insignificant, the accumulated impact of all minor changes should be considered all together to account for potential interactions with interfacing systems (e.g., core neutronic and thermal hydraulic designs, reactor core components, reactor control systems). The impact of mixed cores on fuel reliability and performance requires specific attention [7].

The fuel design changes are validated via a design qualification process that includes lead test rods, lead test assemblies and/or demonstration irradiation campaigns. The demonstration irradiation can be initiated with a small number of fuel rods or assemblies with design changes that are loaded at specific locations in the core while considering the thermal hydraulic, neutronic and/or thermal mechanical aspects. The demonstration irradiation continues with gradually increased number of fuel assemblies with design changes to form a transition core to a full core. The operating experiences can be gathered from other nuclear power plants (NPPs) with similar characteristics such as to reduce the qualification process.

3.2 Quantification and Management of Fuel Design and Operating Margins

Margin quantification and management is an important aspect in qualification of fuel design changes. For this purpose, a practical definition of margins and limits has been proposed based on the margin model of the utilities [8–11]. This margin model is described in details in the revised guide. It sub-divides the total margin into safety margin, design margin, licensing margin, analysis margin and operating margin controlled by different players.

Distinguishing the role played by the different fuel assembly components is helpful for better quantifying the available margins each stakeholder can use while performing fuel design and safety evaluations.

All fuel design changes are ensured via confirmation that sufficient margins exist to the limits of the fuel failures (i.e., safety or design margin) and to the limit of reactor operating conditions (i.e., operating margin). The quantification of these margins depends on the advances in the following fields:

- Use of best estimate calculation codes to simulate the plant conditions and core and fuel behaviors,
- Good understanding of the safety and design limits definition, based on relevant experiments and analytical analysis,
- Use of plant surveillance and on-site fuel inspection to verify the accuracy of the predictions for a set of critical physical parameters,
- Definition of operating limit and margin based on an empirical database from analysis of plant operating experience and events,
- Accurate quantification of calculation codes' uncertainties, based on a wide range of experimental data,
- Use of best estimate plus uncertainty analysis (BEPU) design methodology to remove unnecessary conservatisms and reduce uncertainties.

According to IAEA SSG-2 (Rev. 1) [12], different options of deterministic safety analyses can be chosen, with different levels of conservatism associated with the used computer codes, the assumptions on the availability of the reactor core protection and safeguards systems, and on the initial and boundary conditions used in the analysis. Historically, fuel designs and safety analysis were based on semi-empirical and conservative approaches. Following the significant improvements of the available analysis tools, the BEPU approach is more and more used to quantify margins by considering the uncertainties on operating conditions and on analysis methods such as:

- Overall code or individual modelling uncertainties (basic equation and closure laws),
- Representation (nodalization) uncertainties, numerical inadequacies,
- Uncertainties due to the scaling issues,
- User effects, computer / compiler effects,
- Input data uncertainties on the analysis of an individual event.

The BEPU approach requires high-level verification, validation and uncertainty quantification tools for the computer codes which are used and considerable effort to develop the right methodology [12]. For industry applications, a graded BEPU approach needs to be applied to quantify the margins efficiently and cost-effectively in the safety analysis for DBAs conditions [13].

A good balance between safety, design and operating margins should be considered in the fuel design and safety evaluation process. Margin management ensures that the fuel is designed and operated with adequate margins as licensed:

- The fuel designer needs to provide reliable, robust, and high-performance products with adequate design, operating and analytical margins to meet safety, flexibility, and economic requirements.
- The core designer needs to provide flexible and cost-efficient cycle-specific reload core designs that are within analyzed design limits.
- The plant operator needs to anticipate and manage the operation of the fuel within operating limits to ensure safety, reliability, and performance.

In case of insufficient design and operating margins due to fuel or plant modifications, these modifications need to be prioritized commensurate with the potential impacts on plant safety and reliability. This prioritization is typically based on cross-disciplinary reviews. The management should review and approve the prioritization list and provide adequate support. If the consequences of margin loss are acceptable, a comprehensive rationale needs to be documented to close the issue. If the consequences of margin loss are unacceptable, the following actions are required:

- Development of an action plan,
- Evaluation of the potential consequences of exceeding the limits and associated costs,
- Assessment of costs/benefits associated with margin recovery.

3.3 Good Practices for Fuel Design Change Verification

Based on operating experience and the results of relevant research and development programs, fuel design or design change verification is realized quantitatively by verification of the specific fuel design and safety limits for all applicable plant states in fuel design and safety analysis.

Definition and Determination of Fuel Design and Safety Limits. Fuel design and safety limits are defined to ensure that the nuclear fuel performs satisfactorily in the reactor core throughout its design lifetime and for all applicable plant states. Fuel suppliers use validated fuel performance calculations codes, in-pile and out-of-pile experiments and in-reactor performance data to demonstrate compliance with these limits.

Fuel design and safety limits to be verified in the design and safety analysis can be summarized as follows:

- For operational states including normal operation and AOOs, the probability of failure of the fuel cladding, resulting from DNB or from any other failure mechanisms, needs to be insignificant.
- For accident conditions including DBAs and design extension conditions (DECs), the fuel failure needs to be limited for each type of accidents to ensure a coolable geometry.

Details of fuel design and safety limits, as well as their applicability to ATF are described in Appendix I of NF-G-2.1 Revision 1.

Code Verification, Validation, and Uncertainty Quantification. To verify fuel design and safety limits, computer codes need to be used to predict neutronic, thermal hydraulic and thermal mechanical behaviors of fuel rods and fuel assemblies during all applicable plant states. These computer codes need to be verified and validated to ensure that they are suitable for the intended application and need to be approved for use. The uncertainties of the key models need to be quantified based on experimental and operational data.

The technical aspects of code development, verification and validation, uncertainty quantification are provided in NF-G-2.1 Revision 1.

Fuel Design and Safety Evaluation. Fuel design needs to address fuel damage or failure mechanisms to provide acceptable levels of key performance parameters within which such fuel damage or failures are prevented. The fuel performance for all applicable plant states needs to be evaluated to determine whether all design and safety limits are met. Fuel assembly components are reviewed as separate components as well as integrated bodies (i.e., fuel rods and fuel assemblies).

Fuel design evolutions, new operating envelope (e.g., maximum rod burnup and power), and change of material of fuel components require new fuel design and safety evaluation to verify that existing design and safety limits, computer codes, and methods used for such evaluation remain applicable for the new design, under all applicable plant states. In case the established design and safety limits do not apply, new limits need to be established based on appropriate data and theoretical considerations.

Fuel design and safety evaluation needs to account for operational experience, comparisons with experimental data, directly or through analyses (using fuel performance codes), and other information. The calculation codes need to be verified and validated, audited, and approved by the safety authorities. Appropriated provisions need to be provided to treat the mixed core effects.

The good practices for fuel design and safety evaluation are provided for the following main steps:

- Generation of neutronic design input data
- Verification of fuel design and safety limits
- Cycle specific fuel design verification and safety evaluation
- Loading pattern fuel reliability risk assessment

Details on the verification of fuel design and safety limits are described in Appendix II of NF-G-2.1 Revision 1.

3.4 Qualify Assurance, Qualification and Licensing of Fuel Design

Fuel Design Control. In the field of nuclear fuel, the closer the industrialization phase, the more stringent are the quality assurance requirements. Part of the results obtained during the fuel development process will be used later in the licensing process. For that reason, it is important to identify in each phase of the project, all the tests and studies that are needed to prove at a late stage that the proposed solutions meet the design criteria and the safety standards.

The design control should cover the following activities:

- Design input (e.g., design requirements),
- Design process,
- Control of interfaces between different disciplines/entities involved in fuel design,
- Configuration management (e.g., design or computer code version changes management),
- Analysis, Verification, Validation.

Fuel Design Qualification Process. The fuel design qualification and licensing are under the responsibility of the fuel supplier on one side and the fuel customer on the other side. It is important that the responsibilities of each party involved in the design and safety evaluations are clearly defined. The fact that a new fuel product needs to obtain an operating license from the safety authorities must be considered during the entire fuel development process.

In principle, minor design changes are usually implemented directly, while major changes require further qualification including experimental in-pile irradiations, and/or lead test fuel irradiations in commercial power reactors. After the fuel designer develops a new type of nuclear fuel or a fuel with major design changes, which was not previously implemented in a nuclear power plant, it communicates with the fuel customer to start the qualification process.

The typical qualification process in commercial power reactors should include:

- A comprehensive design documentation, including outcomes of the available validation/qualification tests,
- An early involvement of the nuclear power plant operator and the safety authority,
- A well-defined implementation strategy,
- A complete on-site inspections program (if needed, examinations in hot cells) after each cycle,
- A regular reporting of the inspection results, to the plant operator and to the safety authority, to get the authorization for reloading at the following cycle.

Irradiation campaigns in commercial power reactors, on-site inspection and, if needed, examinations in hot cells, are mandatory steps in the development process of new fuel products. They are used to validate the new designs and to quantify their performance in representative irradiation conditions. When lead test fuels are loaded in a commercial power reactor, the fuel designer should provide guidelines for the loading pattern design to minimize risk for detrimental interaction with resident fuel assemblies while providing enough bounding irradiation conditions to enable proper demonstration (e.g., positioning of lead test fuels, restriction on the power peaking factors, compatibility with existing fuels, etc.).

If necessary and approved by the operator and safety authority, the extension of campaigns can be performed beyond normal irradiation durations to provide fuel rod candidates for test reactors that allows the study of fuel behavior beyond the operating limits (e.g., ramp tests to fuel failure, accident test conditions such as LOCA and RIA tests).

The success of these irradiation campaigns ensures that future operation with full batch reloads can be performed on an industrial scale without contingency. Furthermore, as the current fuel assembly discharge burnup target is often beyond 60 GWd/t (for LWRs), the qualification process is quite long (around ten years), including irradiation time, cooling time and transportation as well as all post irradiation examinations, and tests to achieve convincing results. Also testing and measuring irradiated material or under irradiation is very costly, and requires significant investments over long periods. These time and cost constraints need to be considered in the qualification and licensing processes.

Fuel Design Licensing Process. After the selection of the fuel design, the operator communicates with the regulator or the technical safety organization to start the licensing process. The fuel designer should provide a complete set of Fuel Design Documentation, which, after the review of the operator or its engineer, is submitted for licensing review, including:

(a) Fuel Design Basis

- Design criteria (design and safety limits),

- Basis for the design criteria (degradation and failure mechanisms),
- Adequacy of the design criteria (analytical and experimental evidence),
- Design conditions (all applicable plant states).

(b) Fuel Design (Change) Description and Qualification

- Description of the fuel design or design change,
- Summary of the qualification process and results,
- Guidelines for use of the fuel (e.g., ramp rate limitation, water chemistry, etc.).

(c) Computer Code Verification, Validation, and Uncertainty Quantification

- Description of the code and models,
- Summary of the verification and validation results and the quantified model uncertainties,
- User's guide on the application domain and detailed guidelines.

(d) Fuel Design Verification and Safety Evaluation

- Topical reports with detailed description and demonstration of the verification methodology (analytical and/or experimental methods),
- Design reports with summary of the results of the verification work demonstrating that the design basis requirements have been met and reference to relevant topical reports for detailed information.

The fuel design verification and safety evaluation documents should include at least the following reports [14]:

- Fuel Compatibility and Safety Evaluation Report (covering geometrical, neutronic, thermal hydraulic, mechanical, and mixed core aspects),
- Fuel Rod Thermal Mechanical Design Report,
- Fuel Rod LOCA Safety Evaluation,
- Fuel Rod non-LOCA Safety Evaluation (if not covered in the Fuel Compatibility and Safety Evaluation Report).

The operator or the engineer should perform independent review and audit of the Fuel Design Documentation [15].

The regulator or the technical safety or support organization reviews this Fuel Design Documentation, requests for additional information, and if deemed necessary, performs audits at the fuel designer and fuel manufacturer sites before approving loading the new fuel. This fuel licensing process may take 6 months to 3 years, depending on the types of fuel design changes and the level of improvement in the codes and methods used for the safety demonstration [14].

4 Good Practices to Improve Fuel Reliability During Operation

Fuel assemblies should be designed and manufactured as appropriate to ensure:

- Fuel assemblies do not fail and release fission products during normal operation and AOOs,

- Suitable geometry of the fuel assembly is maintained during its lifetime so that the insertion of control rods is not impeded,
- Reactor core coolability is maintained during postulated accidents.

For each fuel assembly design, the fuel vendors or fuel designers should determine the design limit on the linear power generation rate (also known as bounding power history or thermal design limit) to ensure that all design criteria are met, and that their operating or economic restrictions are also considered to ensure no PCI induced failures during power ramps. For all fuel assembly designs of the reactor core, the most limiting design limit on the linear power generation rate is identified to establish the core operating limit, which must be respected during normal operation. This should also ensure that the most limiting design limit during AOOs on the linear power generation rate is respected. In addition, the cycle specific fuel design verification is performed by the core designer (either the fuel vendor or the operator) to ensure that the core loading pattern is appropriate and that the core operating limits are not exceeded. Finally, the operator should monitor the core operation to ensure that all fuel assemblies operate within the above defined core operating domain during their entire lifetime.

To ensure trouble-free performance of the fuel, it is crucial that the operator carries out:

- Design audits to ensure the safety analysis is performed according to the standard and contractual requirements. Quality assurance of the contractor is also audited.
- Fuel manufacturing audits to ensure that the qualified manufacturing processes are effectively enforced, that the verifications are performed as required, and that the quality plan is effectively applied.

Good practices aiming at trouble-free fuel operation have been consolidated into guidelines by industrial organizations such as World Association of Nuclear Operators (WANO) [16], Institute of Nuclear Power Operations (INPO) [17]. The Electric Power Research Institute (EPRI) has developed a series of Fuel Reliability Guidelines that describe best practices for eliminating fuel failures at nuclear power plants, within the framework of the Fuel Reliability Program, focusing on the following four areas [18]:

- Fuel surveillance and inspection.
- PWR & BWR fuel cladding corrosion and crud.
- Pellet cladding interaction (PCI).
- Grid-to-rod fretting (GTRF).

The EPRI Fuel Reliability Guidelines involve a continuous learning and improvement process where guidance drives the actions to monitor and assess fuel performance, to obtain operating experience along the way, and then to feed that operating experience back into the guidance.

5 Conclusions

Many stakeholders are essential to improvements in fuel reliability and performance: the operator sets the high expectation, the fuel designer and manufacturer improves fuel designs, qualification and manufacturing practices to provide fuel products with high performance and quality, and the operator including engineering, maintenance and operating staff implements policies and best practices to improve reliability.

The IAEA Nuclear Energy Series (NES) Guide on the quality and reliability aspects in nuclear power reactor fuel engineering (NF-G-2.1) has been updated:

- to reflect state of the art information on fuel reliability and performance issues,
- to maintain consistency with the recommendations stated in the new IAEA Safety Standards on the design of the reactor core (SSG-52), and
- to extend applicability to newly developed ATFs [19, 20] and the future SMR fuels.

The NF-G-2.1 revision 1 provides comprehensive recommendations on fuel design changes, manufacturing, qualification, in-reactor operation, and on-site services, with the objective to assist the IAEA member states to achieve excellence in fuel reliability and performance.

Acknowledgements. The authors are grateful to J. Alonso Pacheco (ENUSA, Spain), M. Aullo Chaves (ENUSA, Spain), J. Armstrong (CNL, Canada), A. Gusev (TVEL, Russian federation), K. Inagaki (CRIEPI, Japan), J. Judah (J. Judah and Associates, Inc., Canada) and E. Mader (EPRI, USA) for their contributions to drafting and review of the IAEA guide NF-G-2.1 Revision 1.

References

1. International Atomic Energy Agency: Quality and Reliability Aspects in Nuclear Power Reactor Fuel Engineering, IAEA Nuclear Energy Series No. NF-G-2.1, IAEA, Vienna (2015).
2. International Atomic Energy Agency: Design of Reactor Core for Nuclear Power Plants, IAEA Safety Standards Series No. SSG-52, IAEA, Vienna (2019).
3. Organisation for Economic Co-Operation and Development/Nuclear Energy Agency: State-of-the-Art Report on Light Water Reactor Accident-Tolerant Fuels, NEA No. 7317, OECD (2018).
4. International Atomic Energy Agency: Advances in Small Modular Reactor Technology Developments - 2020 Edition: A Supplement to IAEA Advanced Reactors Information System (ARIS), IAEA, Vienna (2020).
5. International Atomic Energy Agency: Review of Fuel Failures in Water Cooled Rectors (2006–2015), IAEA Nuclear Energy Series No. NF-T-2.5, IAEA, Vienna (2019).
6. International Atomic Energy Agency: Progress on Pellet-Cladding Interaction and Stress Corrosion Cracking: Experimentation, Modelling and Methodologies Applied to Support the Flexible Operation of Nuclear Power plants, IAEA-TECDOC-1960, IAEA, Vienna (2021).
7. International Atomic Energy Agency: Operation and Licensing of Mixed Cores in Water Cooled Reactors, IAEA-TECDOC-1720, IAEA, Vienna (2013).
8. Druenne, H., Zhang, J., Dalleur, J.P. et al.: The Belgian Approach to In-core Fuel Management with Mixed Fuel Assemblies, Proc. TopFuel, Manchester, UK (2012).
9. Zhang, J., Waeckel, N., Yueh, K.: "Utility Perspective on Needs Related to Nuclear Fuel Operation," Proc. of OECD/NEA Workshop on Nuclear Fuel Modelling to Support Safety and Performance Enhancement for Water-Cooled Reactors, Paris, 7–9 March 2017, OECD/NEA (2021).
10. Institute Of Nuclear Power Operations: Excellence in the Management of Design and Operating Margins, Report 09–003, INPO (2009).

11. Zhang, J. and Schneidesch, C.: "Application of the BEPU safety analysis method to quantify margins in nuclear power plants," Nuclear Engineering and Design Volume 406, 112233, May (2023).
12. International Atomic Energy Agency, Deterministic Safety Analysis for Nuclear Power Plants, Safety Standards Series No. SSG-2 Rev 1, IAEA, Vienna (2019).
13. Zhang, J., Kovtonyuk, A., Schneidesch, C., "Towards a Graded Application of Best Estimate Plus Uncertainty Methodology for Non-LOCA Transient Analysis," Nuclear Engineering and Design 354 (2019).
14. Zhang, J., Van Parys, R., Dalleur, J-P., Schneidesch, C., Schinazi, C., Meert, A.: "An Improved Fuel Design Specification and Evaluation Process for Belgian Nuclear Power Plants," Proceedings of the 2017 Water Reactor Fuel Performance Meeting (WRFPM/TOPFUEL2017), Jeju lsland, Korea, September 10–14 (2017).
15. Strasser, A., Epperson, K., Holm, J., Rudling, P.: "Design reviews for reliable fuel performance", WRFPM Top Fuel 2010 (Proc. LWR Fuel Perf. Mtg Orlando, 2010), 1–10, ANS, LaGrange Park, IL (2010).
16. World Association Of Nuclear Operators, Managing Core Design Changes, SOER 2004–1, WANO (2004).
17. Institute Of Nuclear Power Operations: Guidelines for Achieving Excellence in Nuclear Fuel Performance, Report 07–004, INPO (2007).
18. Deshon, J., Whiteside, K., Burnham, R.: "Establishing and Sustaining a Technical Program to Achieve Zero Fuel Failures," 11th International Conference on WWER Fuel Performance, Modeling and Experimental Support, Bolero, Bulgaria, from 26 September to 3 October, (2015).
19. Organisation for Economic Co-Operation and Development/Nuclear Energy Agency: CSNI Technical Opinion Paper No. 19 - Applicability of Nuclear Fuel Safety Criteria to Accident-tolerant Fuel Designs, Nuclear Safety NEA No. 7576, OECD (2022).
20. International Atomic Energy Agency, Status of Knowledge for the Qualification and Licensing of Advanced Nuclear Fuels for Water Cooled Reactors, IAEA-TECDOC-xxx, IAEA, Vienna (2023).

Parametric Study of Phenomena Influencing Secondary Hydriding During LOCA Transients

A. M. Kpemou[1(✉)], J. Desquines[1], T. Taurines[1], S. Guilbert[1], M. C. Baietto[2], B. Normand[2], J. Soulacroix[3], A. Ambard[4], and F. Bourlier[5]

[1] IRSN, Cadarache, Saint Paul-Lez-Durance, France
Apou-martial.kpemou@irsn.fr
[2] INSA Lyon, Université de Lyon, Lyon, France
[3] EDF, DIPNN – DIRECTION TECHNIQUE, 69363 Lyon CEDEX 07, France
[4] MMC Department, EDF R&D, Avenue des Renardières, 77818 Moret-sur-Loing, France
[5] Framatome (Fuel Business Unit), 69007 Lyon, France

Abstract. Secondary hydriding may occur during Loss of Coolant Accidents (LOCA) when nuclear fuel cladding is exposed to steam at high temperature. Indeed, during a LOCA, the cladding may burst, and the steam ingress results in cladding inner surface oxidation. The hydrogen released during this oxidation can be partly absorbed by the cladding, leading to a localized high hydrogen content. This study aims at characterizing the influence of different parameters on the maximum hydrogen content, localization and distribution regarding secondary hydriding. The oxidation duration has been studied using an axisymmetric testing procedure reproducing LOCA conditions. Metallographic analysis and hydrogen measurements have been performed. EPMA and μ-LIBS micro-analysis have been used to map oxygen and the hydrogen local content.

Keywords: Secondary hydriding · LOCA · High temperature oxidation · Microanalysis

1 Introduction

LOCA, which stands for Loss of Coolant Accident, is a hypothetical accident scenario resulting from a break in the main coolant line between the coolant pump and the reactor pressure vessel. This scenario is considered for the design of the Emergency Core Cooling System (ECCS). During a LOCA transient, the cladding tubes made of zirconium alloy may burst, and the steam ingress leads to the oxidation of cladding inner surface in the vicinity of the burst balloon. The primary work on the subject by Chung and Kassner [1] and Uetsuka [2] has evidenced a phenomenon called secondary hydriding. Indeed, during the oxidation stage of a LOCA, the inner gaseous environment of the cladding is progressively enriched with hydrogen released by the oxidation reaction. At a certain distance from the burst opening, the internal gaseous environment is highly enriched in hydrogen which can be partly absorbed by the cladding leading to a localized high hydrogen content of a few hundred up to several thousand wppm at the peak [1, 2].

J. Liu and Y. Jiao (Eds.): WRFPM 2023, SPPHY 299, pp. 68–79, 2024.
https://doi.org/10.1007/978-981-99-7157-2_7

Over the last twenty years, several semi-integral LOCA tests have been performed by many nuclear research institutions such as JAEA [2–5], ANL [6], NRC [7], KIT [8, 9], EDF R&D [10], and CEA [11]. These semi-integral tests aim to experimentally reproduce LOCA transients, on rods with inserted pellets, and then study the behavior of fuel claddings. These tests have been performed on fuel claddings, including unirradiated and irradiated claddings, using irradiated fuel pellets or mock-up pellets. The complex nature of these tests combined to the uneven deformed cladding geometry obtained after the burst, impedes the direct study of the various parameters influencing secondary hydriding.

In the present work, a dedicated experimental approach has been used to investigate the influence of oxidation duration on secondary hydriding. An axisymmetric testing procedure was developed, reproducing LOCA conditions, including cladding oxidation and the final quench of the sample. Each sample was characterized by a set of post-test analysis. Post-quench material microstructure, including inner and outer oxide layer thickness axial profiles, was characterized using metallography. Hydrogen axial profiles were obtained mainly by Hot Vacuum Extraction (HVE). A sample was analyzed using Electron Probe Micro-Analysis (EPMA) and Micro Laser Induced Breakdown Spectroscopy (μ-LIBS) to map simultaneously oxygen and hydrogen local content in an extended region. The analytical test results can be used to support secondary hydriding modeling using simple and well-defined boundary conditions.

2 Materials and Methods

2.1 Materials

The specimens used in this study were cut from a $M5_{\mathbf{Framatome}}$ [1] recrystallized cladding with nominal chemical composition described in Table 1. The outer diameter of the cladding tube is 9.5mm and the thickness 0.57 mm.

Table 1. Chemical composition of the studied $M5_{\mathbf{Framatome}}$ cladding.

Nb (wt%)	O (wt%)	S (ppm)	Fe (wt%)
1	0.14	20	0.035

2.2 Sample Geometry and Constitutive Materials

The samples are cut from as-received $M5_{\mathbf{Framatome}}$ cladding tubes and filled with ceramic pellets. The use of undeformed cladding allows separate effect tests to perform parametric studies with a uniform gap between the pellets and the inner cladding surface. The mock-up pellets diameter was varied to study the fuel cladding-gap impact on secondary

[1] M5 and $M5_{\mathrm{Framatome}}$ are trademarks or registered trademarks of Framatome or its affiliates, in the USA or other countries.

hydriding. Eight mm in diameter porcelain pellets (in average) have been inserted, resulting in a gap of 160 μm. The sample is left open on its upper end, and tightly closed on the lower end with Swagelok cap (see Fig. 1a). The oxidation temperature profile is monitored by a B-type wire thermocouple (TC) welded on the outer surface of the cladding at approximately 10 mm from the opening.

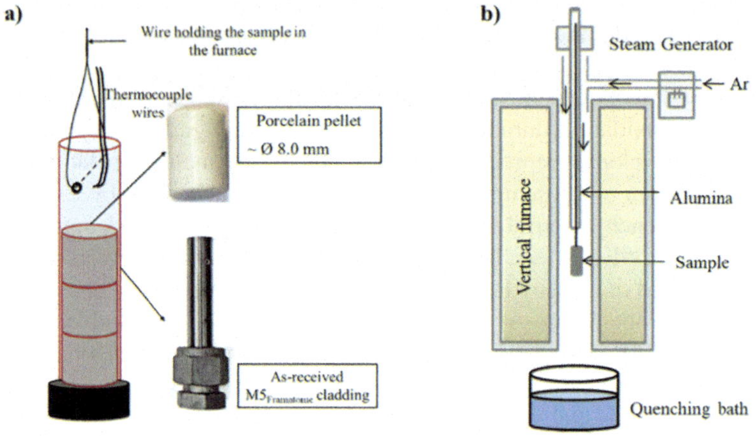

Fig. 1. (**a**) Sample geometry and constitutive materials and (**b**) the experimental device and protocol

2.3 High Temperature (HT) Oxidation

To study secondary hydriding, oxidation tests were performed in steam plus argon environment at approximatively 1200 °C. Each sample is inserted upwards into a vertical furnace under pure argon. After a first fast heating phase, steam is then added to the downstream gas flow to achieve a 50–50% mixture when reaching the target temperature leading to inner and outer cladding surfaces oxidation. The sample is hold for a certain time before quenching (see Fig. 1b). The steam flow rate normalized to the cross-sectional area of the furnace alumina tube was 19.6 mg.cm^{-2}.s^{-1} which is several times above the necessary flow to avoid steam starvation at the outer surface [12].

During the steam oxidation, an oxide layer of zirconia (ZrO_2) is formed on the outer surface of the cladding. On the inner side, the steam ingress, within the narrow pellet-cladding gap, oxidizes the inner surface. Due to the limited space left and gas flow occlusion by argon and hydrogen in the gap, steam starvation occurs in this region. The steam partial pressure progressively disappears replaced by hydrogen partial pressure increase while moving away from the opening. A high hydrogen uptake can then occur at a few millimeters away from the opening, where the inner oxide layer is very thin or absent also corresponding to a high hydrogen partial pressure and probably negligible remaining steam content.

This hydrogen uptake combined with the diffusion process in the material lead to a hydrogen axial profile in the cladding which is called secondary hydriding.

2.4 Post-Test Characterization

Hydrogen content measurements

After LOCA quenching, nanoscale δ hydride precipitates are located in the samples requiring high resolution microscopy to be observed [1, 13]. Consequently, destructive measurement of hydrogen content was used in the present study.

Hydrogen content measurements were then performed by HVE with a LECO ONH 836 analyzer after each oxidation test. The HVE method is based on the extraction of hydrogen gas from a metallic sample by fusion. In this study, the HVE double pic method developed by PSI[2] allows measuring hydrogen content in oxide layers and in the base metal of the sample was used [14]. Hydrogen axial profiles were obtained by measuring hydrogen content in well-localized small cladding pieces cut at different axial positions (see Fig. 2). Calibration of the analyzer is always made before a set of measurements by means of hydrogen-certified standard samples. The average hydrogen content uncertainty is 8% which is the uncertainty of the hydrogen standard content.

Metallographic analysis

Optical metallography is used to measure both inner and outer oxide layer thicknesses after high temperature oxidation. It also allows the identification of the different phases of the material due to oxidation and quenching (oxide layer, α(O)-phase and prior β-phase). These measurements were performed using a Keyence VHX 5000 optical microscope (see Fig. 2).

EPMA & μ-LIBS micro-analysis

The local oxygen content in the different phases of the material was measured by Electron Probe Micro-Analysis (EPMA) with a CAMECA SX-100 Castaing microprobe. This analysis was performed on one sample. Additionally, the local content of iron and niobium was also measured. A surface area of 6000 μm × 600 μm was mapped with a resolution of 2 μm.

Micro-Laser Induced Breakdown Spectroscopy (μ-LIBS) analysis was also carried out by LASALYS. Oxygen and hydrogen maps were obtained on the sample used for EPMA analysis in order to have a complete characterization of this sample. The μ-LIBS analysis is based on the interaction of a pulsed laser beam onto the surface sample [15]. A surface of 3000 μm × 600 μm was scanned with a 3 μm lateral displacement step.

2.5 Test Matrix

Eight oxidation tests were conducted for different oxidation duration. These tests allowed studying the influence of oxidation duration on secondary hydriding. The oxidation temperature was set at 1200 °C. The temperature measured by the welded TC ranged from 1160 °C to 1210 °C. The tests parameters are summarized in Table 2.

[2] Paul Scherrer Institute (PSI).

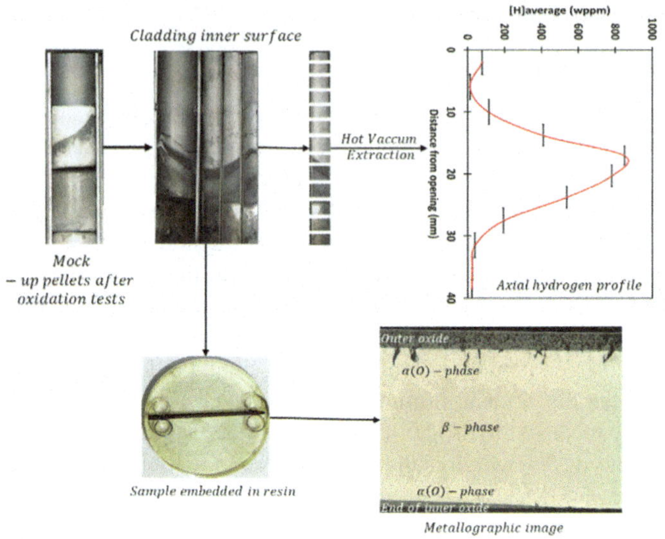

Fig. 2. Analysis methods applied on each sample after oxidation tests

Table 2. Secondary hydriding analytical tests matrix

Sample ID	Average oxidation temperature at TC position (°C)	Oxidation duration (s)	ECR – CP (%)[3] [16]
M5–100	1160	106	8
M5–200	1190	191	13
M5–400-1	1193	387	19
M5–400-2	1197	378	19
M5–600	1189	646	24
M5–1000	1210	979	33
M5–1400	1189	1430	36

3 Experimental Results and Discussion

3.1 Metallographic Analysis

To determine the distribution of different material phases in the cladding, each oxidized sample underwent metallographic analysis. During the oxidation process, the inner and outer surfaces of the cladding were both oxidized, resulting in the formation of a zirconia layer on the surface, as well as α(O)-phase and β-phase in the metal. However, due to steam consumption along the gap by inner surface cladding oxidation, and also due

[3] ECR values are calculated with Cathcart-Pawel correlation in doubled-sided oxidation configuration.

the steam starvation occurring in the gap (limited space left and gas flow occlusion by argon and hydrogen in the gap), the inner zirconia layer growth process is progressively interrupted after only a few tens of millimeters away from the opening (see Fig. 3).

Figure 4 shows the evolution of the outer zirconia layer with oxidation duration at the thermocouple position. At high temperature, as expected according to [17], the oxide growth rate follows a parabolic law.

3.2 Influence of Oxidation Duration on Secondary Hydriding

The main results related to hydrogen content following oxidation tests are summarized in Table 3. Hydrogen axial profiles were obtained for each sample by means of HVE. For some samples, two profiles were obtained using two parts of the sample at different azimuthal positions as shown in Fig. 2. All the hydrogen axial profiles obtained are summarized in Fig. 5.

With respect to oxidation duration, the peak hydrogen content increases with a parabolic like kinetics, indicating a fast uptake rate in the first few minutes, followed by a slower uptake rate over time. The total amount of hydrogen uptake by the sample follows the same trend (see Fig. 5b). Since the main source of hydrogen comes from the oxidation of the internal surface of the cladding, and this oxidation follows a parabolic law, one can conclude that the hydrogen supply decreases with time. Also, the internal oxide layer growth may reduce and close the gap and then impede the ingress of steam and the transport of hydrogen in the gap interrupting this way the hydrogen uptake process. This phenomenon related to oxide formation may also explain the decrease of the hydrogen uptake for longer oxidation duration. Hydrogen peak position also appears to shift away from the opening with oxidation duration.

These analytical tests also showed that hydrogen uptake occurs in regions without protective oxide layer, since the hydrogen peak is located after the maximum progress of the internal oxide in the gap (see Table 3).

After long durations, the diffusion process in the material takes over on the hydrogen uptake process. This leads to a broadening of the axial hydrogen profile over time and then reduces the maximum hydrogen content at the peak while the total hydrogen uptake is higher (see Fig. 5b).

3.3 EPMA & μ-LIBS Micro-Analysis

Oxygen and hydrogen maps were obtained using EPMA and μ-LIBS microanalysis on a sample (Sample ID: M5-400-1) that underwent approximately 400 s exposure to steam at 1200 °C (see Fig. 6). The analysis was performed at the vicinity of the end the inner oxide layer.

The EPMA analysis was used to measure the local oxygen content in the sample, and this data was used to calibrate the oxygen map obtained from the μ-LIBS analysis. The measured oxygen content in the different material phases is illustrated in Fig. 6a and b. The oxygen maps evidence the decrease up to vanishing of the inner oxide layer and a comparable trend is observed for the internal α(O)-phase layer.

Figure 7 shows the oxygen and hydrogen radial mean contents in the Region Of Interest (ROI, see Fig. 6a) across the cladding thickness. Oxygen content regions above

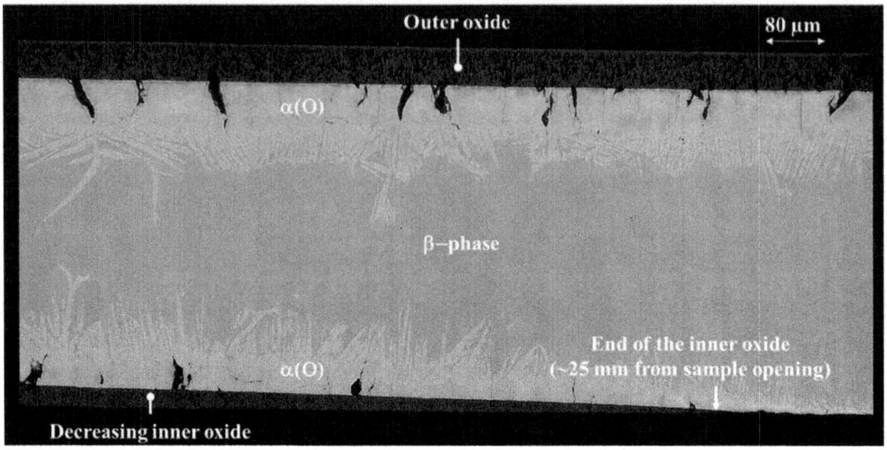

Fig. 3. Metallographic image showing the decreasing and the end of the inner oxide from a sample that underwent oxidation test in LOCA conditions.

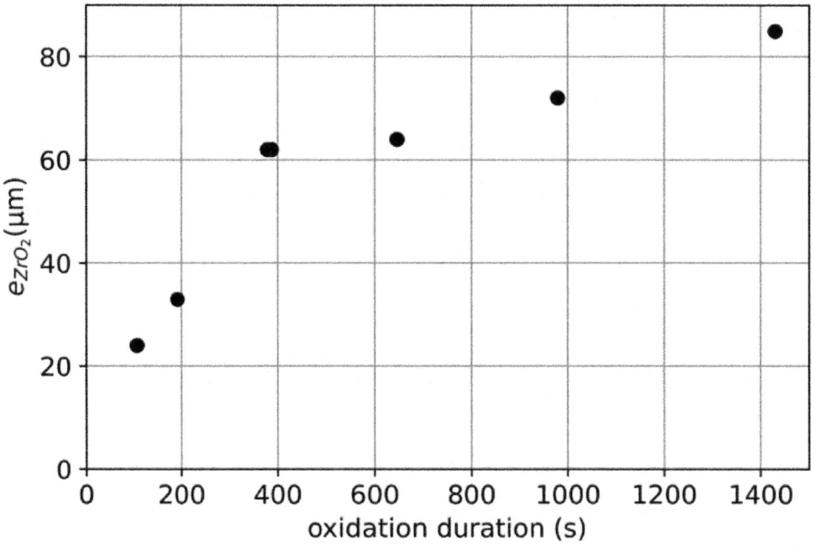

Fig. 4. Outer oxide layer thickness evolution over oxidation duration

25 wt.%, mainly covering surfaces exposed to steam, correspond to the oxide layer, the oxygen diffusion from oxide layers stabilizes α(O)-phase at contents below 8% and above 2–3 wt.% and lower contents are associated to the prior β-phase far away from oxygen sources. Higher hydrogen content is observed in the prior β-phase and at the boundary between this phase and the α(O)-phase on the associated profile in Fig. 7. This last concentration increases result from low hydrogen solubility of the α(O) layer. The measured hydrogen content in the oxide layer is probably an artefact, since the

Table 3. Secondary hydriding results after steam oxidation tests.

Sample ID	Peak hydrogen content [H] (wppm)	Integrated hydrogen picked-up (wppm*mm)	Hydrogen peak position from opening (mm ± 2)	Inner oxide layer end position from opening (mm)
M5–100	212 ± 19	2316	21	19
M5–200	641 ± 57	6225	16.5	17
M5–400-1	952 ± 82	11066	21	21
M5–400-2	891 ± 73	10869	20	21
M5–600	911 ± 79	14001	22	21
M5–1000	1098 ± 95	15305	25	25
M5–1400	1024 ± 88	18299	22	23

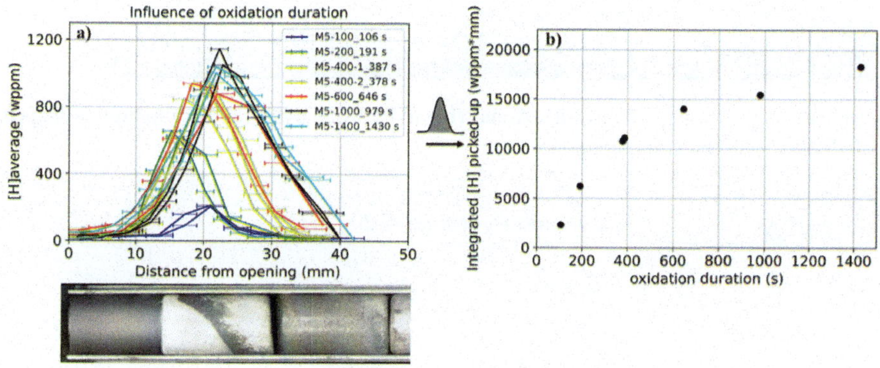

Fig. 5. (**a**) Axial hydrogen profiles and (**b**) total hydrogen content uptake by the cladding at different oxidation duration.

measurement method wasn't calibrated in oxide layer and the laboratory experience is that very low hydrogen content is expected in this layer (see Fig. 7).

The hydrogen maps when averaged provide consistent measurements with the hydrogen axial profile from HVE (see Fig. 8).

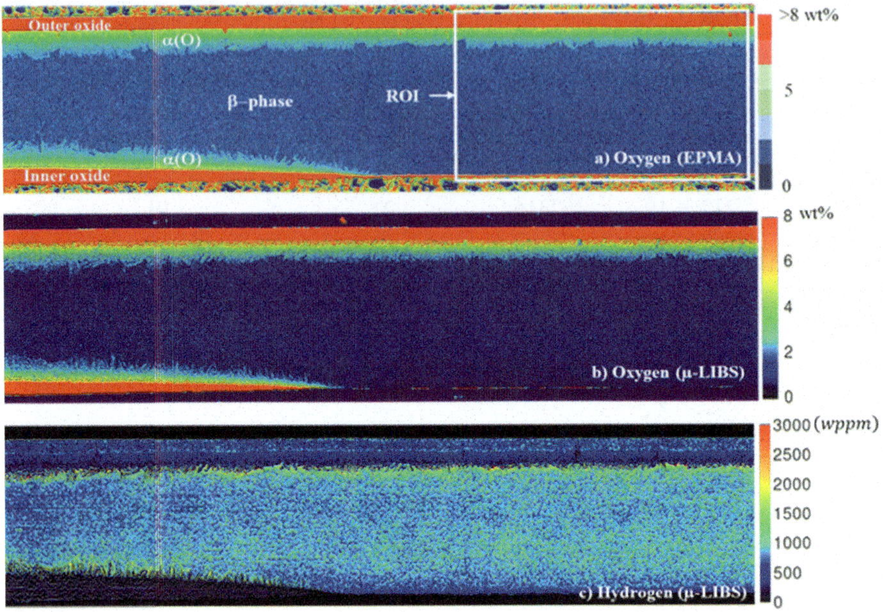

Fig. 6. (**a**)- Oxygen maps from EPMA and (**b**)- μ-LIBS analysis, and (**c**)-hydrogen map from μ-LIBS after ~400s steam oxidation.

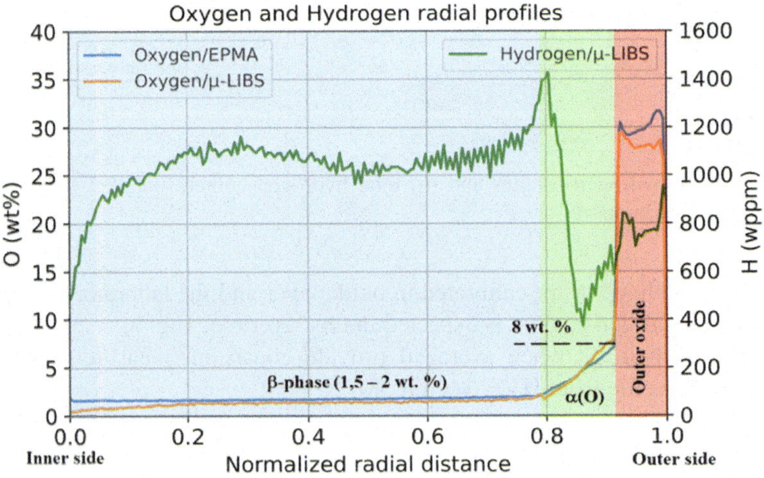

Fig. 7. Oxygen and hydrogen radial distribution in the different phases of the material (in the ROI).

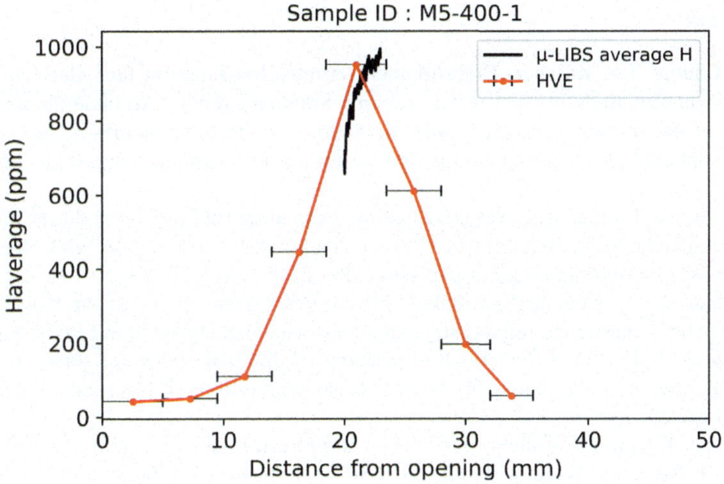

Fig. 8. Comparison between μ-LIBS average hydrogen content and HVE axial hydrogen profile.

4 Conclusion

The purpose of this work was to present an experimental approach for studying the parameters influencing secondary hydriding during a LOCA. The influence of oxidation duration was the focus of the paper. It was investigated through oxidation tests performed under conditions comparable to the ones expected during LOCA. The results show that the hydrogen uptake increases with oxidation duration but with a decreasing rate at longer duration.

In addition, micro-analysis using EPMA, and μ-LIBS were carried out to map oxygen and hydrogen on a sample that underwent a ~400 s long oxidation at 1200 °C. The analysis revealed that hydrogen is mostly concentrated in the prior β-phase of the cladding. These analyses can also help identify any relationship between oxygen and hydrogen distribution in the material and investigate pseudo ternary $M5_{\mathbf{Framatome}}$-O-H phase diagram.

Influences of gap size, opening size and oxide layer effects on secondary hydriding are currently under investigation. Analytical segmented semi-integral tests are also planned. Simulations of these tests are underway with the SHOWBIZ code developed at IRSN (a new version of the code DIFFOX developed earlier at IRSN [18]).

Aknowledgement. The μ-LIBS measurements were performed at LASALYS (1 Avenue du Champ de mars, 45074 Orléans Cedex 2, France). The authors are grateful to the GGP – CC project for supporting this work. Additionally, the authors would like to thank Mrs. Gaelle VILLEVIEILLE for her contribution in preparing and performing the EPMA analysis.

References

1. H.M. Chung, T.F. Kassner, Embrittlement criteria for Zircaloy fuel cladding applicable to accident situations in light-water reactors. Summary report, Argonne National Lab., IL (USA), 1980. https://www.osti.gov/biblio/5496633-embrittlement-criteria-zircaloy-fuel-cladding-applicable-accident-situations-light-water-reactors-summary-report (accessed July 18, 2022).
2. H. Uetsuka, T. Furuta, S. Kawasaki, Zircaloy-4 Cladding Embrittlement due to Inner Surface Oxidation under Simulated Loss-of-Coolant Condition, J. Nucl. Sci. Technol. 18 (1981) 705–717. https://doi.org/https://doi.org/10.1080/18811248.1981.9733309.
3. H. Uetsuka, T. Furuta, S. Kawasaki, Fracture-bearing capability of zircaloy cladding on thermal-shock during the reflood stage of a loss-of-coolant accident in LWRs, Japan, 1983.
4. F. Nagase, T. Fuketa, Effect of pre-hydriding on thermal shock resistance of zircaloy-4 cladding under simulated loss-of-coolant accident conditions, J. Nucl. Sci. Technol. Tokyo. 41 (2004) 723–730.
5. F. Nagase, T. Chuto, T. Fuketa, Behavior of High Burn-up Fuel Cladding under LOCA Conditions, J. Nucl. Sci. Technol. 46 (2009) 763–769. https://doi.org/https://doi.org/10.1080/18811248.2007.9711583.
6. M. Billone, Y. Yan, T. Burtseva, R. Daum, Cladding embrittlement during postulated loss-of-coolant accidents., Argonne National Lab. (ANL), Argonne, IL (United States), 2008. https://doi.org/10.2172/946677.
7. M. Flanagan, P. Askeijung, A. Puranen, Post-Test Examination Results from Integral, High-Burnup, Fueled LOCA Tests at Studsvik Nuclear Labo, U.S. Nuclear Regulatory Commission's, 2013. https://www.nrc.gov/reading-rm/doc-collections/nuregs/staff/sr2160/index.html (accessed March 20, 2023).
8. J. Stuckert, M. Große, C. Rössger, M. Klimenkov, M. Steinbrück, M. Walter, QUENCH-LOCA program at KIT on secondary hydriding and results of the commissioning bundle test QUENCH-L0, Nucl. Eng. Des. 255 (2013) 185–201. https://doi.org/https://doi.org/10.1016/j.nucengdes.2012.10.024.
9. J. Stuckert, M. Grosse, M. Steinbrueck, M. Walter, A. Wensauer, Results of the QUENCH-LOCA experimental program at KIT, J. Nucl. Mater. 534 (2020) 152143. https://doi.org/https://doi.org/10.1016/j.jnucmat.2020.152143.
10. R. Thieurmel, J. Besson, E. Pouillier, A. Parrot, A. Ambard, A.-F. Gourgues-Lorenzon, Contribution to the understanding of brittle fracture conditions of zirconium alloy fuel cladding tubes during LOCA transient, J. Nucl. Mater. 527 (2019) 151815. https://doi.org/https://doi.org/10.1016/j.jnucmat.2019.151815.
11. J.-C. Brachet, D. Hamon, M. Le Saux, V. Vandenberghe, C. Toffolon-Masclet, E. Rouesne, S. Urvoy, J.-L. Béchade, C. Raepsaet, J.-L. Lacour, G. Bayon, F. Ott, "Study of secondary hydriding at high temperature in zirconium based nuclear fuel cladding tubes by coupling information from neutron radiography/tomography, electron probe micro analysis, micro elastic recoil detection analysis and laser induced breakdown spectroscopy microprobe, J. Nucl. Mater. 488 (2017) 267–286. https://doi.org/https://doi.org/10.1016/j.jnucmat.2017.03.009.
12. S. Leistikow, G. Schanz, Oxidation kinetics and related phenomena of zircaloy-4 fuel cladding exposed to high temperature steam and hydrogen-steam mixtures under PWR accident conditions, Nucl. Eng. Des. 103 (1987) 65–84. https://doi.org/https://doi.org/10.1016/0029-5493(87)90286-X.
13. L. Legras, E. Torres, M.C. Baietto, J. Desquine, A. Cabrera-Salcédo, M.B. Yriex, MLLS fitting on plasmon pic for mapping hydrides in a Zr alloy with a complex αZr+βZr acicular microstructure obtained by water quenching, in: Eur. Microsc. Congr. 2016 Proc., John Wiley & Sons, Ltd, 2016: pp. 1024–1025. https://doi.org/10.1002/9783527808465.EMC2016.5904.

14. A. Hermann, H. Wiese, R. Bûhner, M. Steinemann, G. Bart, Hydrogen Distribution Between Fuel Cladding Metal and Overlaying Corrosion Layers, in: Park City, Utah, 2000.
15. J.C. Brachet, D. Hamon, M.L. Saux, V. Vandenberghe, C.C. Toffolon-Masclet, E. Rouesne, S. Urvoy, J.L. Béchade, C.C. Raepsaet, J.L. Lacour, G. Bayon, F. Ott, Study of secondary hydriding at high temperature in zirconium based nuclear fuel cladding tubes by coupling information from neutron radiography/tomography, electron probe micro analysis, micro elastic recoil detection analysis and laser induced breakdown spectroscopy microprobe, J. Nucl. Mater. 488 (2017) 267. https://doi.org/https://doi.org/10.1016/j.jnucmat.2017.03.009.
16. R.E. Pawel, J.V. Cathcart, J.J. Campbell, S.H. Jury, Zirconium Metal-Water Oxidation Kinetics V. Oxidation of Zircaloy in High Pressure Steam, n.d.
17. J. Desquines, C. Duriez, S. Guilbert, T. Taurines, High temperature oxidation and room temperature axial strength of pre-oxidized zircaloy-4 cladding after a simulated LOCA, J. Nucl. Mater. 543 (2021) 152559. https://doi.org/https://doi.org/10.1016/j.jnucmat.2020.152559.
18. C. Duriez, S. Guilbert, S. A., G. C., L. Belovsky, J. Desquines, Characterization of Oxygen Distribution in LOCA Situations, in: 2010. https://doi.org/10.1520/JAI103156.

Radial Hydride Precipitation in Fuel Cladding During Back-End Cooling Transient Under Decreasing Pressure

J. Desquines[1], C. Sartoris[1](✉), M. Guémas[1], and A. Gérard[2]

[1] IRSN, Cadarache, France
{jean.desquines,christine.sartoris,marine.guemas}@irsn.fr
[2] IRSN, Les Angles, France
alain.gerard@irsn.fr

Abstract. During spent fuel assembly transportation or dry cask storage, filling gas and fission gas release pressurize the fuel rods contained in the cask. These rods are heated by fuel assembly residual power resulting in internal pressurization stress increase in the rods. During the subsequent cooling associated to removal of rods from the transportation cask or during the decrease of residual power after a long storage period, the cladding hoop stress is expected to decrease. Concomitant cladding temperature and stress decrease may influence the morphology of precipitated hydrides in corroded claddings. Radial hydride precipitation has a detrimental role on the mechanical resistance of the cladding and is promoted by large hoop stresses during the cooling of the cladding. There is, consequently, a need to predict the radial hydride precipitation in corroded cladding tubes during dry storage or transportation cooling transients. Existing models usually address the radial hydride precipitation conditions under constant stress leading to over-estimated radial hydride content within the cladding.

Keywords: Fuel transportation · Back-end · Radial hydrides · Zircaloy-4

1 Introduction

In the present study a model describing the radial hydride precipitation in a pre-hydrided unirradiated (SRA Zircaloy-4) cladding alloy under time varying stress and temperature transient. The model is first described and, and afterwards validated using pressurized tests with constant depressurization rate during the cool down at 24 °C/h. This model is implemented in IRSN SHOWBIZ [1] simulation code including back-end cycle modeling capabilities.

J. Liu and Y. Jiao (Eds.): WRFPM 2023, SPPHY 299, pp. 80–86, 2024.
https://doi.org/10.1007/978-981-99-7157-2_8

2 Modeling the Influence of Decreasing Stress on Radial Hydride Precipitation

2.1 Constant Stress Model

IRSN already developed a model to predict the radial hydride precipitation under constant stress at 24 °C/h cooling rate [2]. Based on metallographs performed at room temperature, two tangential stress thresholds have been identified, $\sigma_{t,0\%}$ below which only circumferential hydrides can precipitate and $\sigma_{t,100\%}$ above which 100% of newly precipitated hydrides are formed in the radial direction.

$$\sigma_{t,0\%}(MPa) = 0.02[H](wppm) + 186.9 - 0.3862Min\left(T_{max}\left(^{\circ}C\right); T_{SSD}\left(^{\circ}C\right)\right) \quad (1)$$

$$\sigma_{t,100\%}(MPa) = 110 + 65\left(1 - e^{-\frac{[H](wppm)}{65}}\right) \pm 20 \quad (2)$$

T_{SSD} is, the temperature required to dissolve [H](wppm) hydrogen while heating from room temperature, derived from the Une and Ishimoto correlation [3]:

$$T_{SSD}\left(^{\circ}C\right) = -\frac{4395}{ln\left(\frac{[H](wppm)}{12.8\times10^{4}}\right)} - 273.15 \quad (3)$$

The data from Slattery [4] on Zircaloy-4 was used to adjust the terminal solubility while cooling (Fig. 1):

$$C_{SP}(\text{wppm}) = 69100.e^{-\frac{3596}{T(K)}} \quad (4)$$

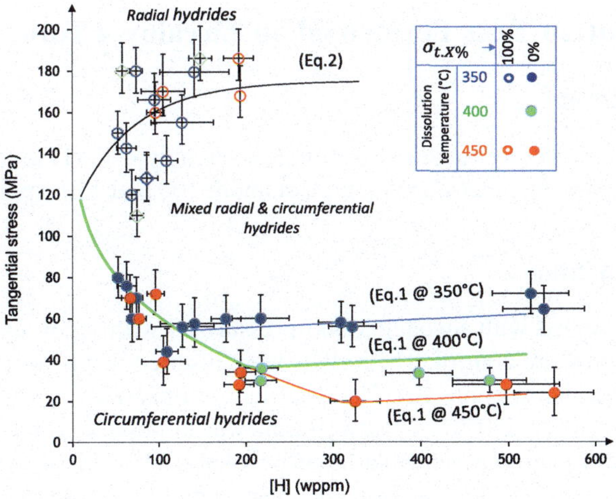

Fig. 1. Hydride orientation/precipitation map between 350 and 450 °C maximum temperature.

The fraction of radial hydrides precipitated at room temperature under constant hoop stress, σ_t, is modeled using a simple lever rule [2]:

$$F_R([H], \sigma_t, T_{max}, T_{SSD}) = \begin{cases} 0\, for\ \sigma_t \leq \sigma_{t,0\%} \\ \frac{\sigma_t - \sigma_{t,0\%}}{\sigma_{t,100\%} - \sigma_{t,0\%}}\, for\ \sigma_{t,0\%} \leq \sigma_t \leq \sigma_{t,100\%} \\ 1\, for\ \sigma_{t,100\%} \leq \sigma_t \end{cases} \quad (5)$$

2.2 Decreasing Stress Model

The above-described model was interpretated and transformed to extend its applicability to the instantaneous fraction of radial hydride precipitated in the cladding at each cooling step under applied hoop stress [5, 6]. A very close approach was developed by Rashid and Machiels [7]. The fraction of hydrogen precipitated in the radial direction ($d[H]_R$) formed when precipitating $d[H]_P$, at each time step is defined as: $f_R = \frac{d[H]_R}{d[H]_P}$. . This parameter can be easily derived from the constant stress model by comparing the radial hydride fraction at room temperature between two very close precipitation temperatures (~T):

$$f_R([H], \sigma_t, T, T_{max}) = \frac{\partial (F_R([H], \sigma_t, T_{max}, T_{SSD}) C_{SP}(T))}{\partial C_{SP}} \quad (6)$$

In the above equation, the T_{SSD} temperature is deduced from $T_{SSD}(\boldsymbol{C_{SP}(T)})$. To illustrate the model, the obtained function is represented in Fig. 2. The plots show that under constant stress, radial hydrides rather form during early stage of precipitation below $\sigma_{t,100\%}$. And above this value, radial hydride precipitation is expected all along the cooling transient.

3 Pressurization Tests Performed on Zircaloy-4 Tubes

3.1 Tested Material

Six 90 mm long tubes constituted of low tin SRA Zircaloy-4 17x17 cladding tubes were uniformly pre-hydrided under pure vacuum at about 100 and 150 wppm (Table 1).

3.2 Testing Conditions

The tubes were sealed with Swagelok and pressurized with argon using a GE Druck PACE 5000 for pressure control. The rodlets were inserted in a Instron heating furnace set-up with a Eurotherm controller. The transient is illustrated in Fig. 3. The sample was heated during 1h to 350 °C, a 2h dwell time was imposed to achieve homogeneous temperature within the furnace and then slowly cooled at 24 °C/h. When reaching 315 °C, about 25 °C above precipitation solvus, the internal pressurization was established and controlled at different cooling rates.

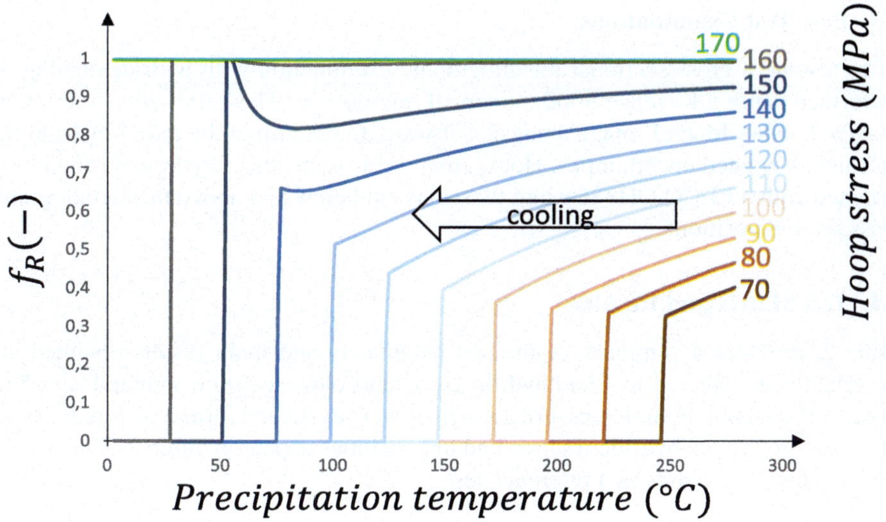

Fig. 2. Predicted fraction of radial hydrides formed during cooling from 350 °C modeling 100 wppm hydrogen content.

Table 1. Zircaloy-4 composition.

Sn (wt%)	Fe (wt%)	Cr(wt%)	O(wt%)	H(wtppm)
1.30	0.21	0.11	0.14	7

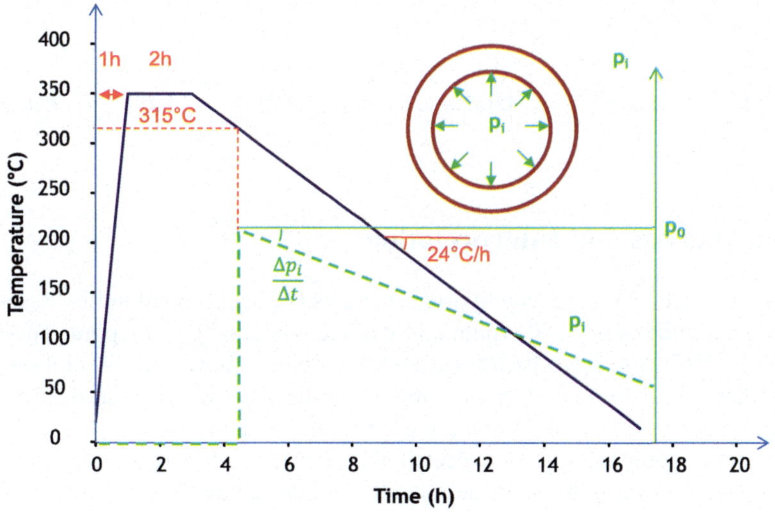

Fig. 3. Heating and pressurization transients imposed to the samples.

3.3 Post-Test Examinations

After testing, a cross sectional metallographic examination with hydride etching was performed using a Keyence 5000 numerical microscope. The entire cross-section was analyzed, using ImageJ image analysis software, to determine the radial hydride fraction and associated uncertainties. Hot vacuum measurements of hydrogen content were obtained with a LECO OH836 using two rings cut below and above the metallographic cross section position.

3.4 Test Matrix and Results

Table 2, provides a summary of the test parameters and main results obtained after material testing. Tests 1 to 4 had hydrogen contents close to 100 wppm and tests 5 to 7 close to 150 wppm. Four depressurization rates were evaluated form 0, 1, 5 and 10 bar/h. The measured radial hydride fraction and uncertainties appear in Table 2. The test #7 at constant pressure stands as a reference test.

Table 2. Test matrix and main results.

Test	$[H]$	$p_{i,max}$	$p_{i,min}$	dp_i/dt	$\sigma_{tt,max}$	$\sigma_{tt,min}$	F_R
(#)	(wppm)	(bar)	(bar)	(bar/h)	(MPa	MPa)	(%)
1	82	150.3	138.3	1.0	110.2	101.4	31.9 ± 4.6
2	96	150.4	89.8	5.0	110.3	65.9	26.4 ± 4.7
3	98	150.3	25.4	10.0	110.2	18.6	19.9 ± 2.4
4	101	150.1	137.6	1.0	110.0	100.9	30.1 ± 4.6
5	148	150.4	29.5	10.0	110.3	21.7	14.1 ± 4.7
6	155	150.5	88.9	5.1	110.4	65.2	18.6 ± 4.3
7	156	150.2	150.2	0.0	110.1	110.1	25.0 ± 4.7

4 Test Analysis and Interpretation

The experimental post-test radial hydride fractions (F_R) are plotted in Fig. 4, separating samples containing about 100 wppm and the one containing 150 wppm. The samples containing 150 wppm have hydrogen content slightly above the dissolution limit at 350 °C (there are consequently remaining circumferential hydrides at maximum temperature) whereas the samples containing 100 wppm experience full dissolution at the same maximum temperature. As expected, the samples with remaining circumferential hydrides lead to lower radial hydride fraction than the samples experiencing full dissolution (100wppm). For both hydrogen contents, a depressurization during the cooling transient has a limiting effect on radial hydride precipitation. This is consistent with the modeling results of Fig. 2 showing that the instantaneous radial hydride fraction precipitated decreases with decreasing stress.

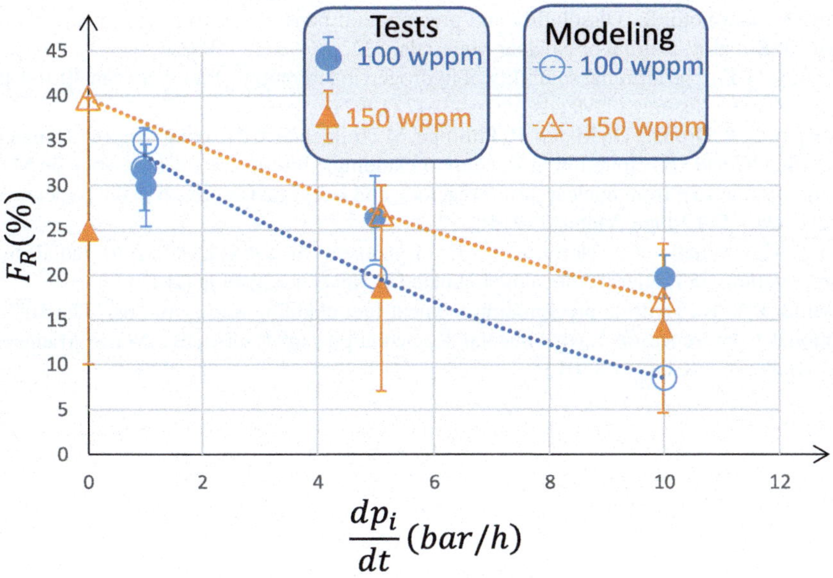

Fig. 4. Test and modeling results of moderate cooling transients at various depressurization rates.

The tests have been modeled using the above-described modeling and the results are compared to experimental data in Fig. 4. The models acceptably predict the influence of depressurization however the prediction accuracy has some sensitivity to hydrogen content.

There is a clear benefit in modeling the influence of decreasing stress on radial hydride precipitation and simple models provide acceptably accurate predictions.

5 Conclusion

This paper summarizes long-term objectives of the testing campaign performed at IRSN on Zircaloy-4. Past results [2] showed that there was a complex influence of maximum temperature and hydrogen content on radial hydride precipitation under constant stress.

These past results were used to build a model to assess the influence of decreasing stress on radial hydride precipitation. The model provides a reasonably good assessment of the influence of a decreasing pressure. This model is now implemented in the SHOW-BIZ [1] multiphysics code to add a modeling capability dedicated to back-end cycle to the already existing modeling developed for oxygen diffusion under LOCA conditions.

References

1. https://www.irsn.fr/recherche/plateforme-logicielle-fuel.
2. Desquines, J., Drouan, D., Billone, M., Puls M.P., March, P., Fourgeaud, S., Getrey, C., Elbaz, V., Philippe, M.: Influence of temperature and hydrogen content on stress-induced radial hydride precipitation in Zircaloy-4 cladding, Journal of Nuclear Materials 453 131–150 (2014).

3. Une, K., Ishimoto, S., Dissolution and precipitation behavior of hydrides in Zircaloy-2 and high Fe Zircaloy, Journal of Nuclear Materials, 322, pp. 66–72 (2003).
4. Slattery, G. F., The terminal solubility of hydrogen in zirconium alloys between 30 and 400°C, Journal of the Institute of Metals, 95, p.43–47 (1967).
5. Desquines, J., Drouan, D., March, P., Philippe, M.: Influence of decaying stress and temperature on radial hydride precipitation in Zircaloy-4 claddings, International Conference on Management of spent fuel from nuclear power reactors: and integrated approach to the back end of the fuel cycle, 15–19 June, Vienna (2015).
6. Feria, C., Aguado, C., Herranz, L. E.: Extension of FRAPCON-xt to radial hydride reorientation, 45 Reunion Annual de la Sociedad Nuclear Espanola (2019).
7. Rashid, J. Y. R., Hydride precipitation in spent fuel cladding during storage, The 10th International Conference on Environmental Remediation and Radioactive Waste Management, September 4–8, Glasgow (2015).

Analysis and Assessment of BEO-Doped Fuel with Fuel Rod Performance Code Jasmine

Kaiyuan Wang(✉), Yanan Zhu, and Xin Jin

China Nuclear Power Technology Research Institute Co. Ltd, Shen-Zhen, China
wky20080795@126.com

Abstract. UO_2-BeO fuel, a quite potential accident tolerant fuel, is used to increase the temperature margin of UO_2 fuel and improve the safety performance of reactor operation. With the advantage of higher thermal conductivity than that of the UO_2 pellet, the heat transfer performance of the UO_2-BeO pellet is significantly improved, and the pellet temperature is much lower. The development of the JASMINE code based on the models of UO_2-BeO pellets had been completed to analyze and evaluate the irradiation performance of UO_2-BeO pellets. In this paper, the performance analysis of UO_2 and UO_2-10 vol% BeO fuel is performed respectively. Based on the experiment data from the Halden reactor, the results show that the predictions calculated by updated JASMINE are in good agreement with the measurements, demonstrating an excellent ability to analyze the in-pile behavior of UO_2-BeO fuel.

Keywords: UO_2-BeO · Fuel performance · JASMINE

1 Introduction

The traditional uranium dioxide (UO_2) fuel has been used for nearly 70 years in nuclear power stations, with advantages including high melting point, stability, and neutron economy under irradiation. However, due to the low thermal conductivity of UO_2, the pellet temperature is high, which greatly limits the improvement of reactor performance. After Fukushima nuclear accident, the concept of accident tolerant fuel (ATF) was proposed and employed for expanding the safety margins and strengthening nuclear safety [1]. Uranium oxide doped beryllium oxide fuel (UO_2-BeO), one of the most potential accident tolerant fuels, has high thermal conductivity, high melting point, low chemical reactivity at high temperatures, and low thermal neutron absorption rate [2], compared with the UO_2 fuel.

Up to now, a great deal of research has been carried out on the thermodynamics performance of UO_2-BeO fuel [3–5]. According to the comparative results of UO_2-BeO fuels with different BeO volume fractions made by Kevin McCoy [6], it was shown that even a slight increase in thermal conductivity will lead to a significant improvement in fuel performance. McDeavitt [7] found that the thermal conductivity of UO_2-10vol%BeO pellets was 60 to 75% higher than that of UO_2 pellets.

J. Liu and Y. Jiao (Eds.): WRFPM 2023, SPPHY 299, pp. 87–96, 2024.
https://doi.org/10.1007/978-981-99-7157-2_9

Over the past years, China General Nuclear Power Corporation (CGN) has done a lot of work with accident-tolerant fuels and obtained fruitful results [8, 9]. The present work aims to evaluate UO_2-BeO fuel performance using the updated JASMINE code and optimize the pellet design in the future. The JASMINE [10] code, developed by CNPRI, was updated to support the evaluation of UO_2-BeO (10vol%) fuel. Furthermore, the updated JASMINE code was used to perform a preliminary assessment of the Halden test rod IFA-716R3.

2 Theoretical Model

2.1 Non-Irradiation Model

Density

Based on the uranium dioxide matrix, the composite ceramic UO_2-BeO fuel is doped with different proportions of beryllium. In general, the overall performance of composite ceramic depends directly on the amount of each component. Meanwhile, the volume and mass fraction of UO_2-BeO can be converted to each other.

$$w_{UO_2} = \frac{v_{UO_2}\rho_{UO_2}}{v_{BeO}\rho_{BeO} + v_{UO_2}\rho_{UO_2}} \tag{1}$$

where w_{uo2} represents the weight fraction of UO_2, ρ_{uo2} and ρ_{BeO} are the densities of UO_2 and BeO in g/cm^3, ν_{uo2} and ν_{BeO} are the volume fraction of UO_2 and BeO.

The UO_2-BeO fuel density could be calculated by a volume weighting function.

$$\rho_{UO_2-BeO} = v_{BeO}\rho_{BeO} + (1 - v_{BeO})\rho_{UO_2} \tag{2}$$

In general, the density of BeO [11] and UO_2 [12] is 2.987 g/cm^3 and 10.96 g/cm^3, respectively.

Heat capacity

The specific heat capacity of UO_2-BeO fuels can be calculated using the quality-weighted average method [13].

$$c_{UO_2-BeO} = w_{UO_2}c_{UO_2} + \left(1 - w_{UO_2}\right)c_{BeO} \tag{3}$$

where c_{uo2} and c_{BeO} are the specific heat of UO_2 and BeO in J/(kg·K), respectively.

Thermal expansion

Given the low thermal expansion coefficient of BeO, the UO_2-BeO fuel has a distinguishing feature that delays the gap closures. The thermal expansion coefficient for UO_2-BeO fuel can be calculated using the volume weighting function [14].

$$\alpha_{UO_2-BeO} = (1 - v_{BeO})\alpha_{UO_2} + v_{BeO}\alpha_{BeO} \tag{4}$$

where α_{uo2} and α_{BeO} are the coefficients of linear thermal expansion for UO_2 and BeO in 1/K, respectively.

Emissivity

The emissivity of the UO_2-BeO pellet is a function of temperature, given as follows [13]

$$\varepsilon_{UO_2-BeO}= 0.676 + 8.93 \times 10^{-5}T-1.76 \times 10^{-8}T^2 \quad (5)$$

where T is the pellet surface temperature in K.

Elasticity modulus

The mechanical responses of the UO_2-BeO composite fuel depend on the metallurgical processes but exist as slight variations. Theoretically, a composite has a higher elastic modulus and lower Poisson ratio. For the UO_2-BeO, its elasticity modulus can be expressed as:

$$E_{UO_2-BeO}=(1 - v_{BeO})E_{UO_2} + v_{BeO}E_{BeO} \quad (6)$$

where Young's modulus and Poisson's ratio of BeO recommended by Soga [15] are used in this study.

2.2 Irradiation Model

Thermal conductivity

The thermal conductivity of the pellet is a key parameter in the design of PWR rods. Intensive research showed that the thermal conductivity of UO_2 doped with BeO would be significantly increased [9, 16]. Additionally, the pores in ceramic materials will also have a significant impact on the thermal conductivity. In general, the thermal conductivity of UO_2-BeO can be normalized by the following equation [16, 17].

$$\lambda_{UO_2-BeO}=\lambda_{100\%}(1 - \eta p) \quad (7)$$

$$\eta=2.6 - 5 \times 10^{-4}T \quad (8)$$

where p is the porosity, T is the temperature in °C, η is the experimental parameter, $\lambda_{100\%}$ is the pure UO_2–BeO thermal conductivity in W/(m·K), and is calculated using the modified Halden thermal conductivity model [18]:

$$\lambda_{100\%}=\frac{1+0.03v_{BeO}}{0.1148+0.004Bu + 2.475 \times 10^{-4}(1 - 0.00333Bu)T}+0.0132e^{0.00188T} \quad (9)$$

where ν_{BeO} is BeO volume fraction in % and Bu is the fuel burnup in MW·d/kg Oxide.

Relocation

The pellet radial total relocation strain can be defined by the following equation.

$$\varepsilon_{\theta}^{re} = \frac{\Delta u}{r} \cdot k_{re} \quad (10)$$

where $\varepsilon_{\theta}^{re}$ is pellet radial total relocation strain (–), Δu is the variation of pellet radius (m), r is pellet radius (m), k_{re} is the relocation recovery factor (–).

The variation of pellet radius can be calculated by FRAPCON 3.4 relocation model [19]. According to the previous study [20], the k_{re} can be obtained by the following method.

$$k_{re} = a \cdot \exp\left(b\Delta t^c P_{con}^d\right) \tag{11}$$

$$a = \begin{cases} 1, \ q \le 40kw/m \\ 1 + 3(q-40)/40, \ q > 40kw/m \end{cases} \tag{12}$$

where P_{con} is contact pressure (MPa), q is fuel rod power (kW/m), Δt is time increment (h), a,b,c,d are model coefficients.

Densification

Based on the Halden experimental data [13], the modified ESCORE densification model [21] for the UO_2-BeO pellet was adopted in this study.

$$\Delta\rho = \Delta\rho_{max}\left(1 - e^{(-Bu/0.593)}\right) \tag{13}$$

where $\Delta\rho$ is densification strain(%), $\Delta\rho_{max}$ is the maximum density variation of the sintering experiment (%).

Irradiation swelling

As reported in the reference [22], the swelling rate of UO_2-BeO is similar to the UO_2 fuel. Therefore, the MATPRO swelling model [23] for UO_2 fuel is adopted in this study.

$$S_s = 7.435 \times 10^{-13}\rho_0\Delta Bu \tag{14}$$

where S_s is fission swelling strain(-), ρ_0 is theoretical density (kg/m^3), ΔBu is increment burnup (MW·s/kg(U)).

Fission gas release

The fission gas release process is strongly influenced by temperature, burnup, grain size, and so on. However, because of the lack of UO_2-BeO fuel experimental data, the fission gas model of pure UO_2 is still a good choice for qualitative evaluation, such as the Bernard model [24].

3 Verification and Evaluation

3.1 Experimental Verification

In this paper, the reliability of the updated JASMINE code is preliminarily verified by using the experimental data of the IFA-716R3 rod containing 10 vol%BeO. The objective of the IFA-716R3 test is to study the effect of the dopant concentration on thermal conductivity and fission gas release [25], etc. The IFA-716R3 rod was irradiated under the steady-state condition in the Halden reactor for 15 cycles [22] and was monitored using the pressure transducers and fuel thermocouples in the upper ends of the fuel rod. The power history is shown in Fig. 1.

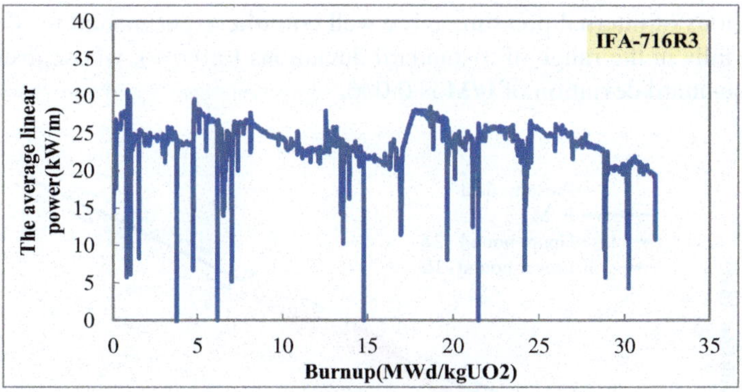

Fig. 1. The power history of the IFA-716R3 fuel rod.

Temperature verification

Fuel temperature is very important for the safe operation of fuel rods. The temperature measurements of the IFA-716R3 rod are used to evaluate the temperature prediction capability of JASMINE, as illustrated in Fig. 2.

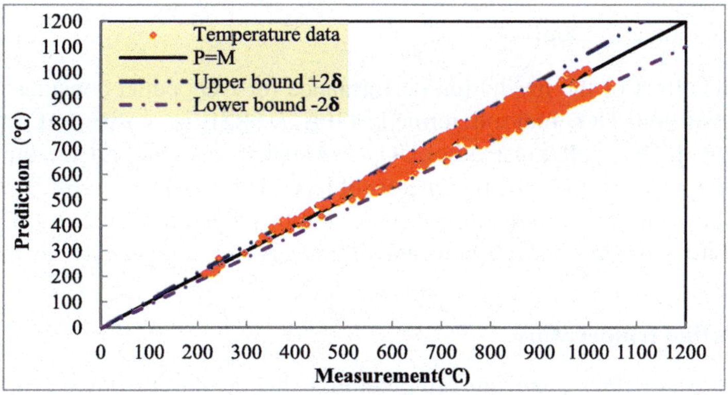

Fig. 2. Comparison of fuel temperature predictions and measurements.

During the 15 cycles, the pellet temperature was varied around 200°C - 1000 °C irradiated with all reactor states. The JASMINE prediction agrees well with the experimental data, whose average P/M is 1.003 and the standard deviation of P/M is 0.040. The data points fall within the interval of 2 standard deviations (±4%). As a result, the JASMINE code based on the modified model, especially the thermal conductivity model, shows a good ability to predict the fuel rod temperature.

Rod pressure verification

The rod's internal pressure is also an important parameter in the fuel rod performance code. The measurements and the calculated values are compared and shown in Fig. 3.

The predictions of internal pressure agree well with the experimental results, and most of the data falls in the range of 3 standard deviations (±0.6%), whose average P/M is 0.998 and standard deviation of P/M is 0.006.

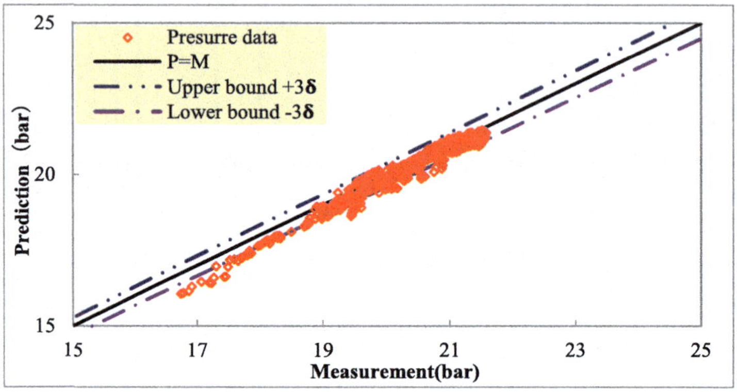

Fig. 3. Comparison of internal pressure predictions and measurements.

3.2 Performance Evaluation

To study the effect of the irradiation performance for UO_2 pellet considering the BeO additive, a fictitious UO_2 fuel rod, termed as IFA-716R3F, is constructed from the IFA-716R3 rod with UO_2 pellet instead of UO_2-BeO pellet. For comparison, the power history, enrichment, and geometric parameters of IFA-716R3 and IFA-716R3F are assumed to be the same. Therefore, some important irradiation parameters, such as the fuel centerline temperature, rod internal pressure, and fission gas release, are analyzed and evaluated in detail.

Fuel centerline temperature

Figure 4 shows the fuel centerline temperature using the modified and original version of the JASMINE code. It is shown that the temperature of the UO_2 fuel is about 95°C higher than that of UO_2-BeO, which mainly results from the higher thermal conductivity of UO_2-BeO fuel. The difference of about 95 °C measured fuel temperature in the IFA-716R3 rod at this power is consistent with a fuel thermal conductivity increase of 30%, which was reported in the literature [4, 22]. Many studies have reported that BeO, as an optimized addition, can significantly reduce the peak pellet temperature by 200 ~ 300°C [9, 26]. The temperature change is also affected by effective thermal conductivity (BeO distribution in the pellet) and the power level.

As we know, the melting point of UO_2-10vol%BeO fuel is 2600°C, which is lower than the melting point of pure UO_2 pellets 2800°C [27]. However, thanks to the high thermal conductivity of doped UO_2 fuel, the peak temperature of the pellets is significantly reduced, especially at high temperatures where the thermal gradient is smoother, which can reduce the risk of pellet failure. In addition, although the doped BeO can effectively reduce the operating temperature of UO_2 pellets, the effect of the eutectic interaction between UO_2 and BeO on the pellet melting temperature should be paid attention to in the practical application, which needs further continuous research.

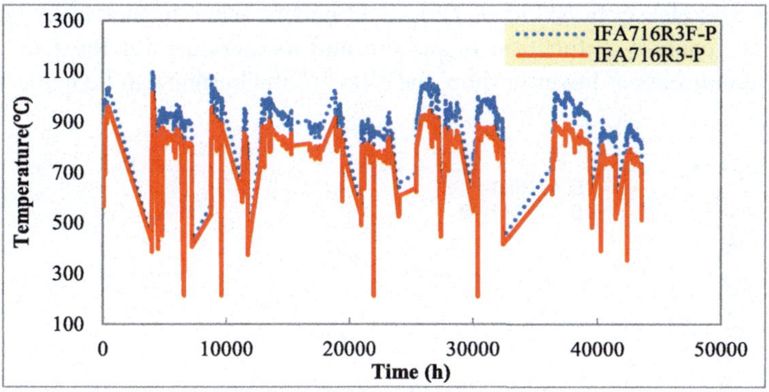

Fig. 4. The fuel temperature prediction comparison for IFA-716R3 and IFA-716R3F.

Internal pressure

The fuel rod's internal pressure is closely related to fuel temperature, fission gas release, and free volume. Generally, increasing the fuel rod temperature leads to a higher internal pressure. Figure 5 shows the calculated pressure history for the IFA-716R3F rod and IFA-716R3 rod. The comparison of the calculations indicates that the internal pressure of the IFA-716R3F rod is significantly higher than that of the IFA-716R3 rods, mainly because of the higher temperature. Besides, it can be also seen that the internal pressure of the IFA-716R3F rod increases considerably after 25,900 h (at a burnup of approximately 17 MWd/kgUO_2) because a large amount of fission gas is released when the estimated peak fuel temperatures for UO_2 fuel reach the levels close to the FGR model threshold.

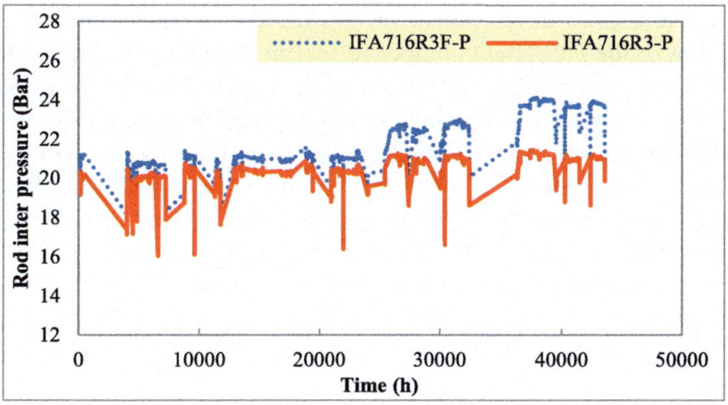

Fig. 5. The internal pressure comparison for IFA-716R3 and IFA-716R3F.

Fission gas release

Another purpose of the IFA-716R3 test is to study the fission gas of the doped beryllium oxide fuel. Since the fuel temperature is smaller than the release threshold, there is no significant fission gas release during the IFA-716R3 experiment [22]. Therefore, the fission gas release threshold of UO_2-BeO pellets needs further investigation. The threshold is generally a function of burnup and temperature for the UO_2 fuel [28]. Under the conditions of lower burnup, the effect of the burnup can be neglected.

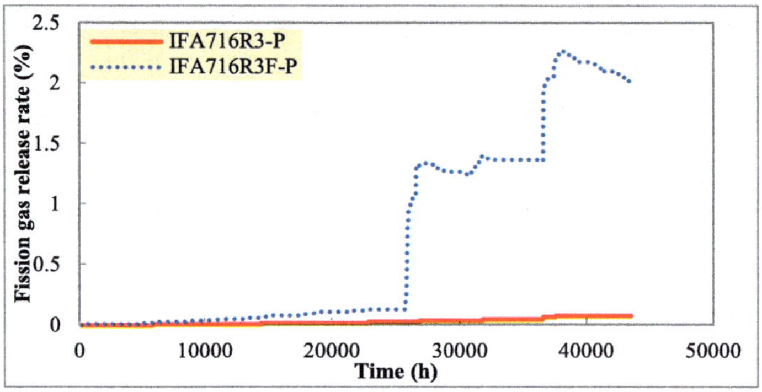

Fig. 6. The fission gas release comparison for IFA-716R3 and IFA-716R3F.

Based on the UO_2 FGR model, an obvious fission gas release phenomenon is observed when UO_2 fuel rods reach a certain release threshold (at a burnup of approximately 16 MWd/kgUO_2), as shown in Fig. 6. The fission gas release fraction of beryllium oxide pellets calculated by JASMINE is less than 0.1% during the whole life, which is also in good agreement with the experimental results.

4 Conclusions

In this study, the second development of JASMINE has been completed, which can be used to analyze the UO_2-BeO composite fuel in the future. Firstly, compared with the experimental data, it can be confirmed that the updated JASMINE code has a good predictive ability in both fuel temperature and internal pressure during the irradiation history. Then, simulation results show that the fuel centerline temperature of UO_2-BeO(10 vol%) fuel is reduced by about 95°C during normal conditions with the power history of IFA-716R3 when using the UO_2-BeO pellet instead of UO_2 pellet. Due to the low temperature, the internal pressure and fission gas release of UO_2-BeO pellet fuel rods are significantly lower than those of UO_2 fuel rods, which shows the excellent performance of the UO_2-BeO pellet. However, to improve the prediction capability, there are still some experiments and theoretical work to be studied, such as fission gas release mechanism and fuel densification research, etc.

References

1. Sato Y, Tanifuji Y, Terasaka Y, et al. Radiation imaging using a compact Compton camera inside the Fukushima Daiichi Nuclear Power Station building[J]. Journal of Nuclear Science and Technology, 2018, 55(9): 965–970.
2. Manly W D. Utilization of BeO in reactors[J]. Journal of Nuclear Materials, 1964, 14: 3–18.
3. Zhou W, Revankar S T, Liu R, et al. Microstructure-based thermal conductivity and thermal behavior modeling of nuclear fuel UO_2-BeO[J]. Heat Transfer Engineering, 2018, 39(9): 760–774.
4. Y. Russin, Fuel pellets produced at JSC ULBA for the IFA 716 experiment, Paper F1.5, EHPG, Storefjell, Norway, March 2010.
5. Garcia C. Sintering and thermal behavior of uranium dioxide in beryllium oxide matrix[D]. , 2014.
6. McCoy K, Mays C. Enhanced thermal conductivity oxide nuclear fuels by co-sintering with BeO: II. Fuel performance and neutronics[J]. Journal of Nuclear Materials, 2008, 375(2): 157–167.
7. Sean M. McDeavitt, Chad Garcia, Jean C. Ragusa1, Joshua Smith, James Malone, "Behavior Assessments For UO_2-BeO Enhanced Conductivity Fuel In A PWR". ANS LWR Fuel Performance Meeting, Top Fuel September 15–19, 2013, North Caroline, USA. p. 1161–1166.
8. Liu T, Xue J, Li R, et al. The Research on Accident Tolerant Fuel in CGN[J].
9. Li B, Yang Z, Jia J, et al. High temperature thermal physical performance of BeO/UO_2 composites prepared by spark plasma sintering (SPS)[J]. Scripta Materialia, 2018, 142: 70–73.
10. Jin X, Wei X, Liu X, et al. JASMINE: A Fuel Rod Thermal-Mechanical Performance's Code[R]. American Nuclear Society-ANS, 555 North Kensington Avenue, La Grange Park, IL 60526 (United States), 2016.
11. Bobkov V, Fokin L, Petrov E, et al. Thermophysical properties of materials for nuclear engineering: a tutorial and collection of data[J]. IAEA, Vienna, 2008.
12. Dan Gabriel Cacuci. Nuclear engineering handbook[M]. CRC Press, 2010.
13. He Y, Wu Y, Wang S, et al. Analysis of UO_2-BeO Fuel Performance During Normal Conditions and RIA[C]//International Conference on Nuclear Engineering. American Society of Mechanical Engineers, 2018, 51531: V009T16A070.
14. Liu R, Zhou W. Multiphysics modeling of novel UO_2-BeO sandwich fuel performance in a light water reactor[J]. Annals of Nuclear Energy, 2017, 109: 298–309.
15. Soga N. Elastic constants of polycrystalline BeO as a function of pressure and temperature[J]. Journal of the American Ceramic Society, 1969, 52(5): 246–249.
16. Ishimoto S, Hirai M, Ito K, et al. Thermal conductivity of UO_2-BeO pellet[J]. Journal of Nuclear Science and Technology, 1996, 33(2): 134–140.
17. Loeb A L. Thermal conductivity: VIII, a theory of thermal conductivity of porous materials[J]. Journal of the American Ceramic Society, 1954, 37(2): 96–99.
18. Mcgrath M A, VOLKOV B Y, RUSSIN Y. In-reactor Investigation of the composite UO_2-BeO fuel: background, results and perspectives[C]//The Nuclear Materials Conference. 2016.
19. Geelhood, K. J., Luscher, W. G., Beyer, C. E., Flanagan, M. E., 2011. FRAPCON-3.4: a computer code for the calculation of steady state thermal-mechanical behavior of oxide fuel rods for high burnup. Richland, WA: US Nuclear Regulatory Commission, Office of Nuclear Regulatory Research.
20. Wang K Y, Wang Y, Jin X, et al. A new relocation recovery model for LWR fuel[J]. Annals of Nuclear Energy, 2020, 145: 107528.
21. Rashid Y, Dunham R, Montgomery R. Fuel analysis and licensing code: FALCON MOD01[J]. EPRI Report, 2004, 1011308.

22. Tverberg T. Update on the in-pile results from the fission gas release mechanisms study in IFA 716[M]//Technical Report HWR-1090. Organization for Economic Co-operation and Development Halden Reactor Project, Halden, Norway, 2014.
23. Allison C M, Berna G A. SCDAP/RELAP5/MOD 3.1 code manual: MATPRO, A library of materials properties for Light-Water-Reactor accident analysis. Volume 4[J]. Office of Scientific & Technical Information Technical Reports, 1995.
24. Bernard L C, Jacoud J L, Vesco P. An efficient model for the analysis of fission gas release[J]. Journal of Nuclear Materials, 2002, 302(2–3): 125–134.
25. Brémond O. IFA 716.1 fission gas release mechanisms[R]. Technical Report HWR 1008, Institutt for Energiteknik, Halden, Norway, 2011.
26. Solomon A, Revankar S, McCoy J K. Enhanced thermal conductivity oxide fuels[R]. Purdue University School of Nuclear Engineering, 2006.
27. Ray S, Lahoda E, Franceschini F. Assessment of different materials for meeting the requirement of future fuel designs[C]//2012 Reactor Fuel Performance Meeting. 2012, 115: 2–6.
28. Bernard L C, Bonnaud E. Finite volume method for fission gas release modeling[J]. Journal of nuclear materials, 1997, 244(1): 75–84.

Reassessment of FRAPTRAN's Cladding Failure Criteria in LOCA Within R2CA H2020 Project

B. Dif(✉), A. Arkoma, and J. Heikinheimo

VTT Technical Research Centre of Finland, Espoo, Finland
{brahim.dif,asko.arkoma,janne.heikinheimo}@vtt.fi

Abstract. In this study, various cladding deformation models and burst criteria were implemented in U.S. NRC's transient fuel performance code FRAPTRAN to examine the effect on the prediction of cladding burst under loss-of-coolant accidents (LOCAs). The work is carried out within EU H2020 R2CA project (Reduction of Radiological Consequences of design basis and design extension Accidents) with the aim to improve the modelling of radiological consequences in LOCA and steam generator tube rupture (SGTR) accidents. Cladding plastic deformation in FRAPTRAN's ballooning model BALON2 was modified by employing two high-temperature creep laws [1, 2] in combination with a dynamic crystallographic phase transformation model [3]. Additionally, new LOCA burst criteria based on true stress limit [4] developed by IRSN were tested to recalculate FRAPTRAN validation cases. The modified code modules were further incorporated in VTT's statistical method for full core analysis to evaluate the number of failing rods in LOCA simulations of EPR type plant. The results demonstrate that the prediction of burst could be improved depending on the selected deformation model and burst criterion. Overall, the cladding creep was found to have more impact on the results than the burst criterion.

Keywords: LOCA · Ballooning · Burst criterion · FRAPTRAN

1 Introduction

During a LOCA, fuel rods are subjected to high temperatures and internal pressures that can lead to cladding deformation, ballooning, and ultimately, rupture. To predict fuel behavior accurately, various failure criteria have been proposed over the years. The adequate assessment of failing rods is crucial to demonstrate compliance with regulatory acceptance criteria under postulated design basis accidents. In Finnish context, the Regulatory Guides on nuclear safety require that the "number of fuel rod failures in a Class 2 postulated accident shall not exceed 10% of the total number of fuel rods in the reactor" [5]. The fulfilment of this criterion is studied in EPR type reactor with the FRAPTRAN code using various creep models and burst criteria.

J. Liu and Y. Jiao (Eds.): WRFPM 2023, SPPHY 299, pp. 97–107, 2024.
https://doi.org/10.1007/978-981-99-7157-2_10

2 FRAPTRAN Description

FRAPTRAN is a widely used fuel performance code developed for analyzing the transient behavior of nuclear fuel rods at high burnups [6]. To enhance its thermal hydraulics modeling capabilities, FRAPTRAN has been coupled in the past with the subchannel thermal hydraulics code GENFLO [7], an in-house code developed by VTT. The coupled version is used by VTT to assess the number of failing fuel rods in the calculation chain of the full core statistical simulations.

2.1 Cladding Mechanical Deformation Model

Total mechanical deformation in FRAPTRAN is modelled by considering the elastic strain (ε_θ), thermal expansion ($\varepsilon_{\mathrm{th}}$), and plastic strain ($\varepsilon_{\mathrm{plastic}}$):

$$\varepsilon_{\mathrm{total}} = \varepsilon_{\mathrm{th}} + \varepsilon_\theta + \varepsilon_{\mathrm{plastic}} \tag{1}$$

The plastic stress-strain behavior in the code is modelled using the modified Hollomon's equation [8]. FRAPTRAN calculates the plastic strain using the mechanical model FRACAS-I for all nodes before ballooning. When plastic strain exceeds the instability limit calculated by the code, FRACAS-I is no longer used. After that, the BALON2 model is called, and plastic strain is calculated only for the ballooning node.

$$\varepsilon_f = \left[\left(\frac{n}{m}+1\right)10^{-3}\left(\frac{\sigma}{\mathrm{K}}\right)^{1/m}\Delta t + \varepsilon_{\mathrm{i}}^{\left(\frac{\mathrm{n}}{\mathrm{m}}+1\right)}\right]^{\frac{\mathrm{m}}{\mathrm{n+m}}} \tag{2}$$

Equation (2) is the time-dependent solution of the modified Hollomon equation solving for the strain [8]. Where ε_f is the true effective strain at the end of a time interval, ε_i is the true effective strain at the start of a time interval, Δt is the time interval, and σ is the true effective stress. The strength coefficient K, strain rate exponent m, and strain hardening exponent n are material-dependent parameters that describe the metallurgical state of the cladding [8].

2.2 LOCA Failure Criteria

In FRAPTRAN, LOCA failure criterion has been established based on empirical stress and strain limits [9]. In this context, the term "fuel failure" refers to a situation in which the fuel rod loses its integrity, so that the fuel rod cladding is breached and is no longer hermetically sealed. Under LOCA conditions, the code predicts cladding failure in the ballooning node given the following conditions:

- The primary criterion is triggered if the local cladding true hoop stress exceeds an empirical limit. There are essentially two different cladding type-specific stress criteria: E110 cladding having its own model, while all other cladding types use a common correlation (Eq. 3) which is a function of temperature, fast neutron fluence, and cold work. Heating rate and strain rate do not affect this criterion [8].

$$\sigma_{\theta B} = \begin{cases} 1.36K_A, & for \quad T \le 750K \\ 46.861429K_A \exp\left(\frac{-1.9901087\times10^6}{T^2}\right) & for \quad 750 < T \le 1050K \\ 7.7K_A & for \quad T > 1050K \end{cases} \tag{3}$$

$\sigma_{\theta B}$ is the tangential component of true stress at burst
K_A is the strength coefficient for annealed cladding as determined by MATPRO.

- The second condition is met if the predicted cladding permanent hoop strain exceeds FRAPTRAN's strain limit as defined in [9].

It is worth noting that the two criteria have been established based on experimental data available by 1980 [8]. Since then, the development of new cladding materials, changes in fuel design, and the trend to go to higher burnups, highlight the need to examine the applicability of these criteria.

3 Modifications in the Code

For LOCA modelling, the relevant models were identified and updated accordingly. The updates were conducted on FRAPTRAN 1.4 as that version had an established coupling with the subchannel code GENFLO. The developments were validated by re-calculating various validation cases. For the validation of the changes, the modified stand-alone FRAPTRAN (referred to as FRAPTRAN-VTT1.4) was used without the GENFLO coupling. It is worth noting that similar modifications to introduce high-temperature creep and phase transformation models were previously implemented in the finite element (FE) based solution module (mechan = 1) in FRAPTRAN-QT1.4b [10]. The developments in this work on the other hand are applied to the BALON2 module that is used with the default mechanical model FRACAS-I (mechan = 2). The implementations in the FEA model were applied for calculations based on the hoop stress and strain, whilst the modifications in FRAPTRAN-VTT1.4 concern mechanical deformation based on the effective stress and strain.

3.1 High-Temperature Creep Deformation

Firstly, the ballooning model in FRAPTRAN was updated to consider high-temperature creep strain deformation, instead of the plastic strain, which was previously used. The total deformation hence becomes as.

$$\varepsilon_{\text{total}} = \varepsilon_{\text{th}} + \varepsilon_{\theta} + \varepsilon_{\text{creep}} \tag{4}$$

Equation (4) takes into account two different high-temperature creep models: Rosinger 1984 [1] and Kaddour et al. 2004 [2]. The former is based on Zircaloy-4 (Zry-4) data and is suitable for generic zircaloy claddings, while the latter provides different parameters for both Zry-4 and Zr1%NbO (M5) types. As highlighted in [11], although both creep laws are developed for steady state conditions, plastic deformation during ballooning in LOCA conditions is assumed to follow the same law. Consistently, the same approach has been adopted in this work. This assumption is particularly relevant at large strains approaching the burst strain and is consistent with the original work in developing the burst failure criteria by Rosinger [12]. The two models report creep strain rates in Norton creep formulation:

$$\dot{\varepsilon}_{creep} = \frac{A}{T}\sigma^{n}\exp\left(-\frac{Q}{RT}\right) \tag{5}$$

with A being strength coefficient, temperature T, activation energy Q, gas constant R, and stress exponent n. To implement the models consistently in FRAPTRAN, the strength coefficients A_f was obtained using the Hill theory [12].

$$A_f = \frac{A_z}{(F+G)^{(n+1)/2}} \tag{6}$$

The approach in Eq. (6) accounts for the fact that the strength coefficients A_z in the original work were obtained from uniaxial testing, whereas FRAPTRAN calculates effective isotropic strain and stress. F and G are the anisotropic factors defined by Hill. These parameters depend on the phase present in the cladding material. Under normal operation (Zr-α) is anisotropic and therefore F = 0.956, G = 0.304. Whilst under accident scenarios in the mixed α + β phase and the Zr-β phase, the cladding exhibits isotropic behavior so F = G = 0.5 [12].

Table 1 provides the original parameters of the Kaddour et al. model (A_Z, n, Q) [2], in addition to the recalculated A_f values as implemented in FRAPTRAN-VTT1.4.

Table 1. Kaddour et al. (2004) creep law parameters, with A_Z, n, Q from [2]

Material phase		Parameter			
		$A_z[MPa^{-n}s^{-1}]$	n	$A_f[MPa^{-n}s^{-1}]$	$Q/R[K]$
		Zry-4			
α-phase	$\sigma \leq 15MPa$	1.00E + 06	1.3	7.67E + 05	22852
	$\sigma > 15MPa$	1.63E + 08	5	8.15E + 07	38006
β-phase		1.00E + 04	4.25	1.00E + 04	18041
		Zr-1%NbO (M5)			
α-phase	$\sigma \leq 15MPa$	6.78E + 03	1	5.38E + 03	15996
	$\sigma > 15MPa$	2.28E + 03	4	1.28E + 03	23333
β-phase		1.23E + 03	4.3	1.23E + 03	17079

In the mixed phase region where both alpha and beta phases co-exist, the total creep rate ($\dot{\varepsilon}_{\alpha\beta}$) is calculated using linear interpolation with respect to the phase composition.

$$\dot{\varepsilon}_{\alpha\beta} = \dot{\varepsilon}_{\alpha}(1-y) + \dot{\varepsilon}_{\beta} y \tag{7}$$

$\dot{\varepsilon}_{\alpha}$ represents the strain rate from the α phase, $\dot{\varepsilon}_{\beta}$ is the strain rate from the β phase, and y is the volume fraction of β phase present in the cladding.

3.2 Phase Transformation

Inherently, FRAPTRAN does not have a specific model for phase transformation to calculate the phase composition in the cladding which limits its applicability with the creep described above. To address this limitation, a dynamic phase transformation model developed by Massih (2009) [3] has been implemented. The model solves the differential Eq. (8) which gives the time evolution of the volume fraction of β phase (y) with respect to time (t) and temperature (T):

$$\frac{dy}{dt} = k(T)\left[y_s(T) - y\right] \tag{8}$$

The term $k(T)$ represents the temperature-dependent reaction rate constant while $y_s(T)$ is the equilibrium volume fraction of β phase at temperature T. This information is then used to calculate the total strain rate during ballooning as in Eq. (7). The full description of model and solution methodology can be found in [3].

3.3 IRSN New Burst Criteria

Following a thorough literature analysis conducted by IRSN as part of the R2CA project [4], the experimental database for LOCA tests was updated, and new stress limit envelopes for failure criteria were fitted as a function of temperature as shown in formula (9). The coefficients for the minimum, maximum and best estimate (BE) envelopes are documented in Table 2.

$$\sigma_{\theta\ \mathrm{burst}}(T) = k \times e^{-qT}. \tag{9}$$

Table 2. Parameters for new IRSN true stress burst criteria [4].

Coefficients	Stress limit		
	σ_{min}	σ_{max}	σ_{BE}
k(MPa)	11015.44	75541.69	9080.44
$q\left(°\mathrm{C}^{-1}\right)$	0.007472	0.007022	0.005817

4 Results of the Validation Cases

In total, eight validation tests were recalculated with FRAPTRAN-VTT.1.4 updated models and criteria: 7 selected LOCA validation cases from FRAPTRAN integral assessment database listed in Table 3 [13], in addition to one case for M5 cladding, IFA-650.15 (PWR 17 × 17, 65 MWd/kgU) from Halden.

The original plasticity model and Kaddour et al. (2004) creep were used for all 8 validation cases. Each test against five different failure criteria: FRAPTRAN original criterion, IRSN BE, IRSN min., IRSN max., and by Meyer and Wiesenack (2022) [14] (8 × 5 calculations each). Rosinger model, on the other hand, is not suitable for Zr1%Nb cladding type and therefore only six validation cases were tested against the five criteria. When using the newly implemented criteria (IRSN, Meyer and Wiesenack), the secondary strain limit was disabled.

Table 3. Simulated LOCA tests from the FRAPTRAN integral assessment database [12].

LOCA test	Rod type	Rod average burnup, [MWd/kgU]
IFA-650.5	PWR 15x15	83
IFA-650.6	VVER-1000	56
IFA-650.7	BWR 10 × 10	44
LOC-11C rods 1 and 4	PWR 15 × 15	0
LOC-11C rod 2	PWR 15 × 15	0
LOC-11C rod 3	PWR 15 × 15	0
TREAT rod 16 and 17	BWR 7 × 7	0

The number of correctly and incorrectly predicted cases can be summarized as follows. FRAPTRAN original plasticity was found to be the most conservative deformation model; rod failure was falsely predicted for one test (LOC-11C rod 2) with all 5 burst criteria. The Kaddour model had one false occurrence of burst for the same test when selecting the IRSN min. criterion. Rosinger model, on the other hand, exhibits less accurate predictions, with two false occurrences of burst and one case where the model fails to predict rod rupture. The missed burst occurrence was registered for TREAT test when selecting Rosinger creep and the original FRAPTRAN burst criterion. With the original plastic deformation model, the failure in this rod was triggered by the strain limit. However, the strain evolution from Rosinger creep is low in all cases as can be observed in Fig. 2. Overall, the model is generally quite rigid compared to plastic deformation from the other two and therefore the strain limit was not reached.

In Figs. 1 and 2, the calculations results for all different deformation models and burst criteria are presented and compared to the experimental values of burst temperature and maximum residual hoop strain, respectively. In terms of burst temperature, the FRAPTRAN original plasticity and the Kaddour model provide more conservative forecasts. The IRSN min. is the most stringent stress limit among the employed burst criteria. Moreover, IRSN BE and IRSN max. produce comparable results to the original criterion in the code whereas IRSN min. exhibit a more conservative boundary. Overall, the choice of creep model and failure criterion for application depends on the level of conservatism needed in the analysis. Due to the limited number of tests and the inherent uncertainty included in developing the failure criteria, statistical estimation of probability for each model is not performed here.

Figure 3 illustrates an example of FRAPTRAN stress calculations for IFA 650.15. Ballooning starts at 166s and BALON2 module is called. At that time, the local stress in the node is already over the IRSN min. criterion and since the code is structured to check for failure only after BALON2 is called, the cladding is failed instantly. Other criteria trigger failure at different times as shown along the calculation. Given the fact that the cladding failed after 245 s in the experiment, Kaddour et al. (2004) creep model produces better predictions with most criteria compared to the original plastic strain deformation. Kaddour model was developed for M5 cladding whilst FRAPTRAN original strain calculation uses generic zircaloy correlations.

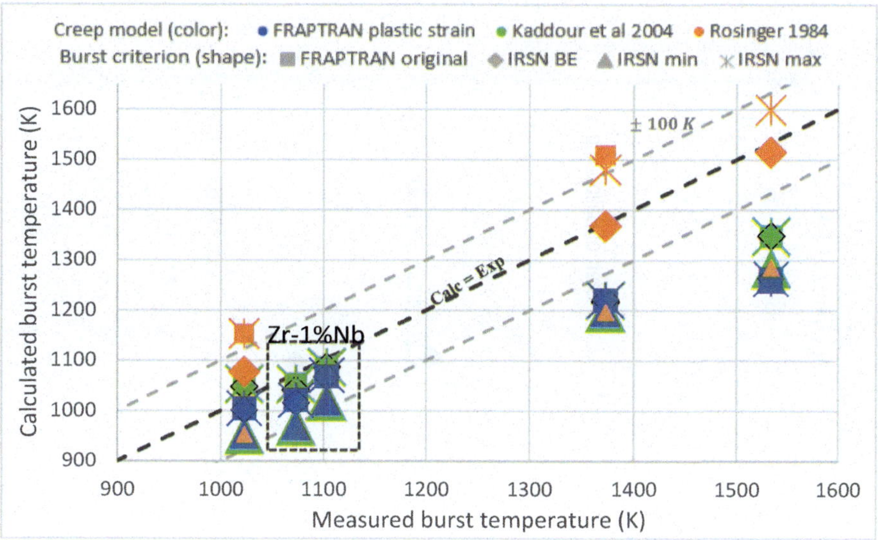

Fig. 1. Calculated vs experimental burst temperature for simulated validation cases.

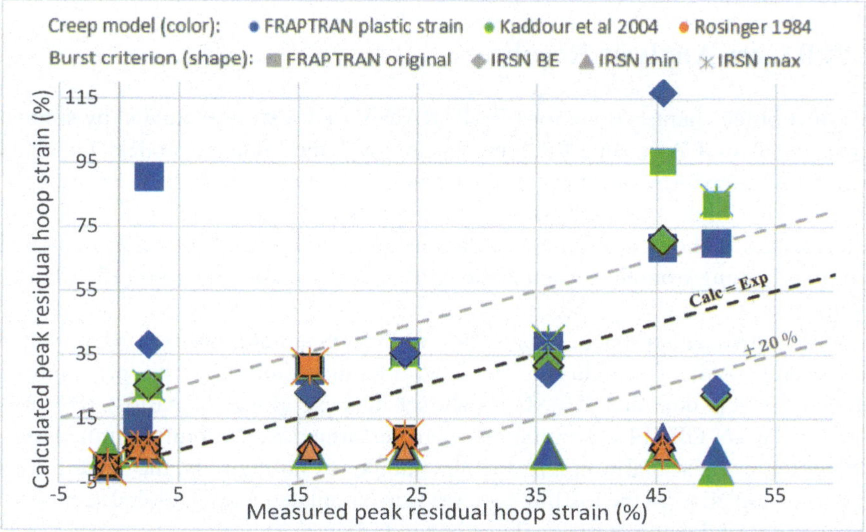

Fig. 2. Calculated vs experimental max. residual hoop strain for simulated validation cases.

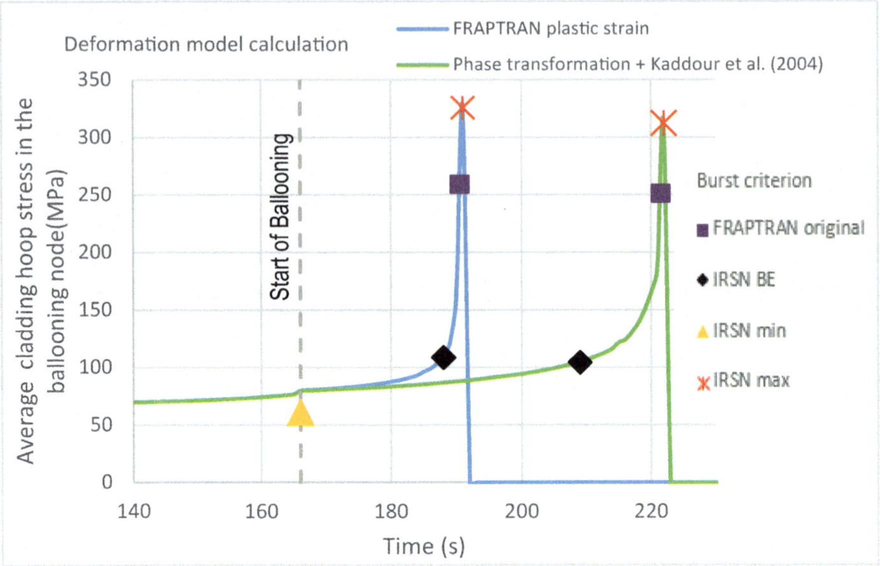

Fig. 3. FRAPTRAN-VTT1.4 stress calculations with various deformation models and burst criteria for IFA-650.15.

5 Full Core Analysis Results

The effect of the changes done for FRAPTRAN-VTT.1.4 were evaluated by simulating a large break LOCA in an EPR type reactor with the updated models for cladding creep, phase transformation laws and new fuel failure criteria. Full description of the original EPR simulations methodology and results are reported in [15, 16]. There are no differences in the steady state calculations done with FRAPCON nor the boundary conditions coming from the system code APROS between the previous [15, 16] and the current analysis.

In the statistical approach, the number of APROS simulations was 59, as stated by the first order Wilks' formula for one-sided tolerance limit when the probability and confidence levels both are 95%. For each set of these global scenarios, 1000 single-rod FRAPTRAN-GENFLO simulations were performed on randomly sampled rods. To pinpoint the worst scenario in terms of the number of failing rods, the same rods (that were first sampled from the loading pattern) were simulated in each global scenario. The statistically varied parameters are described in [15].

Various combinations of models for cladding plastic deformation and fuel failure criteria were tested with FRAPTRAN-GENFLO full core analysis:

- Kaddour et al. creep model + IRSN BE failure criterion
- Kaddour et al. creep model + IRSN min. failure criterion
- Kaddour et al. creep model + Temperature limit by Meyer and Wiesenack
- Original plastic deformation model + IRSN min. failure criterion
- Original plastic deformation model + original stress-based failure criterion, as a reference

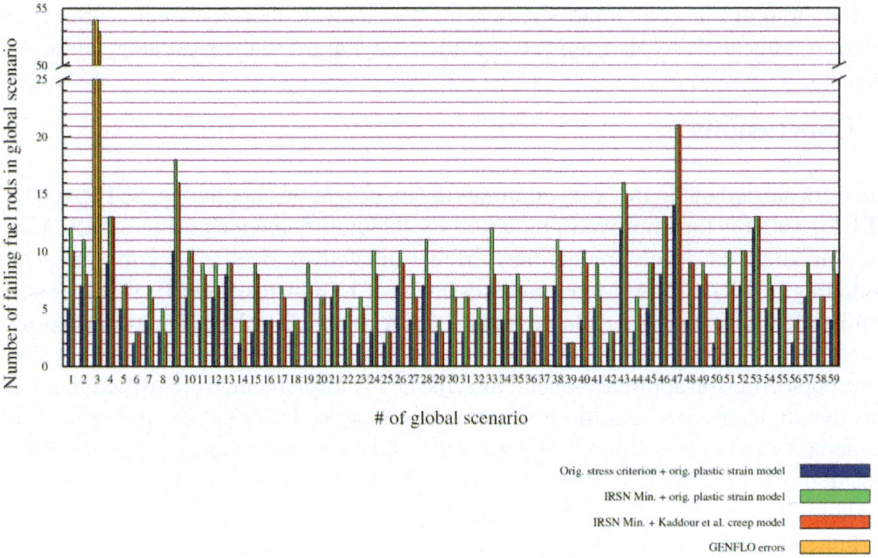

Fig. 4. Number of failing fuel rods in each global scenario according to FRAPTRAN-GENFLO; 1000 rods were simulated per scenario (same rods in each scenario).

From Fig. 4, it is seen that the most stringent combination in terms of number of fuel failures is to use the original plasticity model combined with the IRSN minimum criterion. However, in the worst scenario #47, the number of fuel failures is the same regardless of which plasticity model is used when the IRSN min. criterion is applied. In that scenario, the same rods fail regardless of the plasticity model. Compared to the maximum number of fuel failures (14) when using the as-delivered models, the new maximum number is 21, i.e., 0.7 percentage point difference, which can be considered as small. Thus, there is plenty of margin for the 10% limit set by the regulations in Finland. There were also a number of calculation errors in global scenario #3 due to the subchannel code GENFLO, and if all the crashed simulations would be considered as failed rods, the total fraction of failed rods in that scenario would be 6%.

Finally, all the rods that had been in reactor for two cycles before the LOCA event (about half of the assemblies in the core) were simulated with FRAPTRAN-GENFLO using the boundary conditions of the worst global case (#47). There were only one- and two-cycle rods in the loading, and from the data of the previous analysis it could be seen that the hoop strains in the one-cycle rods were rather low (<2%). For that reason, and to be able to compare the results with the previous analysis [16], one-cycle rods were not simulated here. The applied models were the Kaddour et al. creep model (+ phase transformation) and IRSN min. stress limit. The Kaddour model was chosen as the cladding material in the analysis was M5 and the model has parameters fitted for that cladding material. The number of fuel failures increased by 1.4 percentage points to 3.4% (compared to the analysis in [16]). There were no erroneous simulations. The IRSN min. failure criterion evaluated the cladding to fail right after the start of the ballooning so that the highest strains and stresses are much lower than with the original failure criterion. This analysis also showed considerable margin to the regulatory limit.

6 Conclusions

The work carried out in this study outlines the modifications applied to the FRAPTRAN-VTT.1.4 version for improved prediction of cladding behavior during a LOCA event. The implemented developments have been validated by detailed thermal mechanical modelling of FRAPTRAN integral assessment of LOCA tests and further extended to simulate full core analysis of large break LOCA in an EPR plant. The produced results show that the original plastic deformation model dictates failure in most cases regardless of the burst criterion applied. Nonetheless, the newly implemented creep models are more sensitive to burst criteria as more variation is observed. In terms of failure time, failure temperature and failure strain, Kaddour et al. (2004) model produced better results when considering the overall differences between calculated and measured results amongst the validation cases especially for M5 cladding type.

For full core analyses, it can be concluded that even with the most stringent combinations of plastic deformation models and fuel failure criteria (i.e., original plasticity model combined with IRSN min. criterion, and Kaddour et al. creep model combined with IRSN min. criteria), there is plenty of margin left for the regulatory safety limit (10%) concerning the number of failing fuel rods. Further developments needs were distinguished in GENFLO subchannel code so that it could handle all situations with long-lasting reversed core flows without calculation failures.

Acknowledgements.

This paper has been possible thanks to the work of all the partners of the project who are thanked here for their contribution. This project has received funding from the Euratom research and training programme 2014–2018 under grant agreement n° 847656.

References

1. Rosinger, H. E., 1984. A Model to predict the failure of Zircaloy-4 fuel sheating during postulated LOCA conditions," *Journal of Nuclear Materials*, vol. 120
2. Kaddour, D., Frechinet, S., Gourgues, A. F., Brachet, J. C., Portier, L., Pineau, A., 2004. Experimental determination of creep properties of zirconium alloys together with phase transformation. *Scripta Materialia*, vol. 51, no. 6
3. Massih, A. R., 2009. Transformation kinetics of zirconium alloys under non-isothermal conditions. *Journal of Nuclear Materials*, vol. 384, no. 3
4. Taurines, T., Belon, S., Dif, B., Arkoma, A., Kaliatka, T., Kulacsy, K., Jobst, M., Ovdiienko, I., Van Uffelen, P., Klouzal, J., Calabrese, R., 2023. Rod cladding failure during LOCA—Final report on experimental database reassessment and model/code improvements. R2CA project deliverable report D3.4

5. STUK, 2019. Regulatory Guides on nuclear safety (YVL), B. Plant and system design, YVL B.4 Nuclear fuel and reactor, 15.3.2019. Radiation and Nuclear Safety Authority.
6. Geelhood, K. J., Luscher, W.G., Beyer, C. E., Cuta, J.M., 2011. FRAPTRAN-1.4: A computer code for the transient analysis of oxide fuel rods. PNNL-19400, Vol.1, NUREG/CR-7023 Vol.1
7. Miettinen, J., Hämäläinen, A., 2002. GENFLO - A general thermal hydraulic solution for accident simulation. VTT Research notes 2163, ISBN 951–38–6083–3, ISSN 1455–0865
8. Siefken, L. J., Coryell, E. W., Harvego, E. A., Hohorst, J. K., 2001. SCDAP/RELAP5/MOD 3.3 Code Manual, MATPRO - A Library of Materials Properties for Light-Water-Reactor Accident Analysis.
9. Powers, D.A., Meyer, R.O., 1980. Cladding swelling and rupture models for LOCA analysis. U.S. NRC, NUREG-0630
10. Manngård, Tero, Lars Olof Jernkvist, and Ali Massih. "Evaluation of loss-of-coolant accident simulation tests with the fuel rod analysis code FRAPTRAN-1.4." *HI TEMP, Boston, MA, USA (2011).* Quantum Technologies AB, 2011, Technical report TR11–008V1.
11. Pastore, G., Williamson, R.L., Gardner, R.J., Novascone, S.R., Tompkins, J.B., Gamble, K.A., Hales, J.D., 2021. Analysis of fuel rod behavior during loss-of-coolant accidents using the BISON code: Cladding modeling developments and simulation of separate-effects experiments. Journal of Nuclear Materials, vol. 543
12. Neitzel, H. J., Rosinger, H. E., 1980. The Development of a Burst Criterion for Zircaloy Fuel Cladding under LOCA Conditions. Technical Report KfK 4343, Kernforschungszentrum Karlsruhe, Germany.
13. Geelhood, K. J., Luscher, W.G., Beyer, 2011. FRAPTRAN-1.4: Integral Assessment. PNNL-19400, Vol.2, NUREG/CR-7023 Vol.2
14. Meyer, R. O., Wiesenack, W., 2022. A critique of fuel behavior in LOCA safety analyses and a proposed alternative. Nuclear Engineering and Design, vol. 394
15. Arkoma, A., Hänninen, M., Rantamäki, K., Kurki, J., Hämäläinen, A., 2015. Statistical analysis of fuel failures in large break loss-of-coolant accident (LBLOCA) in EPR type nuclear power plant. Nuclear Engineering and Design, Vol. 285
16. Arkoma, A., Ikonen, T., 2016a. Statistical and sensitivity analysis of failing rods in EPR LB-LOCA. In proceedings of: TopFuel 2016, Boise, Idaho, USA, September 11–15, 2016, Paper 17570

Numerical Calculation on Thermal Expansion Of UO_2 – 3 Vol% Mo Microplate Pellet

H. S. Lee(✉), D. S. Kim, D. J. Kim, J. H. Yang, J. H. Yoon, and H. K. Kim

Korea Atomic Energy Research Institute, Daejeon, Republic of Korea
{hslee88,dskim86,djkim,yangjh,yjh1,hkkim1}@kaeri.re.kr

Abstract. UO_2 – 3 vol% Mo microplate pellets in which micro-scale metal Mo plates radially arrange in the pellet have been successfully fabricated with high radial thermal conductivity. As the Mo microplates are arranged with the radial direction, the radial thermal conductivity may be further increased, but the anisotropic properties of thermal expansion may be revealed. Herein, we numerically investigated the thermal expansion of UO_2 – 3 vol% Mo microplate pellets. After we calculated that the stress distribution of the UO_2 – 3 vol% Mo microplate pellets under LHGR of 200 W/cm. As a results, enhanced radial thermal conductivity of the fuel pellet decreases the temperature gradient of the pellets and mitigates maximum hoop stress by thermal expansion.

Keywords: Microplate fuel pellet · ATF Fuel · Thermal conductivity

1 Introduction

Improving the thermal conductivity of light water reactor (LWR) fuels could provide a great advantage in improving the reactor safety [1]. Although UO_2 has a high melting temperature, a steep temperature gradient appears in the fuel pellet due to its low thermal conductivity. Therefore, many studies of considering composite material nuclear fuel to improve the thermal conductivity with high thermal conductive material have being conducted. Since Mo has high thermal conductivity and high melting temperature [2], it is one of the attractive additives for UO_2 nuclear fuel. For this reason, there are various studies of improving the thermal conductivity of nuclear fuel employing Mo. If Mo particles array an effective heat transfer path in the UO_2 matrix, the amount of the additives to improve thermal conductivity could be minimized. We evaluated the UO_2 – Mo microplate fuel pellet in which a penny shaped Mo microplate was distributed, and confirmed that the thermal conductivity was improved by 47% at 1000 °C compared to UO_2 pellet [3]. The Fig. 1a shows the anisotropic characteristics of thermal conductivity in the UO_2 – 3 vol% Mo microplate pellet. The UO_2 – Mo microplate pellets to which the Mo microplate particles were applied had more improved radial thermal conductivity than the axial direction. When the ratio of the diameter and thickness of the added Mo microplate is defined as the shape factor, as the shape factor increase, the radial thermal conductivity was increased (Fig. 1b). Based on the anisotropy of the thermal conductivity of the microplate, it could be expected that the thermal expansion characteristics would also have anisotropy.

J. Liu and Y. Jiao (Eds.): WRFPM 2023, SPPHY 299, pp. 108–111, 2024.
https://doi.org/10.1007/978-981-99-7157-2_11

In this study, the thermal expansion behavior of UO_2 – 3 vol% Mo microplate pellet was evaluated by numerical calculations.

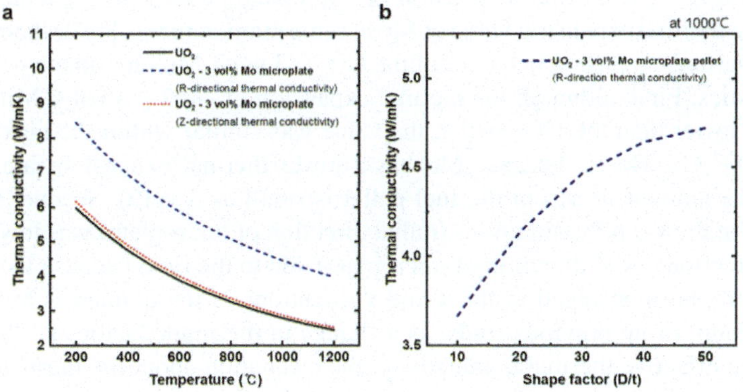

Fig. 1. Thermal conductivities of the UO_2 – 3 vol% Mo microplate pellet [3]. (a) Comparison of directional thermal conductivity of UO_2 – 3 vol% Mo microplate pellets in terms of temperature. (b) Thermal conductivity changes according to Mo microplate shape factor.

2 Numerical Calculation Model

To calculate the thermal expansion of UO_2 – 3 vol% Mo microplate pellet, the unit cube model was designed (Fig. 2). The amount of Mo microplates is 3 vol% in the model, the shape factor of the Mo microplate is configured as 30. And the xy-, yz-, and zx-surface were set as symmetric condition, and the remaining surfaces were set to the free condition. The deformation change of the model by thermal expansion at 300, 600 and 900 °C was calculated. In this study, it was assumed that the Mo microplates were distributed uniformly and parallel in the fuel pellet, and the interface between UO_2 and Mo was in a perfect contact.

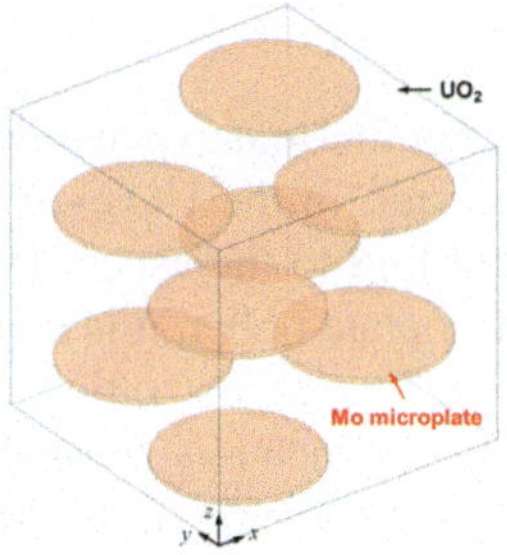

Fig. 2. Schematic illustration to calculate thermal expansion of UO_2 – 3 vol% Mo microplate pellet.

3 Results

The calculation results of thermal expansion in UO_2 – 3 vol% Mo microplate pellet is shown in Fig. 3. The thermal expansion of microplate pellet is lower than UO_2 pellet because the thermal expansion value of Mo at same temperature is lower than UO_2 material. The thermal expansion behavior of the UO_2 – 3 vol% Mo microplate pellet has two characteristics. First, although the thermal expansion of UO2 – 3 vol% Mo microplate pellet was lower than the UO_2 pellet, the value was similar within the maximum error 1.8% (1200 °C). This is because Mo has a lower thermal expansion rate than UO_2, however the amount of Mo in the fuel pellet is small as 3 vol%. Second, the thermal expansion in the x- and y-directions (radial direction of microplate) was lower than that in the z-direction (axial direction of microplate) due to the ratio because the proportion of Mo microplates arranged in the x and y directions in the domain is higher than in the z direction. From previous study, as a change in the shape factor of Mo microplate affected the effective thermal conductivity, the evaluation about the thermal expansion in accordance with the shape factor was conducted.

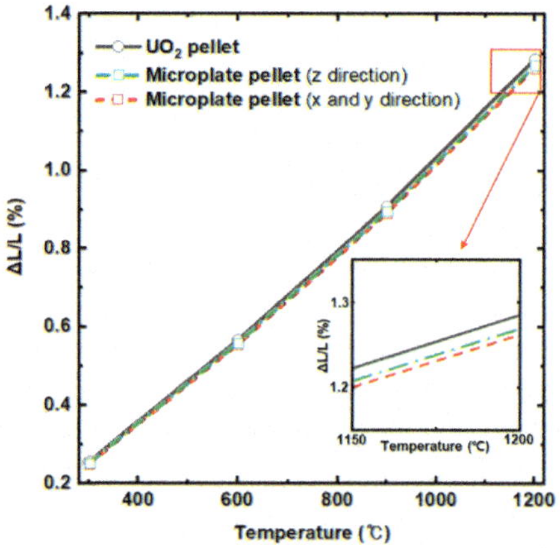

Fig. 3. Thermal expansion of UO_2 – 3 vol% Mo Microplate pellet. The inset plot highlights the thermal expansion value around 1175 °C.

Figure 4 shows the coefficient of thermal expansion of microplate pellets with shape factors of 10, 20, and 30. As a result, in the UO_2 – 3 vol% Mo microplate pellet, the thermal expansion behavior according to the aspect ratio was similar. Due to low Mo content in UO_2 – 3 vol% Mo microplate pellet, it is expected that there is no significant difference in thermal expansion at the aspect ratio of 10 to 30. However, if the Mo particles have an extreme shape factor such as spherical (low aspect ratio) or micro wire (high aspect ratio), there will be a difference in thermal expansion behavior from the results of this study even with the same amount.

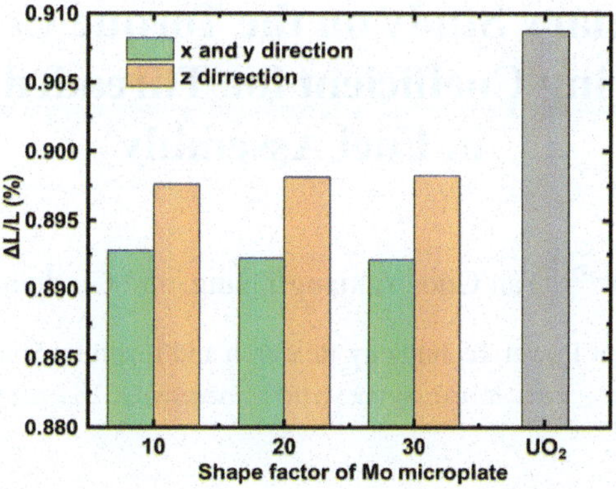

Fig. 4. Comparison of thermal expansion results in terms of Mo microplate shape factor.

4 Conclusion

In order to evaluate the thermal expansion behavior of the UO_2 – 3 vol% Mo microplate pellet, numerical calculations using a cubic model were performed. UO_2 – 3 vol% Mo microplate pellet showed anisotropic thermal expansion behavior, but it was calculated to be similar to UO_2 pellet. To improve this thermal expansion model, further research on thermal expansion measurement of UO_2 – Mo microplate pellets is needed.

References

1. Terrani, K.A., Wang, D., Ott, L.J., Montgomery, R.O.: The effect of fuel thermal conductivity on the behavior of LWR cores during loss-of coolant accidents. Journal of Nuclear Materials 448 (512) (2014).
2. Kim, D.J., Rhee, Y.W., Kim, J.H., Kim, K.S., Oh, J.S., Yang, J.H., Koo, Y.H., Song, K.W.: Fabrication of micro-cell UO_2 – Mo pellet with enhanced thermal conductivity. Journal of Nuclear Materials 462 (289) (2015).
3. Lee, H.S., Kim, D.S., Kim, D.J., Yang, J.H., Yoon, J.H., Koo, Y.H., Song, K.W.: Numerical investigation of the thermal conductivity of UO_2 – Mo microplate pellets to realize enhanced heat transfer in the fuel radial direction. Journal of Nuclear Materials 554 (2021).

Preliminary Study on the Torque Coefficient and Filtering Coefficient for Threaded Fasteners in Fuel Assembly

Hai Wu(✉), Yan Guo, Yuxiang Zhang, and Guoliang Zhang

China Nuclear Power Technology Research and Institute, Shenzhen, China
{wuhai2011,guo_yan,zhangyuxiang,zhangguoliang}@cgnpc.com.cn

Abstract. As an important element of non-permanent connection, threaded fasteners are commonly used in Pressurized Water Reactor (PWR) fuel assembly connections. The relationship between applied torque and preload of screw is always estimated based on the engineering experience, because the friction coefficient between screw and nut is complicated. The threaded fasteners in fuel assembly under Nuclear Power Plants (NPPs) operational conditions are subjected to external tensile or compressive loads, and when the external load is applied to the threaded fastener, the load is distributed between the members of fastener, the ratio between increased preload of screw and the external load is difficult to predict because of the complexity among the connection members. In order to understand the behavior of each member in threaded fasteners, it is necessary to investigate the relationship between torque and preload of the screw, and also the axial load distribution of threaded connections. In this paper, the torque coefficient and filtering coefficient of the threaded connection among fuel assembly components are studied. The torque coefficient results are compared with the theoretical calculation, and the fastener members are connected by preload screw, when the external load is applied to the connection, the filtering coefficient are obtained via FEM simulation, which is compared with the reference value. The comparison has shown very good agreements between referenced test and calculation.

Keywords: Torque · Preload · Threaded fasteners · Filtering coefficient · Torque coefficient

1 Introduction

Threaded connections are commonly used in mechanical structures, its reliability has a crucial impact on the safety of system components. When a threaded connection is failed, the whole system, including the connector, will suffer unpredictable damage. Fuel assembly is one of the most important systems in the Pressurized Water Reactor (PWR), it usually consists of top nozzle, bottom nozzle, guide thimbles, grids and fuel rods. The threaded fasteners are always used to connect the top nozzle and upper part of guide thimbles, as well as bottom nozzle and lower part of guide thimbles. Once the threaded connection fails, the fuel assemblies will not be able to withstand and transmit

J. Liu and Y. Jiao (Eds.): WRFPM 2023, SPPHY 299, pp. 112–122, 2024.
https://doi.org/10.1007/978-981-99-7157-2_12

the loads from reactor internals. Therefore, it is very important to study the performance of threaded connections during normal operation and design basis accident condition of Nuclear Power Plants (NPPs). However, the relation between the torque and preload of screw is difficult to predict due to the friction in the threads and under the screw head, which brings complicated nonlinearity.

In order to understand precisely the behavior of thread fasteners, many works have been done to study the relationship between external torque and preload of bolt, and also the loads distribution when the axial or lateral force is applied. Motosh [1] proposed an early research on the relationship between bolts preload and tightening torque, the physical and mathematical models are established by considering the process of applying torque to a bolt fastener, an analytical formula was given based on the study. Saotshi Izumi et al. [2], developed a three-dimensional finite element model to simulate the tightening and loosening process of threaded fastener, the nonlinearity of the relation between preload and tightening torque was predicted and the distribution of load was also analyzed. After applying the tightening torque, the preload is produced in the shaft of screw. However, when the threaded fasteners are subjected to static or dynamic loads such as vibration and thermal loads, the loads are transmitted to the connected components. The distribution of external loads is determined by the stiffness of components of connection. J. G. Williams et al. [3] studied the relationship between the external load and preload of screw based on the computational and experimental results, it is founded that the conventional analytical methods substantially over predict the proportion of the external load carried by the bolt. J. Wileman et al. [4] provides a simple technique for calculating the member stiffness in many types of bolted connections, an exponential expression for the stiffness is developed, and the results are compared with those of some other methods to prove the validity and rationality of the expression. Shi-kun Lu et al. [5] developed a calculation model for the distribution of axial force and stiffness considering the friction factor of the threaded connection, the calculation results, the test results and the finite element analysis (FEA) results were compared respectively, it is founded that the results were in a reasonable range, proving validity of the calculation of the method of this work is verified. Richard G. Budynas et al. [6] presents the mechanics model to provide the theoretical method to calculate the stiffness of threaded members.

2 Mechanical Model

2.1 The Relationship Between the Applied Torque and Preload of Screw

The physics models of tightening and loosening a thread are presented in Fig. 1(a) and Fig. 1(b) respectively [6]. In these two models, the forces applied to the threads to keep balance. To raise the load, a force P_R acts to the right, on the other hand, to lower the load, a force R_L acts to the left. The friction force is the product of the normal force N and the coefficient of friction f, the direction of friction force is opposed to the motion. The threads are in equilibrium under action of these loads.

For raising the load, the loads and their component in both directions are balanced, therefore, we have

$$\Sigma F_x = P_R - N \sin \lambda - fN \cos \lambda = 0 \tag{1}$$

$$`\Sigma F_y = -F - fN \sin\lambda + N \cos\lambda = 0 \tag{2}$$

Similarly, for lowering the load, we have

$$\Sigma F_x = -P_L - N \sin\lambda + fN \cos\lambda = 0 \tag{3}$$

$$`\Sigma F_y = -F + fN \sin\lambda + N \cos\lambda = 0 \tag{4}$$

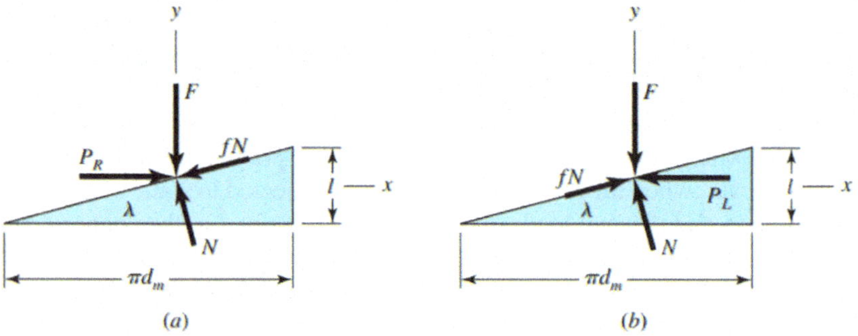

Fig. 1. The physics model of tightening and loosening threads

The normal force N can be eliminated due to it is not interested, then the load P_R and P_L can be solved respectively for raising and lowering the load:

For raising the load, it can be given

$$P_R = \frac{F(\sin\lambda + f\cos\lambda)}{\cos\lambda - f\sin\lambda} \tag{5}$$

And for lowering the load, it gives

$$P_L = \frac{F(f\cos\lambda - \sin\lambda)}{\cos\lambda + f\sin\lambda} \tag{6}$$

Because of the relationship between the half thread angle λ and the pitch diameter d_m and lead l of screw, $\tan\lambda = \frac{l}{\pi d_m}$, then the Eqs. (5) and (6) can be transformed as

$$P_R = \frac{F(l/\pi d_m + f)}{1 - (fl/\pi d_m)} \tag{7}$$

$$P_L = \frac{F(f - l/\pi d_m)}{1 + (fl/\pi d_m)} \tag{8}$$

It must be noted that the applied torque is the product of force (P_R and P_L) and the mean radius $d_m/2$, then Eqs. (7) and (8) can be rewritten as

$$T_R = \frac{Fd_m}{2}\frac{(l + \pi f d_m)}{\pi d_m - fl} \tag{9}$$

$$T_L = \frac{Fd_m}{2}\frac{(\pi f d_m - l)}{\pi d_m + fl} \tag{10}$$

Equations (9) and (10) give the torque to overcome the friction between the internal thread (nuts) and external thread (screws) during tightening and loosening threads. When the screw is tightened or loosened, another part torque is required to overcome the friction between the rotating and stationary members, this torque can be given

$$T_c = \frac{Ff_c d_m}{2} \tag{11}$$

In which f_c is the frictional coefficient between the rotating and stationary members, d_c is the average diameter of the friction. Therefore, the total torque required to tighten or loosen the threads can be given by combining Eqs. (9) and (11), or (10) and (11).

For raising the load, the toque can be expressed as

$$T = \frac{Fd_m}{2}\frac{(l + \pi f d_m)}{\pi d_m - fl} + \frac{Ff_c d_m}{2} \tag{12}$$

For lowering the load, the toque can be given as

$$T = \frac{Fd_m}{2}\frac{(\pi f d_m - l)}{\pi d_m + fl} + \frac{Ff_c d_m}{2} \tag{13}$$

2.2 The Redistribution of External Load Between Screw and Members

When an external load is applied to a fastener or bolted joint, the load is distributed between the screw and the members (components) being fastened together. The distribution of load is affected by several factors, including the stiffness and geometry of the members, the properties of the fastener, and the type and amount of preload applied to the joint.

Ideally, the load should be evenly distributed between the members and the fastener. The preload applied to the joint helps to ensure that the load is distributed more evenly by placing the joint under initial compression. This compression helps to prevent the members from separating and ensures that the load is shared between them. However, in practice, the load distribution may not always be even due to variations in the stiffness and geometry of the members or the properties of the fastener. This can result in uneven loading, which can cause stress concentration and potential failure in the joint. Therefore, it is important to ensure that the connection is properly designed and the fastener is correctly sized and tightened to achieve even load distribution and prevent premature failure.

Considering a bolted connection as shown in Fig. 2, before the external load P is applied, the preload F_p has been correctly produced by tightening, the load between screw and members F_m equals to the screw preload F_p. After applying the external load P, this load can be split into two parts, the first part is P_b, which is carried by bolt, and the second part is P_m, which is carried by the connected members. Then the resultant loads are,

$$\text{For bolt load} \quad F_b = F_p + P_b \tag{14}$$

$$\text{For members load} \quad F_m = P_m - F_p \tag{15}$$

The bolt becomes longer due to the tensile load P, assuming the additional elongation is δ, this elongation can be obtained by Hooke's law.

$$\delta = \frac{P_b}{k_b} \tag{16}$$

And also, the deformation of members is δ based on the geometry compatibility between bolt and members.

$$\delta = \frac{P_m}{k_m} \tag{17}$$

The Eq. (16) can be rewritten as

$$P_b = \delta \cdot k_b = \frac{P_m}{k_m} \cdot k_b \tag{18}$$

Since the external load P is the sum of P_b *and* P_m, thus we have

$$P_b = \delta \cdot k_b = \frac{P_m}{k_m} \cdot k_b = \frac{k_b}{k_m}(P - P_b) \tag{19}$$

Finally, it can be obtained that

$$\frac{P_m}{P} = \frac{k_m}{k_m + k_b} \tag{20}$$

It means that the fraction of external load carried by bolt depend on the relative stiffness of bolt and assembly, the stiffness of bolt k_b can be calculated easily, but the stiffness of connected members k_b cannot be obtained theoretically due to the complexity of structural characteristics. Thus, Finite Element Method (FEM) are usually employed to analyze the preload change in screw, the detailed FEM information is given in §3.3 of this paper.

3 Analysis of Threaded Connection in Fuel Assembly

3.1 The Structural Description of PWR Fuel Connection

Pressurized Water Reactors (PWR) are the most common type of nuclear reactor, accounting for two-thirds of current installed nuclear generating capacity worldwide. A typical PWR fuel assembly consists of fuel rods, top nozzle, bottom nozzle, spacing grids and guide thimble tubes (see Fig. 3). The bottom nozzle and top nozzle end the whole fuel assembly as the whole, there are several spacing grids between the bottom nozzle and top nozzle, these grids provide lateral axial and lateral positioning for fuel rods. The top and bottom nozzle are heavily constructed as they provide much of the mechanical support for the fuel assembly structure [7].

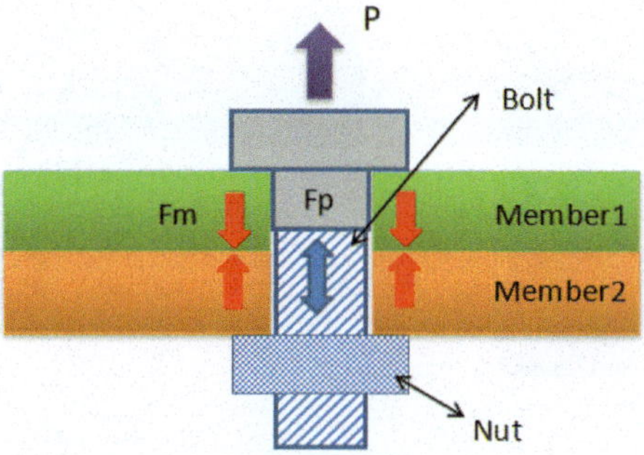

Fig. 2. An external load is applied to the bolted connection

Based on the design criteria of fuel assembly, the top and bottom nozzle should be designed as removable structure [8]. Therefore, the typical PWR top /bottom nozzle and guide thimble tubes are always connected by threaded fasteners or other 'quick disconnect' connections, such as bulge joints, etc. Both the top and bottom nozzles can be removed from the fuel assembly for potential repair by specific tools. However, threaded connections should be designed to maintain enough high preload during the normal operation of nuclear power plant to fulfil the mechanical integrity of the whole fuel assembly.

STEP-12™ fuel assembly is designed and developed by China General Nuclear (CGN), four lead fuel assembly was loaded into Unit 1 of the Ling Ao nuclear power plant in Guangdong province for irradiation tests in February 2016. To date, 3 cycles of irradiation have been completed. At the lower end of STEP-12™ fuel assembly, an lower end plug is weld onto the lower end of guide thimble tube section, the lower end plug is internally threaded to engage with the lower end screw that connects the guide thimble tube to the bottom nozzle (see Fig. 4(a) and Fig. 4(b)), which provides rigid connection of the guide thimble tube to the bottom nozzle with special stainless steel bolt that incorporate a mechanical locking feature.

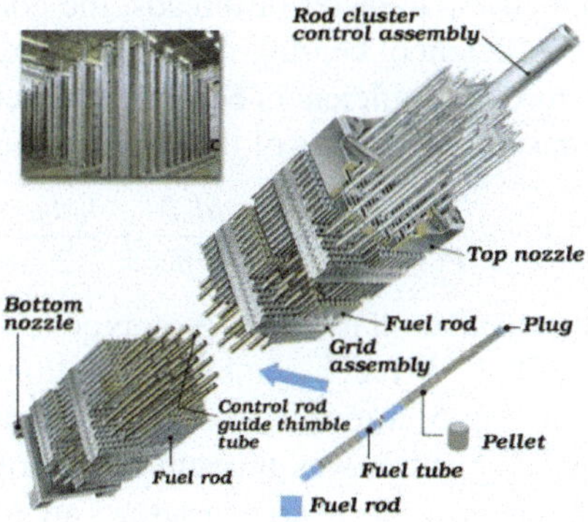

Fig. 3. The schematic view of typical PWR fuel assembly

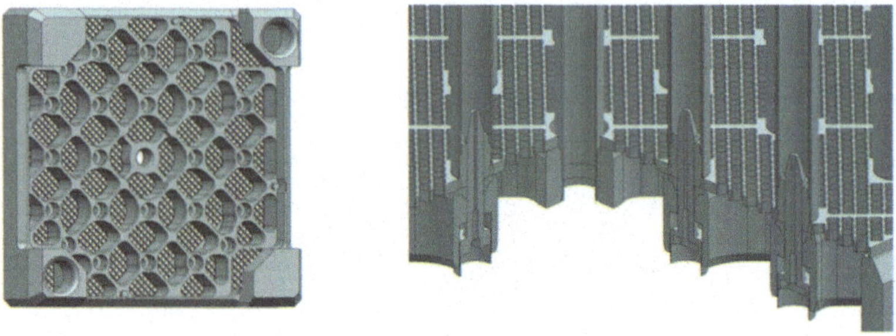

(a) (b)

Fig. 4. The lower end threaded connection between guide thimble tube and bottom nozzle

3.2 Torque Coefficient Analysis

Preload in the shaft of screw plays a key role to resist the external loads, consequently, it's very important to assure the preload during fuel assembly manufacturing. Several bolt fasteners tightening method can be used to develop the preload of screw, such as torque control method, rotation angle method, torque gradient method, elongation measurement method, loading method and thermal heating method, etc. Each method has its own advantages and disadvantages, and the final adopted method depends on the usage environment, convenience and accuracy requirement. The following Table 1 presents a summary of accuracy and relative cost by each method.

For STEP-12™ fuel assembly design, the standard metric screw M6 × 0.75 mm is adopted to connect the guide thimble tube to bottom nozzle, the screws are installed and tightened by torque wrench, the tightening is stopped until a specific torque value is achieved. This is the most common tightening method; it is very simple and quick to used when the bolt diameter does not exceed 30 mm. However, as shown in Table 1, this method suffers from the major intrinsic drawback is the uncertainty is high, in addition to the desired axial tension stress of bolt, this method also simultaneously introduces a torsion stress in the bolt, which results in increasing the total stress of bolt.

As per the mechanical model given in §2.1, the torque coefficient can be estimated by Eq. (13), with the geometric parameter of threads, the coefficient between internal and external threads, the coefficient of bearing-contact surfaces between the nut and the flange, etc. Therefore, the torque coefficient of STEP-12™ fuel design can be calculated by Eq. (21) based on the design parameters of threads and frictional coefficients.

$$K = \frac{T}{F} = \frac{d_m}{2}\frac{(l + \pi f d_m)}{\pi d_m - fl} + \frac{f_c d_m}{2} \tag{21}$$

In order to validate the calculation model, a prototype test was conducted on lower threaded connection of STEP-12™ fuel assembly, the bottom nozzle was fixed on the testing bench, as shown in Fig. 5, a torque was gradually and slowly applied to the shoulder screw, the screw elongation was measured by strain gauge to monitor the preload in the screw during the test. There are ten screws are tested, and the comparison between mechanical model calculation and test are shown in Fig. 6, it can be concluded that the calculation results are within the band of test results due to the complicated friction coefficient in the test condition.

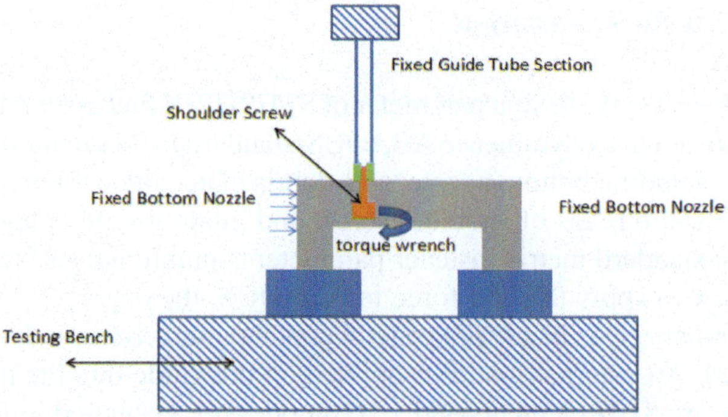

Fig. 5. The schematic diagram of STEP-12™ fuel assembly lower connection test

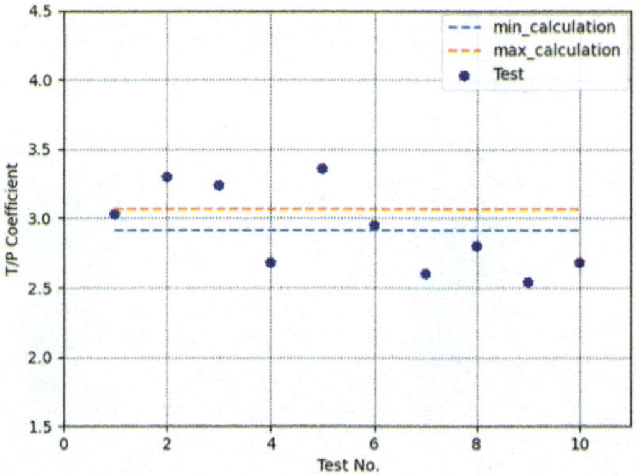

Fig. 6. Comparison between mechanical model and test for torque coefficient

Table 1. Manual preloading method for bolt torque applied accuracy and relative costs. [9]

Preload method	Accuracy	Relative cost
Feel	±35%	1
Torque wrench	±25%	1.5
Turns of nut	±15%	3
Preload indicating washer	±10%	7
Sensor wrench below yield using turn of nut	±15%	8
Sensor wrench yield point sensing	±8%	8
Bolt elongation	±3–5%	15
Strain gauges	±1%	20

3.3 Filtering Coefficient Analysis

FEM Model. Based on the design parameter of STEP-12™ fuel assembly lower connection, a two-dimensional axisymmetric ABAQUS/standard FEM model shown in Fig. 7(a) was developed, including bottom nozzle, anti-debris plate, guide thimble lower plug and shoulder screw. The threads of shoulder screw and guide thimble plug are established according to the standard metric fastener parameters, an internal surface is assigned on the shaft of screw to apply the bolt force in ABAQUS, the degree of freedom of lower surface of bottom nozzle is fixed, and general contact interactions with the whole model itself are defined. An external load P is applied on the guide thimble plug as shown in Fig. 7(a). There are 8107 elements and 10420 nodes are generated in the mesh model, which can significantly reduce the calculation time.

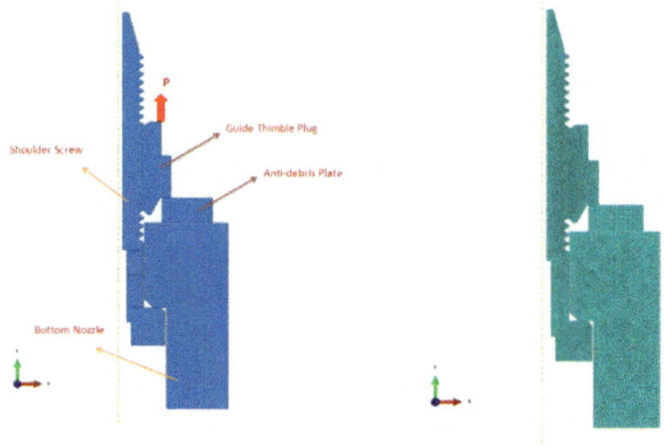

(a) (b)

Fig. 7. The FEM and mesh model of fuel assembly lower connection

FEM Results. Figure 8(a) shows the Von Mises stress results of STEP-12™ fuel assembly lower connection, the maximum stress is about 162 MPa, which is occurred at the first thread position, it is consistent with the practical experience, the maximum stress value is well below the yield stress threshold value of stainless-steel shoulder screw. Figure 8(b) presents the contact status of the lower connection, it can be seen that all the potential contact pairs are well established, especially for the contact between the shoulder screw and guide thimble plug, and among plug, anti-debris plate, bottom nozzle, etc. It can be concluded that the FEM analysis results are reasonable.

The preload in the shoulder screw is monitored during the process of external load applying, and the ratio between the increased shoulder screw preload and total external load, which gives the fraction of external load carried by shoulder screw, it is presented in Fig. 8 and compared with the reference value. The result indicates that the ratio between the creased screw preload and total external is basically the same with reference value, therefore, the FEM can be used to predict this coefficient in some other similar structure or harsh environments, such as hot and irradiation conditions in reactor during normal operation (Fig. 9).

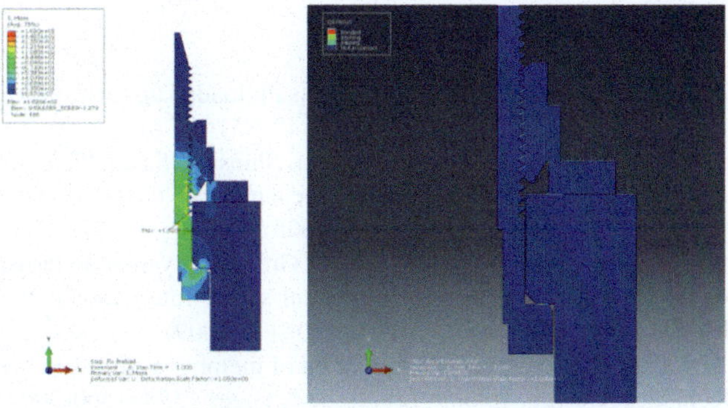

(a) (b)

Fig. 8. The mises and status of contact results of fuel assembly lower connection

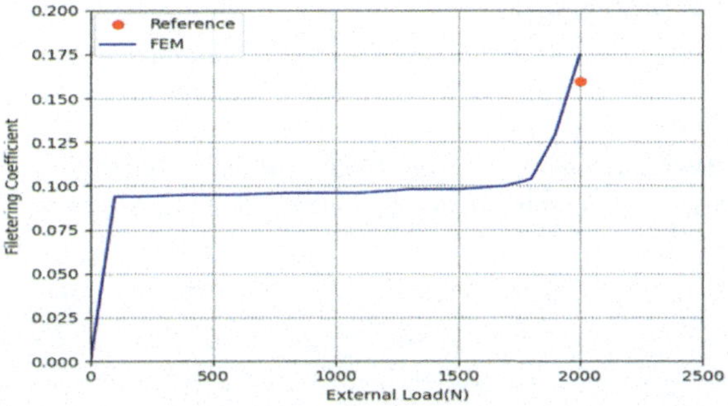

Fig. 9. Comparison between FEM and reference for STEP-12™ fuel design lower connection

4 Conclusion

Threaded fasteners are commonly used in fuel assembly design, the integrity of connection is very important for design evaluation of fuel assembly component, the preload and external load distribution are key factors for evaluation. This paper gives a theoretical method to predict the torque coefficient, which is the ratio between applied torque on screw and resulting preload in screw, and also present a FEM simulation to predict the filtering coefficient, which means the increased preload in screw and total external load.

The results indicate that both these two coefficients can be obtained reliably and easily, such methods can be used to simulate the behavior of fuel assembly connections in harsh environment, such as hot and irradiation conditions, considering additional thermal expansion and irradiation.

References

1. N. Motosh, Development of Design Charts for Bolts Preloaded up to the Plastic Range, Journal of Engineering for Industry, Vol. 98, Issue 3, pp. 849–851(1976)
2. Saoshi Izumi, Takashi Yokoyama, Atsushi Iwasaki, Shinsuke Sakai, Three-dimensional finite element analysis of tightening and loosening mechanism of threaded fastener, Engineering Failure Analysis, Vol. 12, Issue 4, pp. 604–615(2005)
3. J. G. Williams, R. E. Anley, D. H. Nash, T. G. F. Gray, Analysis of externally loaded bolted joints: Analytical, computational and experimental study, International Journal of Pressure Vessels and Piping, Volume 86, Issue 7, Pages 420–427(2009)
4. J. Wileman, M. Choudhury, I. Green, computation of member stiffness in bolted connections, Journal of Mechanical Design, volume 128, Issue 6, Pages 1357–1360(2006)
5. Shi-kun Lu, Deng-xin Hua, Yan Li, Fang-yuan Cui, Peng-yang Li, stiffness calculation model of thread connection considering friction factors, Mathematical Problems in Engineering, Volume 2019, Article ID 8424283(2019)
6. Richard G. Budynas, J. Keith Nisbett, Shigley's Mechanical Engineering Design 10th Edition, ISBN 978–0–07–352928–8(2014)
7. Nuclear Power Homepage, http://www.nuclear-power.com/nuclear-power-plant/nuclear-fuel/
8. National Energy Administration, PWR nuclear power plants system design-core-Part 3: Fuel assembly, NB/T 20057.2(2012)
9. Engineers Edge Bolt Fastener Tightening Method Review, https://www.engineersedge.com/hardware/bolt_fastener_tightening_method_review_15832.htm

ThN's Lattice-Assisted Thermal Conductivity Revisited

Barbara Szpunar(✉)

University of Saskatchewan, Saskatoon, SK, Canada
B.Szpunar@usask.ca

Abstract. The calculated electron density of states of ThN at the Fermi energy is low, which is in agreement with a very low electronic heat capacity coefficient measured experimentally. Therefore, it has been claimed that the electronic thermal conductivity of ThN should be low. However, the total thermal conductivity as measured recently is high. We used Quantum Espresso and EPW codes based on density-functional theory to evaluate the electron density of states, the electronic heat capacity coefficient, and electronic heat conductivity. We confirmed that the calculated electronic heat capacity coefficient was low due to the evaluated low electron density of states of ThN at the Fermi energy while the calculated electronic thermal conductivity of ThN was found to be high. However, the method of evaluating the remaining contribution from phonons is still disputed. The calculated lattice thermal conductivity of ThN is over-predicted unless a large smearing is applied in phono3py code. ShengBTE calculations predict that the lattice thermal conductivity of a pure single crystal may be one order of magnitude higher than the currently measured relatively low conductivity (~20 $Wm^{-1}K^{-1}$) at room temperature.

Keywords: Thermal conductivity · ThN · Enhanced accident tolerant nuclear fuel

1 Introduction

The predicted theoretical electron density of states of ThN at the Fermi energy is low, which is in agreement with a very low electronic heat capacity coefficient (0.001 J mol^{-1} K^{-2}) measured experimentally [1]. Therefore, it has been claimed that the electronic thermal conductivity of ThN should be low. However, the total thermal conductivity is high (55.4 $W(m\cdot K)^{-1}$), as measured by Parker et al. [1]. We used Quantum Espresso [2] and EPW [3] codes based on density-functional theory to evaluate the electron density of states, the electronic heat capacity coefficient, and electronic heat conductivity. We confirmed that the calculated electronic heat capacity coefficient was low [4] due to the evaluated low electron density of states of ThN at the Fermi energy. We found the calculated electronic thermal conductivity of ThN to be high (40.38 $W(m\cdot K)^{-1}$, which is smaller than the total experimental value but higher than the 37.79 $W(m\cdot K)^{-1}$ evaluated using experimental resistivity [5, 6].

J. Liu and Y. Jiao (Eds.): WRFPM 2023, SPPHY 299, pp. 123–130, 2024.
https://doi.org/10.1007/978-981-99-7157-2_13

The phonon density of states of ThN, calculated at a temperature of 0 K, does not agree with the experiment [4, 7] and the calculation of its lattice thermal conductivity is challenging due to predictions of a contribution from acoustic phonons of Th atoms that is too high.

2 Results and Discussion

2.1 Phonon Density of States

We described previously in detail the Density Functional Theory methodology, which is used here with the same parametrization [4] within the QE code [2]. Furthermore, our new 0 K calculation of the phonon density of states (DOS) of ThN using phonopy [8] shows a very high density of states of acoustic phonons of thorium as presented by a medium dashed blue line in Fig. 1, which is not observed experimentally [7]. However the calculated gap between the optical phonons of nitrogen and acoustic phonons of thorium is in agreement with experiment [7] but not in agreement with the gap previously calculated using QE code with second order force constants (FC2) calculated with imposed symmetry (ibrav = 2). On the other hand the DOS of ThN at 300 K calculated by phonopy [8] and ALM [9] codes and indicated by the dashed-dot green line in Fig. 1 does not show such a high density of states of acoustic phonons. The DOS of ThN used in the previous ShengBTE [10] calculations of thermo-physical properties of ThN [4] is similar, as can be compared in Fig. 1, where it is indicated by a short dashed red line. It was calculated using QE [2, 11] and a primitive unit cell without imposing symmetry (ibrav = 0). Furthermore we recalculated by ShengBTE the DOS of phonons of ThN for FC2 evaluated with imposed symmetry (ibrav = 2) and it is shown in Fig. 1 by a solid black line with a noticeably sharp peak emerging again near the slightly larger gap.

2.2 Heat Capacity

ShengBTE code [10] uses only three input files: FC2, which is evaluated by alternative codes like QE, third order force constant (FC3), which is calculated using python script, and the outputs from QE files obtained using input files with incorporated (defined by another python script) displacements of atoms. Additionally a CONTROL file must be provided, which defines lattice vectors and coordinates of base atoms used in the calculations. Although FC2 and FC3 provide these coordinates, they are overwritten by those specified in the CONTROL file. We demonstrated previously [4] that the phonon contribution to heat capacity (C_V), which was calculated **without imposing symmetry** in FC2, is in agreement with experiment [1]. In Fig. 2 we reproduced it, as indicated by dark blue triangles up, and compared it versus the amount calculated C_V with FC2 obtained by QE **with imposed symmetry** (indicated by dark pink empty circles). The latest values are only 0.7% lower at 300 K and the differences gradually become even smaller (0.07%). Both results agree well with experiment [1] (indicated by a solid green line) in Fig. 2 since the electronic contribution to heat capacity is very low (indicated by dotted red line) as evaluated previously [4] using a hybrid functional (B3LYP).

In Fig. 2 we also reproduced the previously [4] calculated total ($C_p + C_e$) heat capacity (dashed-dot red line) where symmetry was not imposed in QE calculations

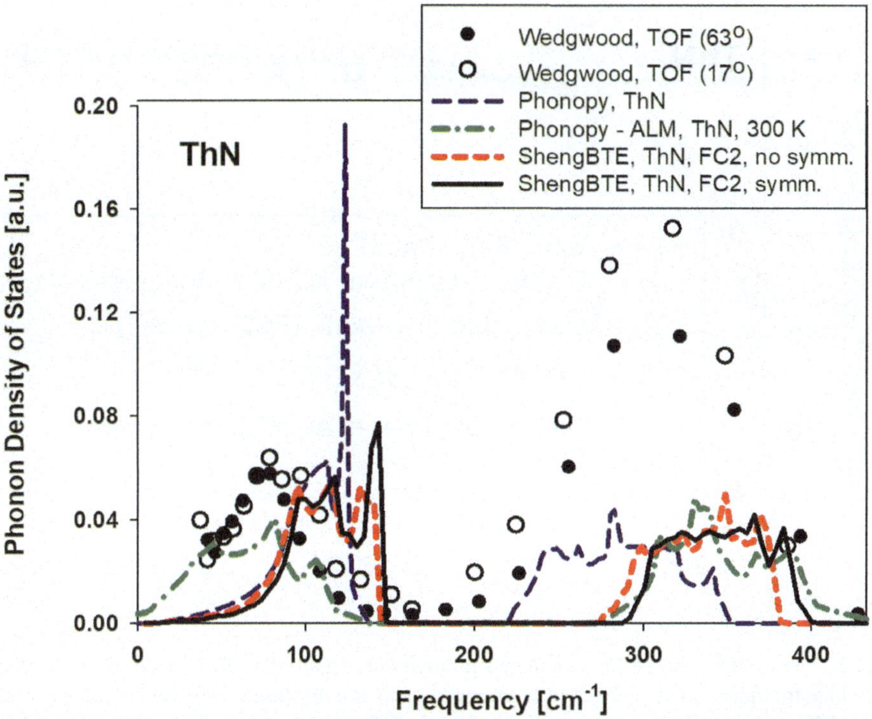

Fig. 1. The TOF spectra of ThN measured with 2 Å neutrons at 63° and 17° scattering angles [7] are indicated by full and open black circles, respectively. The total DOS of ThN calculated at 0 K temperature and 300 K by phonopy and ALM codes are indicated by a medium dashed blue line and a dashed-dot green line, respectively. The DOS of ThN used in the previous ShengBTE calculations [4] is indicated by a short dashed red line. The black solid line represents the respective DOS of ThN calculated in contrast by ShengBTE using FC2 with implementation of symmetry in QE evaluation. The linear tetrahedron method was used in the evaluation of the DOS.

of FC2. The total C_V value using FC2 with symmetry is very close and therefore it is omitted for clarity.

2.3 Lattice Thermal Conductivity

We demonstrate here that the lattice-assisted thermal conductivity, calculated by ShengBTE [10], is very sensitive to the number of symmetry operators used in calculating forces on atoms and any disorientation in the lattice base. The lattice-assisted thermal conductivity of ThN was calculated before [4] using QE [2, 11] and ShengBTE code [10] without imposing a symmetry (ibrav = 0, primitive unit cell) in calculating the second-order force constant (FC2). It is shown in Fig. 3 by a dark pink dashed-dot-dot line. Furthermore, the inconsistently used lattice vectors defining primitive cells in the final input (CONTROL) file to ShengBTE code, which were not updated previously and were rotated by 90 degrees [4], were corrected in the present calculations. Most properties like phonon density of states (indicated by the short dashed red line in Fig. 1),

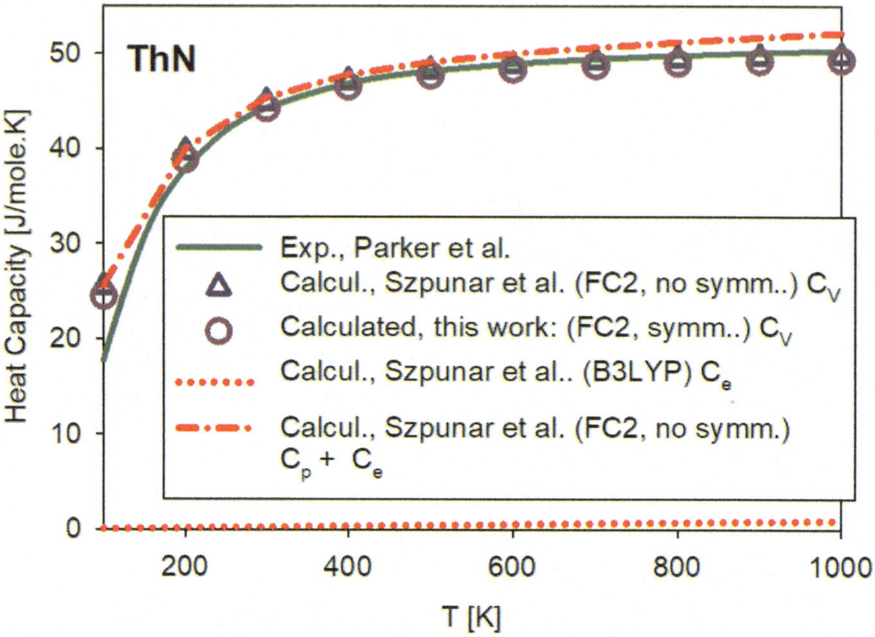

Fig. 2. The comparison of the calculated phonon contributions to heat capacity (indicated by dark blue triangles up) using FC2 without imposing symmetry (ibrav = 0) versus values obtained using FC2 with restricted symmetry (ibrav = 2) in QE calculations, indicated by dark pink open circles. The electronic contribution calculated previously [4] using B3LYP functional is indicated by a dotted red line, while the total heat capacity at constant pressure ($C_p + C_e$) is indicated by a dashed-dot, red line.

specific heat (Fig. 2), and scattering rates were not affected by this artificial misorientation in input of the CONTROL file. However, its overwriting of the correct input from the QE FC2 file resulted in a reduction of thermal conductivity (at 300 K to 19 $Wm^{-1}K^{-1}$ from 27 $Wm^{-1}K^{-1}$), which was in excellent agreement with experiment as described previously [4]. Additionally, we have just identified that nine out of 108 output files of the calculated forces in a simultaneously submitted multiple jobs' run of QE in the array were affected due to a predominant reduction of the number of symmetry operators.

Furthermore, Fig. 3 demonstrates that the lattice-assisted thermal conductivity of ThN calculated using QE and ShengBTE code with the corrected orientation, no symmetry imposed in FC2 evaluation and submitted as an array of 108 multiple jobs' run (pink dashed-dot-dot line) is about 10 times less than the conductivity evaluated using the array limited (via %1) to a single run only (red solid line). The respective results for FC2 calculated with assumed symmetry (ibrav = 2) are more than twice as high (780 $Wm^{-1}K^{-1}$ versus 280 $Wm^{-1}K^{-1}$ at 300 K) and are shown by a dashed black line. This thermal conductivity with a higher order of magnitude than currently measured [1] may eventually be reproduced in a pure single crystal of ThN.

In this work, we repeated calculations of the lattice thermal conductivity of ThN using phono3py [12] code. We found it is over-predicted unless a large smearing (e.g. sigma =

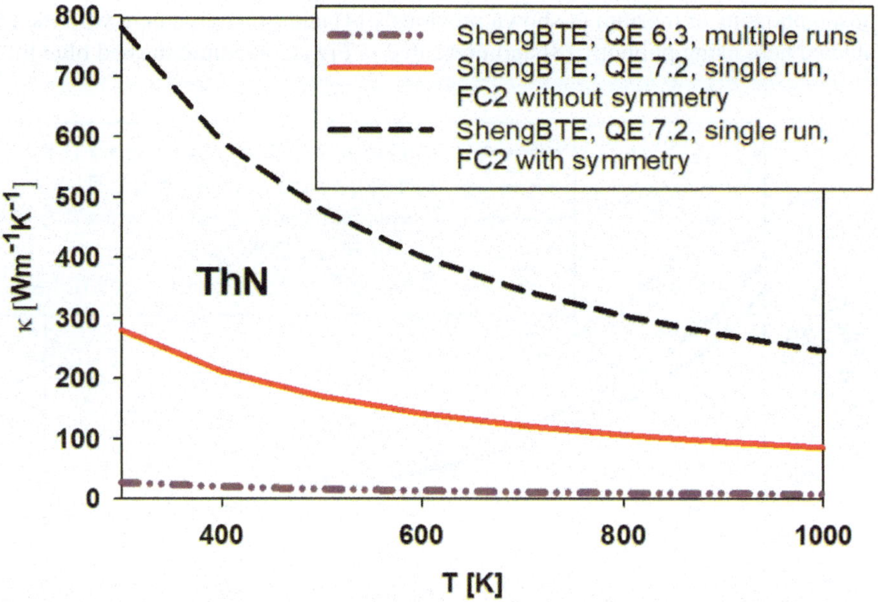

Fig. 3. A comparison of the lattice-assisted thermal conductivity of ThN calculated using QE and ShengBTE code with the corrected orientation, no symmetry assumed for FC2 evaluation and submitted as an array of 108 multiple jobs' run (dark pink dashed-dot-dot line) versus the thermal conductivity evaluated using the array with the restriction of a single run only (red solid line). The respective results for FC2 calculated with assumed symmetry (ibrav = 2) are shown by a dashed black line.

0.9 Thz, (50 × 50 × 50) mesh at 300 K, $\kappa = 23\ \mathrm{W(m\cdot K)^{-1}}$) is applied in phono3py code as shown in Fig. 4. This effective reduction of ThN thermal conductivity by smearing is in agreement with the relaxation time approximation since the phonon linewidths are inversely proportional to the relaxation time. The lattice-assisted thermal conductivity of ThN, as calculated previously [4] using QE [2] and ShengBTE code [10] in a very good agreement with current measurement [1] due to numerical artifacts as discussed above, is reproduced in Fig. 4 by a dotted red line. However most properties (phonon density of states, specific heat, scattering rate) were not affected by the discussed inconsistencies in FC3 and the CONTROL file; therefore we posted online the three input files used there (https://www.barbara-research.ca/ThN-data-thermophysical.htm), which allows to calculate lattice-assisted thermal conductivity and all other thermo-physical properties to be reproduced well using ShengBTE code [10]. The respective lattice thermal conductivity with updated orientation in the CONTROL file by 90 degrees, which resulted in an increase of thermal conductivity (at 300 K from 19 $\mathrm{Wm^{-1}K^{-1}}$ to 27 $\mathrm{Wm^{-1}K^{-1}}$), is indicated by a dark pink dashed-dot-dot line in Figs. 3 and 4 for comparison. ShengBTE also evaluated a much larger thermal conductivity (78 $\mathrm{Wm^{-1}K^{-1}}$) when using the same third-order constant (FC3) with an alternative second-order force constant (FC2) that was calculated with symmetry implemented. This also led to a large density of states of acoustic phonons of thorium as shown previously [4] and is in agreement with the DOS evaluated here using phonopy [8] and presented in Fig. 1, (medium dashed blue line).

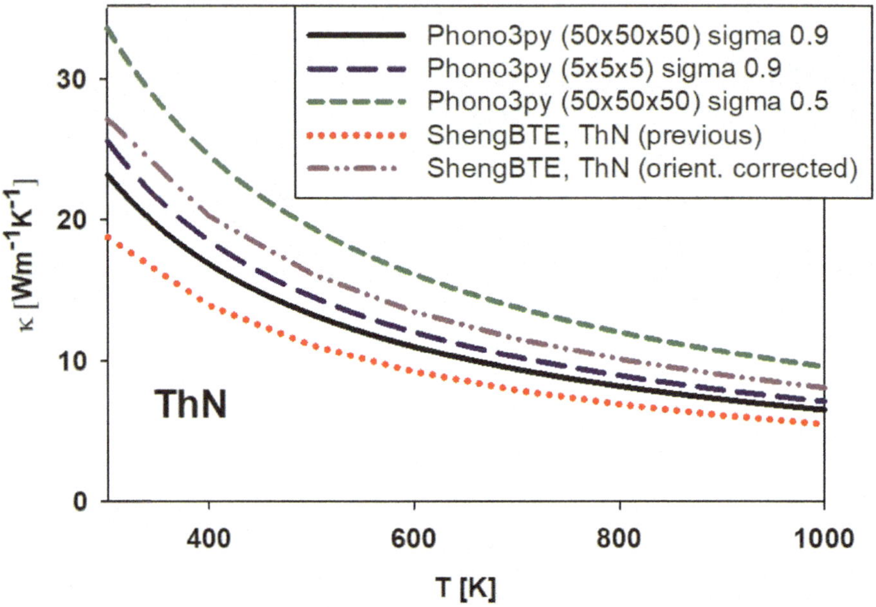

Fig. 4. A comparison of the lattice-assisted thermal conductivity of ThN calculated using QE and ShengBTE code, which is indicated by a dark pink dashed-dot-dot line (the previously evaluated values [4] with artificial misorientation are indicated by dotted red line) versus that evaluated via phono3py for a (50 × 50 × 50) mesh with sigma equal to 0.9 (black solid line) and 0.5 (short dashed green line) and a (5 × 5 × 5) mesh with sigma equal to 0.9 (long dashed blue line).

The other factor that may reduce lattice thermal conductivity in a metallic solid is e-ph scattering. However, the electron density of states at Fermi energy of ThN is very low and, as evaluated previously using EPW code [3], the e-ph coupling is also low [5, 6]. The four-phonon scattering rate may also lower thermal conductivity but it is significant predominantly at high temperatures as demonstrated, for example, for BAs in a recently extended version of ShengBTE code [13]. However, such extensions are not yet planned for the phono3py code [12]. An extension of phono3py to include temperature-dependent random displacements in the evaluation of the thermal conductivity of ThN may be promising as a reduction in the density of states of acoustic phonons of Th was observed with this effect as also discussed elsewhere [14].

3 Summary

It has been confirmed theoretically that ThN has a high lattice thermal conductivity. Additionally, it remains high at high temperatures due to a significant electronic contribution [5, 6]; therefore, it is considered for application as an Enhanced Accident Tolerant

Nuclear Fuel. However, ThN is not stable [15] and therefore more research is required to determine how it can be stabilized and possibly used in encapsulated fuel. An extension of phono3py to include temperature-dependent random displacements in the evaluation of the thermal conductivity of ThN may be promising as a reduced density of states of the acoustic phonons was observed in ThN with this effect.

Acknowledgments. The authors acknowledge access to high-performance supercomputers at Alliance Canada (CalculQuebec, Sharcnet, and WestGrid).

Free access to Quantum Espresso, EPW, ShengBTE, phonopy, phono3py codes with technical support is acknowledged. The very helpful e-mail comments by the developers of phonopy and phono3py (A. Togo) and ShengTE (J. Carrete and A. Karttunen) are also acknowledged.

The author acknowledges a constructive discussion with Dr. S. Poncé and a very helpful 2021 EPW workshop.

This work was supported by a Discovery grant from the National Sciences and Engineering Research Council of Canada.

References

1. Parker S.S., White J.T., Hosemann P., Nelson A.T., Thermophysical properties of thorium mononitride from 298 to 1700 K, J. Nucl. Mater. 526 (2019) 151760,9.
2. Giannozzi *et al.*, QUANTUM ESPRESSO: a modular and open-source software project for quantum simulations of materials, J. Phys. Condens. Matter 21 (2009) 395502–395521.
3. Poncé S., Margine E.R., Verdia C., Giustinoa F., EPW: Electron–phonon coupling, transport and superconducting properties using maximally localized Wannier functions, Comput. Phys. Commun. 209 (2016) 116–133.
4. Szpunar B., Ranasinghe J.I., Malakkal L., and Szpunar J.A., First principles investigation of thermal properties of thorium mononitride, J. Alloys Compd., 879 (2021) 16046, 8.
5. Szpunar B., First principles investigation of the electronic-thermal transport of ThN, UN and ThC, Nucl. Mater. Energy, 32 (2022) 101212.
6. Szpunar B., Ranasinghe J.I., Szpunar J.A., Malakkal L., Comparison of the electronic transport of ThN against ThC, J. Phys. and Chem. Solids, 165 (2022) 110647.
7. Wedgwood F. A., Actinide chalcogenides and pnictides. III. Optical-phonon frequency determination in UX and ThX compounds by neutron scattering, J. Phys. C (1974) 3203–3218.
8. Togo A. and Tanaka I., First principles phonon calculations in materials science, Scr. Mater., **108** (2015) 1–5.
9. Tadano T. Tsuneyuki S, First-Principles Lattice Dynamics Method for Strongly A harmonic Crystals, J. Phys. Soc. Jpn. 87 (2018) 041015, 11.
10. . Li W., Carrete J., Katcho N.A., Mingo N., ShengBTE: A solver of the Boltzmann transport equation for phonons, Comput. Phys. Commun. 185 (2014) 1747–1758.
11. Baroni S., Giannozzi P., Testa A., Green's-function approach to linear response in solids, Phys. Rev. Lett. 58 (1987) 1861–1864.
12. Togo A., First-principles Phonon Calculations with Phonopy and Phono3py, J. Phys. Soc. Jpn., **92** (2023) 012001, 21.
13. Zherui H., Xiaolong Y.; Wu L.; Tianli F.; Xiulin R., FourPhonon: An extension module to ShengBTE for computing four-phonon scattering rates and thermal conductivity, Comput. Phys. Commun. *270 (2022) 108179, 11.*

14. Szpunar B., Ranasinghe J.I., Malakkal L., and Szpunar J.A., First principles investigation of thermal properties of thorium mononitride, J. Alloys Compd., 961 (2023) 71068, p. 4, Corrigendum and extension to:., J. Alloys Compd., 879 (2021) 160467, p. 8.
15. Gouder T., Havela L., Black L., Wastin F., Rebizant J., Boulet P., Bouexiere D., Heathman S., Idir M., Synthesis and electronic properties of Th–N films, J. Alloys Compd., 336 (2002) 73–76.

How to Deal with the Threat of New Energy to the Safe Operation of Nuclear Fuel

Guo Shaosheng(✉), Zhang Xianggui, Zhang Qi, Ou Changgui, Gao Geng, Zhan Zhihe, and Gu Minqiang

China National Nuclear Corporation, Lianyungang, China
{guoss,zhangxg01,zhangqi06,oucg,gaogeng,zhanzh,gumq}@cnnp.com.cn

Abstract. In the last decade, with the carbon peaking and carbon neutrality goals, wind generation and solar photovoltaic generation have been increasing their share of the power grid. Now, it has become normal for nuclear power plants to participate in peak load regulation of the grid. During peak load regulation at the nuclear power plant, the change of reactor power brings the temperature change of fuel cladding and pellet. The pellet-cladding interaction caused by the temperature change of fuel pellets maybe cause the damage of fuel rod cladding, which will greatly threaten the safety of nuclear fuel. In this paper, firstly, the depth of peak load regulation in the nuclear power plant is discussed. Secondly, the power increase/reduction speed in the process of peak load regulation is researched to reduce the thermal stress effect of pellets. In addition, the risk of the local power of the reactor is discussed. At last, some other peak load regulation solutions for nuclear power plants, such as nuclear energy for heat supply, steam supply, pumped storage power plant, and battery storage are being explored.

Keywords: Peak load regulation · Nuclear fuel safety · Fuel cladding · Fuel pellet · Control mode · AO · DNBR · Temperature gradient

1 Energy Situation

Driven by the global energy transition, China, as the largest energy consumer, has been pushing forward its energy revolution in the past decade, and has proposed the strategy of "carbon peaking and carbon neutrality", i.e., "we aim to peak carbon dioxide emissions by 2030 and achieve the goal and vision of carbon neutrality before 2060". Under the guidance of the "carbon peaking and carbon neutrality" strategy, China's energy structure has been continuously optimized, taking data of 2021 as an example: coal consumption accounts for 56.0% of total primary energy consumption, oil for 18.5%, natural gas for 8.9%, and clean energy for 25.5%. Compared to 2012, the proportion of coal consumption in China has decreased by 12.5% points and that of clean energy consumption has increased by 11% points (Fig. 1).

According to the data, by the end of 2022, China's installed capacity of nuclear power has reached 56.89 million kW, nearly 4.5 times of that in 2012. In the past decade, China's nuclear power has stepped into a new stage of safe and efficient development (Fig. 2).

J. Liu and Y. Jiao (Eds.): WRFPM 2023, SPPHY 299, pp. 131–139, 2024.
https://doi.org/10.1007/978-981-99-7157-2_14

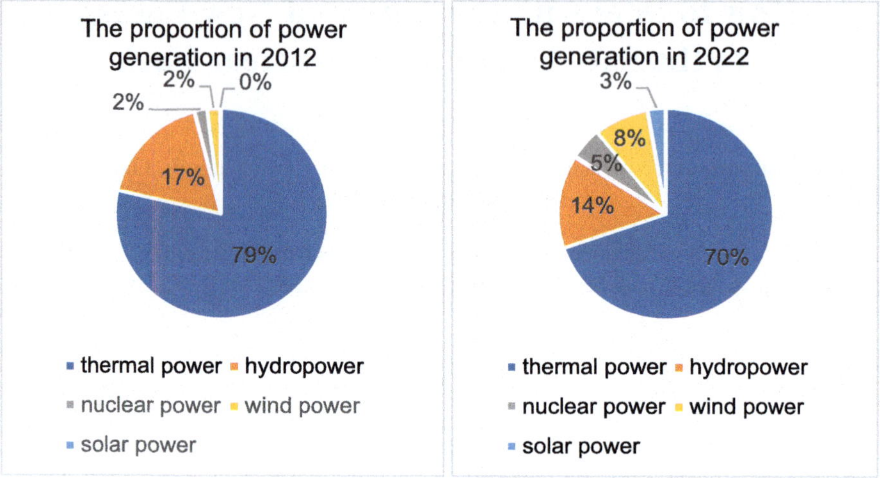

Fig. 1. The proportion of power generation in 2012 & 2022

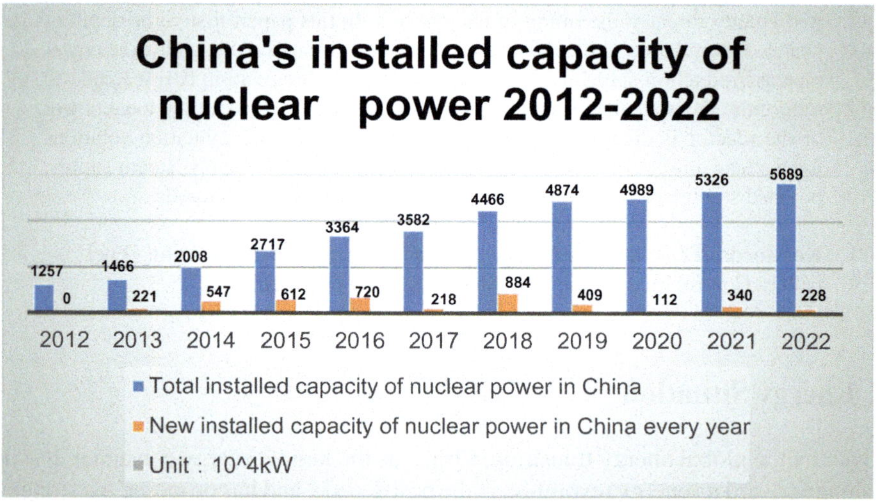

Fig. 2. China's installed capacity of nuclear power in 2012–2022

In terms of renewable energy development, China's newly installed wind and photovoltaic power generation capacity has reached 125 million kW in 2022, exceeding 100 million kW for three consecutive years and created a new record. Wind and photovoltaic power generation capacity has reached 1.19 trillion kWh, an increase of 207.3 billion kWh compared to 2021 with a growth of 21% on year-on-year basis, accounting for 13.8% of the total electricity consumption in China, with an increase of 2% points year-on-year (Fig. 3).

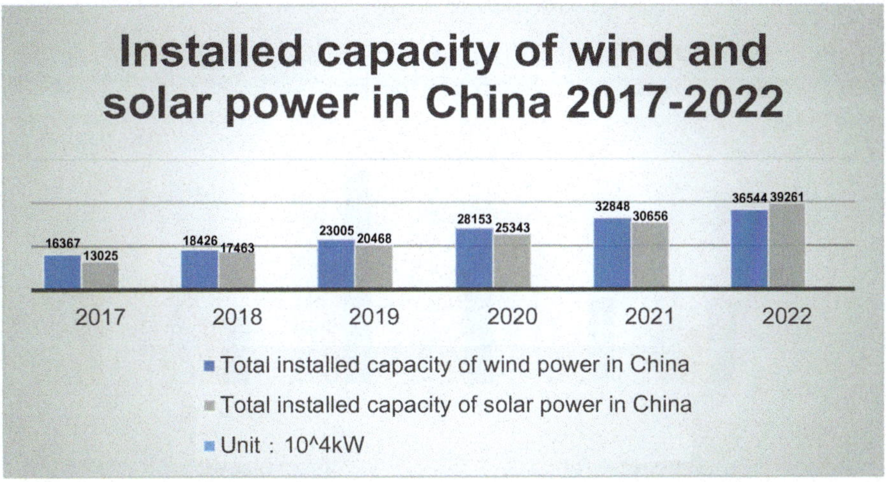

Fig. 3. Installed capacity of wind and solar power generation in China in 2017–2022

2 Analysis of the Situation of Nuclear Power Units Peak Load Regulation

In recent years, with the help of domestic conditions and government policy support, China has taken the lead in the development of wind power, photovoltaic and other renewable energy in the world, and its proportion of installed capacity has surpassed that of nuclear power units. In addition, the intermittent and random features of wind power and photovoltaic power sources dictates that they not only can't provide the peak load regulation ability that the power source should have, but also show the characteristics of reverse peak load regulation during certain periods.

In provinces with large installed nuclear power units and regions with rapid growth of nuclear power installed capacities, the power grid is in conformity with the characteristics of weekends and festivals. Maintaining full power operation of nuclear power units will increase the difficulties in peak load regulation. During low load periods, the problem of insufficient reserve is prominent, and there are risks of water and wind abandonment, which is inconsistent with the concept of efficient utilization of renewable energy. Therefore, for the moment, power grids in provinces such as Zhejiang, Fujian, and Liaoning, etc. need to decrease load to participate in peak load regulation during the Spring Festival, National Day and other special load days [1].

The following Fig. 4 shows the annual average peak load regulation times of a power station in China from 2020 to 2022. It can be seen that the annual number of peak load regulation times of each unit varies according to the situation, but the number of total peak load regulation times of the power station still shows an increasing trend year by year.

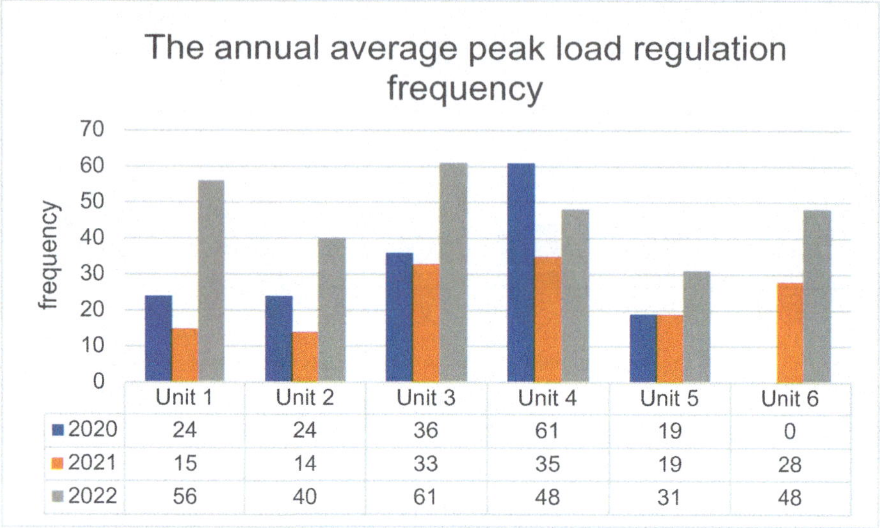

Fig. 4. The annual average peak load regulation frequency

3 Technical Measures to Deal with Nuclear Power Units Peak Load Regulation

The nuclear fuel assembly contains hundreds of fuel rods. Generally the clad is made of zirconium alloy and filled with UO_2 fuel pellets. The zirconium alloy cladding tube and the end plugs at both ends serve to contain the fuel pellets, allowing the radioactive fission products produced by the fission of the fuel pellets to be contained within the fuel cladding.

Frequent participation of nuclear fuel in peak load regulation may increase PCI effect. PCI effect refers to the interaction between fuel pellets and cladding in a reactor. Frequent peak load regulation leads to drastic temperature changes in fuel assembly, and fuel pellets form hourglass structure due to uneven thermal expansion. The fuel cladding generates ring ridges on the shoulder of corresponding pellets, and leads to local stress concentration near the ring ridges, which may lead to local fatigue damage of fuel cladding (Fig. 5).

According to research reports such as EPRI's *Fuel Reliability Guidelines: Pellet-Cladding Interaction, IAEA Nuclear Energy Series–Review of Fuel Failures in Water Cooled Reactors* and relevant practical experience, frequent participation of nuclear power units in peak load regulation, mediation and deep peak load regulation will increase the probability of nuclear fuel failure. Therefore, it is necessary to take corresponding technical measures to control the peak load regulation.

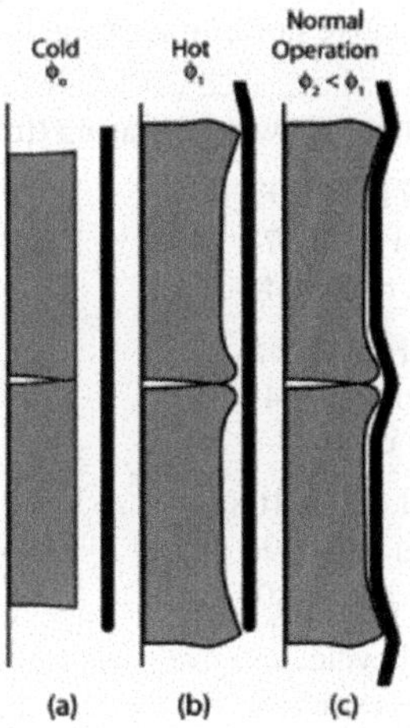

Fig. 5. Diagram of PCI effect

3.1 Balance of the Amplitude and Frequency of Peak Load Regulation with Multi-reactor Management

It can be seen from the above table on peak load regulation data of 6 units that the method of power reduction sharing can maximize the use of multi-reactor management concept, take advantage of units with large capacity in nuclear power base, reduce the range of reactor power variation and the probability of PCI effect occurring.

The core of the VVER reactor is composed of 163 groups of fuel assemblies with a refueling cycle of 18 months. Each refueling cycle adopts 1/3 refueling mode, and there are fuels of the first, the second and the third cycles in the core. According to the technical specifications of fuel assemblies and analysis report *The Ability of Units 1–6 of a Nuclear Power Plant to Participate in Power Grid peak load regulation*, the number of Units 1–6 participating in peak load regulation shall be controlled within the scope of less than 150 times, and the warning value is set to be 120 times.

The table below shows the number of accumulative power increasing times of in-core fuel assemblies of Units 1–4 by a certain day (Table 1).

Operation with frequent peak load regulation will increase the PCI effect of fuel elements, increase the risk of fuel assembly damage, and affect the security of nuclear fuel. Therefore, the unit shall be reasonably planned for peak load regulation according to the accumulated number of peak load regulation of in-core fuel assemblies, and operation with peak load regulation shall not be carried out in the reactor that has reached the specified limit.

3.2 Control the Rate of Change of Reactor Power

According to the reactor technical specifications, there are clear requirements for the speed of reactor raising and lowering power, in order to reduce the impact of PCI effect and prevent the loss of fuel assembly airtightness. When the power of the reactor is raised and lowered frequently, according to years' operating experience of the station,

Table 1. Number of accumulative power increasing times of in-core fuel assemblies

Unit	Number of accumulative power increasing times of in-core fuel assemblies
Unit 1	Fuels of the 3rd fuel cycle: 97 Fuels of the 2nd fuel cycle: 74 Fuels of the 1st fuel cycle: 58
Unit 2	Fuels of the 3rd fuel cycle: 65 Fuels of the 2nd fuel cycle: 44 Fuels of the 1st fuel cycle: 3
Unit 3	Fuels of the 3rd fuel cycle: 108 Fuels of the 2nd fuel cycle: 96 Fuels of the 1st fuel cycle: 80
Unit 4	Fuels of the 3rd fuel cycle: 93 Fuels of the 2nd fuel cycle: 71 Fuels of the 1st fuel cycle: 28

the PCI effect will increase sharply, and the possibility of the fuel assembly losing the airtightness will also increase greatly. Therefore, the reactor peak load regulation speed shall be strictly controlled, when the power grid needs peak load regulation of the unit. According to the fuel performance safety demonstration report, technical specifications and operation experience, the suggestions are as follows.

When the accumulative power rise times of the fuel assembly with the longest running time in the reactor are below 60, the controlled power rise speed shall not exceed 10% NP/h; When the number of times is 60–120, the controlled power rise speed shall not exceed 5% Nnom/h; When the number of times exceeds 120, the controlled power rise speed shall not exceed 3% NP/h.

At present, the duration of reactor low-power operation during peak load regulation is short (less than 48 h). Indeed, it is well known that extended reduced power operation (ERPO) enables the cladding to creep down on the pellet stack. Therefore, when the fuel rod goes back to Nominal Power, the thermal expansion of the fuel pellets generates strong hoop stresses in the cladding which in turn could generate a risk of PCI-SCC failure. So, there are corresponding technical requirements in the technical specification, i.e., when the reactor is in continuous operation below Nominal Power for more than 12 days, the power increase rate is required to be no more than 1% NP/h.

3.3 The Moving Speed of the Control Rod

According to the nuclear physical characteristics of the core, when the control rod is withdrew or inserted down, the neutron flux density distribution at the location of the control rod will change significantly, which leads to the increase of neutron flux density gradient of the nearby fuel rod and the great change of fuel rod power (Fig. 6).

In order to prevent the neutron flux density gradient from leading to a large power gradient and causing the exacerbation of PCI effect, it is necessary to strictly control the change of the neutron flux density distribution in the core and the moving speed of

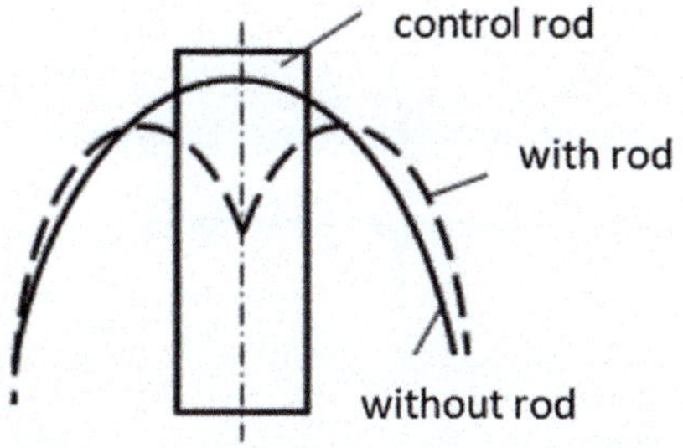

Fig. 6. Neutron flux density distribution during control rod insertion

control rod, so that AO can be controlled within the required range. Therefore, during the unit operation with peak load regulation, Technical Support Branch needs to prepare the reactivity control scheme according to the peak load regulation range, and the operator must strictly implement the reactivity control scheme to ensure that AO, line power density margin and DNBR margin are within the required controllable range.

Taking the peak load regulation data of the third fuel cycle of a power plant unit as an example, there are 93 times of peak load regulations in the third fuel cycle, and the peak load regulation durations are as follows (Fig. 7).

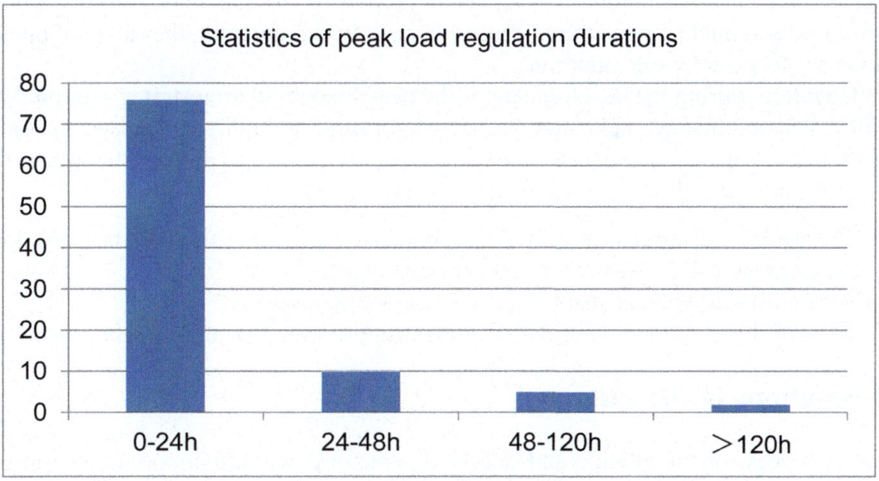

Fig. 7. Statistics of peak load regulation durations

The duration of peak load regulation is different, there are 76 times less than 24 h,10 times between 24 and 48 h, 5 times between 48 and 120 h, 3 times more than 120 h (Fig. 8).

It can be seen from the change trend of primary circuit radioactivity that the material stability of nuclear fuel is well guaranteed by the control methods above in the case of frequent peak load regulation of the unit.

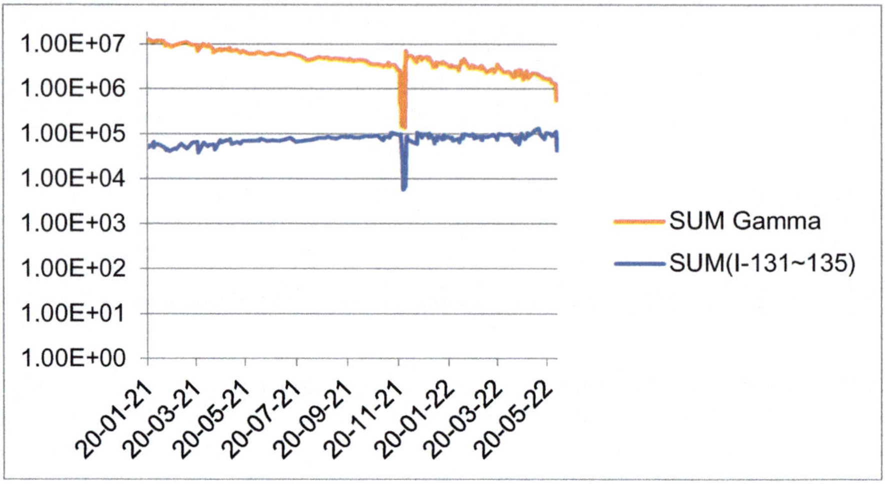

Fig. 8. The trend chart of summary iodine radioactivity and summary Gamma radioactivity in primary circuit of a power plant unit. (Nov.23–29, 2021 was for the unit minor maintenance)

3.4 Conclusion

In face of the current situations of increasing proportion of new energy resources in the power grid and nuclear power units requiring peak load regulation, the safety of nuclear fuel needs to attract more attention.

Therefore, during the development of the peak load regulation strategy for the NPP, deployment and management shall be carried out from the following aspects to control the PCI effect of fuel elements within an acceptable range, and prevent the loss of fuel assembly airtightness or even the damage of fuel assemblies:

(a) The peak load regulation range and frequency of each unit shall be considered comprehensively with multi-reactor management;
(b) To control the speed of reactor raising or lowering power;
(c) To control the range of core parameters during reactor raising or lowering power.

4 Solutions in the Future

Reducing peak load regulation and stabilizing reactor power are important measures to reduce the fuel risk, which can be tried by multiple directions of energy conversion.

4.1 Steam Energy Supply

A steam energy supply project of a nuclear power plant utilizes the main steam in the secondary circuit of the nuclear power unit to produce industrial steam through steam conversion equipment, which is transported to the consumer enterprise in the petrochemical base through the long-distance heat transmission network outside the plant. The design scale of the project is 600 t/h, accounting for approximately 10% of the steam capacity of a unit.

4.2 Pumped Storage Power Station

A certain number of pumped storage power stations can be built in the power grid to meet the needs of peak load regulation, so as to reduce the frequency and range of peak load regulation. As a mountainous area, northern Jiangsu Province can be planned as a suitable place to build reservoirs, which is more suitable for the construction of pumped storage power station.

4.3 Self-built Energy Storage

Utilization of the large-capacity battery packs combining with a variety of new energy generation methods. At present, it is in the experimental stage and far from commercial application stage. Large-scale application is in face of technical bottleneck and cost bottleneck. In recent years, with the breakthrough of sodion energy storage battery technology, the cost of battery will be greatly reduced, and it is expected to be used for power station energy storage in a large scale in the future.

4.4 Grid Construction

The northern Jiangsu region, as an energy base, is relatively backward in economy and the ability to consume electric energy locally is insufficient. Most of the power demands of East China power grid are in northern Zhejiang Province, Shanghai City and southern Jiangsu Province. Therefore, optimizing the construction of power grid and improving the transmission capacity can effectively reduce the peak load regulation condition caused by line power flow limitation.

References

1. 郑宽等. 等十三五"期间核电参与电网调峰前景分析. 中国电力. 2017,1(50):1.
2. Zheng Kuan, et al. Analysis on the Load Following Prospect of Nuclear Power in China during the Period of "13th Five-Year Plan". Electric Power. 2017, 1(50): 1.

Study on Thermal Hydraulic Characteristics of Rod-Type Fuel Irradiation Test Section

Yuanyue Zhang, Wenhua Yang, Liangqian Fu(✉), Wenbin Zhao, Liang Zhang, and Shuai Jin

Nuclear Power Institute of China, Chengdu, China
213163314@seu.edu.cn

Abstract. Irradiation test is the crucial segments to develop new type of nuclear fuel. In order to guide the irradiation test of rod-type fuel from the perspective of design and operation, the thermal hydraulic characteristics of the irradiation device test section were numerically simulated. In this paper, the simulated results are compared with the hydraulic test data to verify the reliability of the numerical simulation method. The thermal hydraulic characteristics of the test section under different linear heat rates (LHR) of fuel and mass flow rates of coolant were analyzed. It was verified whether the temperature, mass flow rate and other parameters met the design requirements and irradiation test criteria. The results show that the numerical simulation method has certain reliability. The LHR that makes the temperature of fuel test rod meet the criteria is 47.9–71.8 kW/m. Due to the outside narrow channel has less shunting, the flow channel design of fuel section is reasonable. When the average LHR of fuel is 60 kW/m, the temperature of the fuel test rod at different mass flow rates can meet the irradiation test criteria and leave a safety margin. The pressure drop of test section is mainly concentrated in the resistance section and fuel section. With the decrease of mass flow rate, the proportion of pressure drop in resistance section significantly increases, and the shunt ratio of outside narrow channel decreases. In summary, the irradiation test is reasonably designed to achieve the expected irradiation goals.

Keywords: Fuel irradiation · Test section · Fuel test rod · Temperature · Pressure drop · Shunt ratio

1 Introduction

HFETR is a research reactor that uses light water as coolant and moderator, and mainly performs irradiation test tasks for power reactor fuels and materials [1]. Fuel irradiation test is the crucial segments to develop new type nuclear fuel. The corresponding irradiation device is mainly composed of a device section and a test section, of which the test section is the main part of receiving neutron irradiation in the reactor. Rod fuel is a common type of fuel. In order to avoid direct contact with the reactor's primary coolant, nuclear fuel is placed in a protective tube to form a fuel test rod for irradiation. The fuel irradiation test puts forward requirements for the temperature of the fuel test rod.

J. Liu and Y. Jiao (Eds.): WRFPM 2023, SPPHY 299, pp. 140–151, 2024.
https://doi.org/10.1007/978-981-99-7157-2_15

However, in order to ensure the integrity of fuel test rod, it is difficult to directly measure its temperature due to the high temperature and neutron flux inside. The heat generated during the irradiation test of the test section is mainly taken away by the primary coolant flowing into the irradiation device, and the flow characteristics of the coolant in the test section determine the corresponding cooling conditions. Therefore, it is essential to obtain the thermal hydraulic characteristics of test section with the help of numerical simulation tools.

At present, the thermal hydraulic characteristics of test section have been studied by computational fluid dynamics (CFD) method simulation. Jin et al. [2] studied the flow resistance characteristics of the rod-type fuel test section through CFD simulation, which provided a reference for the hydraulic design of the irradiation device. They [3] also performed numerical simulations of the flow and heat transfer characteristics of test section and studied the uniformity of the coolant temperature in the test section. But their work was limited to one-way coupling of flow or heat transfer, without assessing the temperature distribution inside fuel test rod. Xing et al. [4] analyzed the thermal hydraulic characteristics of plate-type fuel irradiation test, and studied the influence of the thickness change of the fuel plate based on the bidirectional fluid-heat coupling model. Liao et al. [5] studied the changes of fuel assembly temperature and coolant flow distribution after irradiation swelling of plate fuel assemblies by numerical simulation methods.

Based on the fluid-heat coupling model of rod fuel irradiation test section established by CFX program, this paper studies the thermal hydraulic characteristics of test section under different conditions, evaluates whether the temperature of the fuel test rod meets the requirements of irradiation test, and verifies whether the parameters of coolant reach the design requirements, so as to guide the design and operation of the rod-type fuel irradiation test.

2 Numerical Models

2.1 Geometry

In this paper, the test section of a typical fuel irradiation device for HFETR is studied as the research object, and its structure is shown in Fig. 1. According to the different functions of test section components, it can be divided into four parts: inlet section, fuel section, resistance section and outlet section.

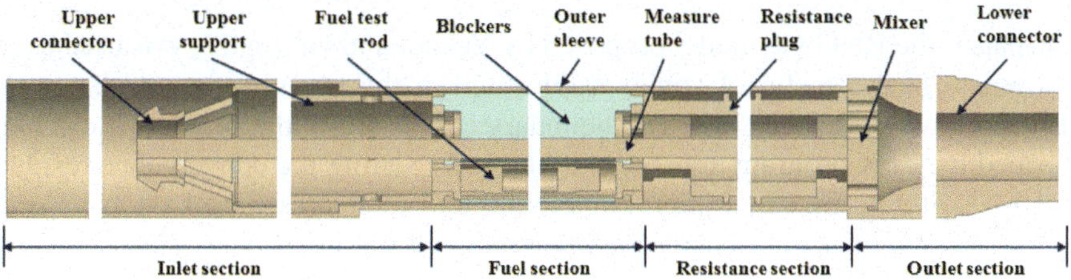

Fig. 1. Schematic diagram of the structure of test section

The fuel test rod is located in the fuel section and consists of fuel, cladding, thermal resistance layer and protective tube, and its structure is shown in Fig. 2. The basic criteria [6] for fuel irradiation test are as follows: (1) The maximum temperature on the outside of protective tube does not exceed 195 °C; (2) The maximum temperature range of the cladding is 490 ± 60 °C; (3) The maximum temperature of fuel does not exceed 820 °C, (4) The range of linear heat rate (LHR) of fuel is 60–90 kW/m.

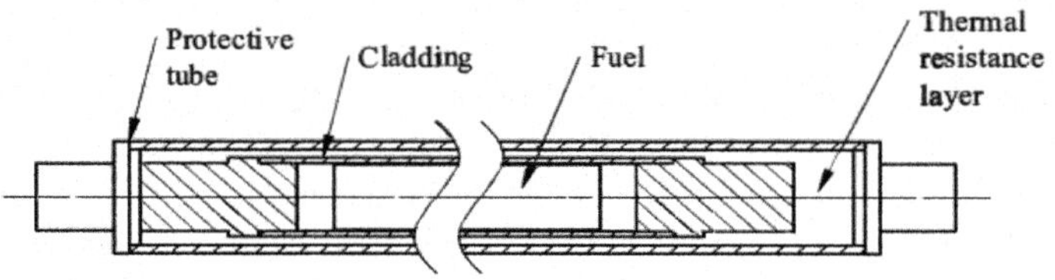

Fig. 2. Schematic diagram of the structure of fuel test rod

The flow channel of the fuel section is shown in Fig. 3, which is mainly divided into two parts, the annular channel and the outside narrow channel. The annular channel is the main flow channel of coolant, while the outside narrow channel is the flow channel formed by the reserved assembly gap.

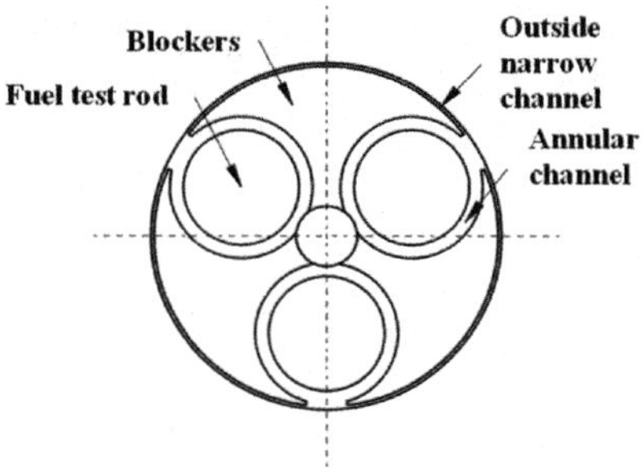

Fig. 3. Schematic diagram of the flow channel of fuel section

2.2 Model Settings

In this paper, the RNG *k-ε* model proposed by Yakhot and Orszag [7] was used to study the flow and heat transfer of compressible fluid in the test section, and the scalable wall function was selected. The main boundary conditions of the simulation calculation performed by CFX program are shown in Table 1. Setting the convergence residual to 1×10^{-4}, monitoring the outlet temperature and inlet pressure of coolant, and the calculation convergence is considered when both are stable.

Table 1. Boundary conditions of simulation calculation

Parameter	Value
Mass flow rate of coolant (kg/s)	1.58
Inlet temperature of coolant (°C)	43
Back pressure of coolant (MPa)	1.4
Gauge pressure of coolant (Pa)	0
LHR of fuel (kW/m)	85

2.3 Mesh Sensitivity Analysis

Due to the complex structure of test section, the hybrid mesh division method is used to mesh the computational fluid and solid domain, and some areas are locally refined. The specific settings are shown in Table 2.

Table 2. Mesh size of test section

Mesh	Mesh size of solid domain (mm)	Local refinement size of solid domain (mm)	Mesh size of fluid domain (mm)	Local refinement size of fluid domain (mm)	Mesh quantity ($\times 10^4$)
Case 1	2.0	2.0	1.5	1.0	628
Case 2	2.0	1.2	1.5	1.0	797
Case 3	2.0	1.0	1.5	1.0	1004
Case 4	2.0	1.0	1.5	0.65	1108
Case 5	2.0	1.0	1.5	0.5	1270
Case 6	1.5	1.0	1.5	0.5	1399
Case 7	1.5	0.9	1.3	0.45	1868

The mesh sensitivity of pressure drop of test section, temperature rise of coolant, maximum temperature of fuel and calculation time was analyzed, and the trend of calculation results with the mesh quantity is shown in Figs. 4 and 5. As the mesh quantity increases, the coolant temperature rise does not fluctuate much, and the changes in other parameters tend to stabilize. The main reasons for the fluctuation of coolant temperature rise are: when the solid domain mesh is refined, resulting in an increase in solid volume, an increase in heat source, and finally an increase in coolant temperature rise. When the fluid domain mesh is refined, resulting in different coolant temperatures at different fluid domain interfaces, and the greater the difference in fluid domain mesh, the greater the difference in coolant temperatures on both sides of the interface, and finally the coolant temperature rise is slightly reduced. Considering the efficiency and accuracy, it can be considered that the Case 5 mesh satisfies the independent solution of mesh, and the mesh setting will be used for subsequent calculation analysis.

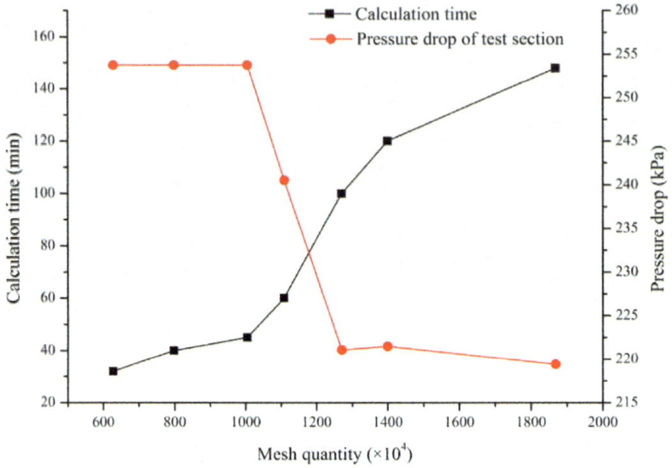

Fig. 4. Mesh sensitivity of calculation time and pressure drop of test section

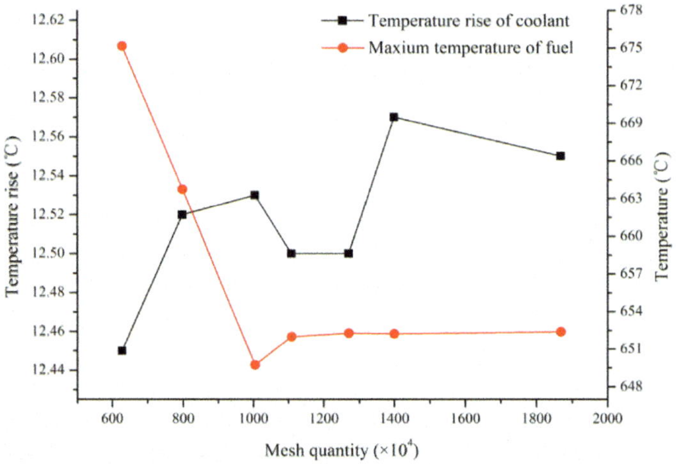

Fig. 5. Mesh sensitivity of temperature rise of coolant and maximum temperature of fuel

2.4 Hydraulic Test Verification

The pressure drop of test section under different mass flow rate of coolant is calculated by CFD simulation, and the results of the hydraulic test outside the reactor are compared, as shown in Fig. 6. As can be seen from the figure, the CFD simulation calculation results are larger. The relative error range of $\pm15\%$ of the hydraulic test value is given in the figure, and with the increase of mass flow rate, the error of simulation calculation also increases, but its relative error is less than 15%, and the maximum deviation is about 12%, indicating that the calculation deviation of flow rate under the same pressure drop is about 6%, and the error is within the acceptable range. Therefore, it can be determined that the RNG k-ε turbulence model is suitable for simulating the thermal hydraulic characteristics of the test section studied in this paper.

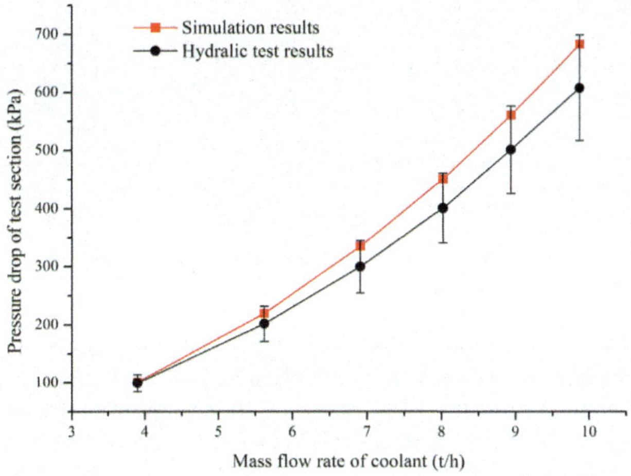

Fig. 6. Pressure drop of test section at different mass flow rate of coolant

3 Calculation Results and Discussion

3.1 Design Condition

Thermal analysis

For the fuel irradiation test, the source of heat is mainly nuclear heat release. There are four groups of fuels in the axial direction of the test section, and the heat release of test section shows the same axial distribution because the neutron flux of the HFETR is an axial cosine distribution [8]. Assume that the axial distribution factors of LHR of four groups of fuel are 1.1, 1.2, 1.05 and 0.65, respectively. When the average LHR of fuel is 85 kW, the temperature distribution of fuel section is shown in Fig. 7.

The peaks of the line 1–3 represent the maximum temperature of fuel, cladding and the outside of protective tube, respectively. It can be seen from Fig. 7 that the trend of maximum temperature also shows a cosine distribution. For the maximum temperature of fuel and the outside of protective tube, they meet the temperature limits below 820 °C and below 195 °C specified in the irradiation test criteria, while the maximum temperature of cladding exceeds the limited range, and the average LHR of fuel needs to be appropriately reduced.

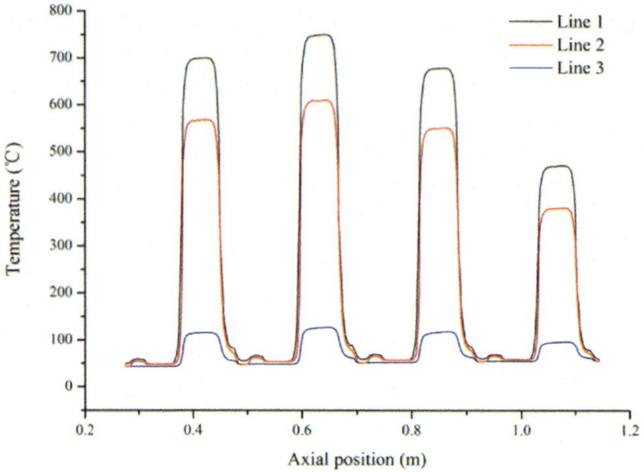

Fig. 7. Axial temperature distribution of fuel section

Figure 8 show the maximum temperature of cladding corresponding to the different average LHR of fuel. According to the existing data points, the maximum temperature of cladding change curve in the LHR range of 40–85 kW/m is fitted. It is concluded that the average LHR of fuel is about 47.9–71.8 kW/m under the limits of the maximum temperature of cladding.

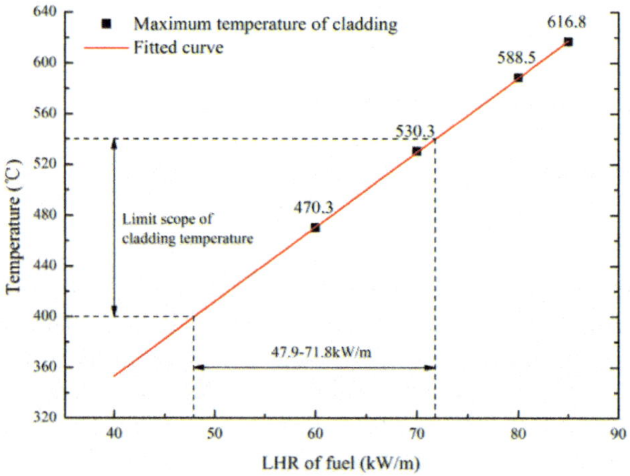

Fig. 8. Curve of the maximum temperature of cladding as a function of LHR

Hydraulic analysis

Figure 9 shows the proportion of pressure drop at different sections to the total pressure drop in the test section. It can be obviously seen from the figure that under the design mass flow rate, the pressure drop of inlet section and outlet section is negligible, although the proportion of length is not small. The pressure drop of test section is mainly concentrated in the fuel section and resistance section, and the pressure drop of resistance section is larger. Due to the simple structure and low flow resistance of outlet section, the pressure drop is negative under the gravity.

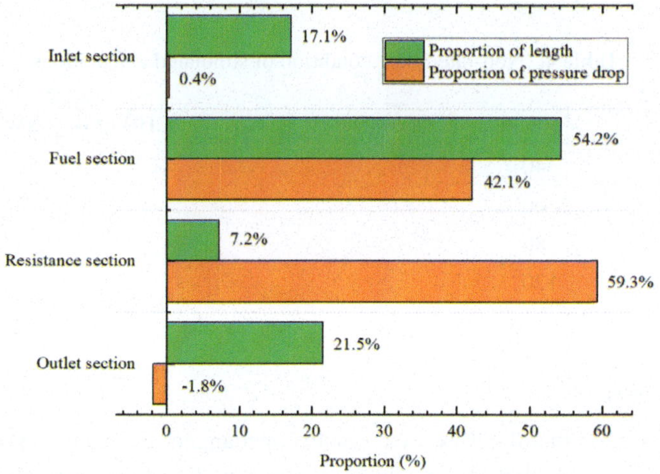

Fig. 9. Proportion of length and pressure drop at different sections

Table 3 lists the hydraulic calculation results of the flow channel of fuel section. It can be seen from the table that the results of the theoretical calculation and CFD simulation are generally consistent. Among them, the calculation results of the outside narrow channel are slightly different between the two methods. Since the coolant in the narrow channel is not directly involved in the heat transfer of fuel test rod, so the smaller the shunt, the better the cooling effect. The proportion of narrow channel flow area to the total flow channel area is about 18%, and the flow distribution calculated by both methods is less than this value. Therefore, it can be considered that the flow channel design is reasonable, and the cooling conditions meet the requirements.

Table 3. Hydraulic calculation results of the flow channel of the fuel section

Method	Mass flow rate (kg/s)	Total average velocity (m/s)	Narrow channel velocity (m/s)	Annular channel velocity (m/s)	Shunt ratio of narrow channel (%)
CFD	1.58	3.58	1.63	4.02	8.2
Theory			1.82	3.97	9.3

3.2 Non-rated Flow Conditions

Non-rated flow conditions are those in which the flow of coolant deviates from the design value. Through the thermal analysis of design condition, the LHR range that meets the irradiation test criteria has been determined, and in order to ensure that the irradiation temperature under non-rated flow conditions meets the criteria, the value of LHR needs to be conservatively selected. The calculation conditions in this section are set as follows.

Table 4. Non-rated flow condition of simulated calculations

Conditions	Relative mass flow rate (%)	LHR of fuel (kW/m)
1	40	60
2	60	
3	80	
4	120	

Thermal analysis

Under the condition that the flow area remains unchanged, the change in the mass flow rate of coolant is equivalent to the change in the velocity of coolant, which will affect the convective heat transfer between the coolant and fuel test rod, thereby causing the temperature change of fuel test rod. Figure 10 shows the maximum temperature of the fuel, cladding and outside of protective tube under different mass flow rates.

With the decrease of mass flow rate, the temperature of the fuel test rod increases, and the increase is increasing. When the relative mass flow rate of coolant changes from 120 to 40%, the temperature rise of fuel test rod decreases from outside to inside, among which the maximum temperature increase on the outside of protective tube is the largest, about 70 °C, the maximum temperature of cladding is about 53 °C, and the maximum temperature of fuel rises by about 49 °C. Under the above conditions, the temperature of the fuel test rod can meet the basic criteria of the irradiation test and leave a safety margin.

Hydraulic analysis

Under the condition that the resistance coefficient of test section remains unchanged, the change in the mass flow rate of coolant will inevitably cause the change in the pressure drop of coolant. The pressure drop of coolant in the test section, resistance section, fuel section and the corresponding proportion of pressure drop under different mass flow rates is shown in Fig. 11.

It can be seen from the figure that the pressure drop of coolant decreases with the decrease of mass flow rate, and the reduction of pressure drop in resistance section is slightly greater than that in fuel section. When the mass flow rate is large, the pressure drop of test section is mainly composed of the pressure drop of resistance section and fuel section. When the mass flow rate is small, the sum of the pressure drop of resistance section and fuel section is greater than the test section, which is because gravity provides

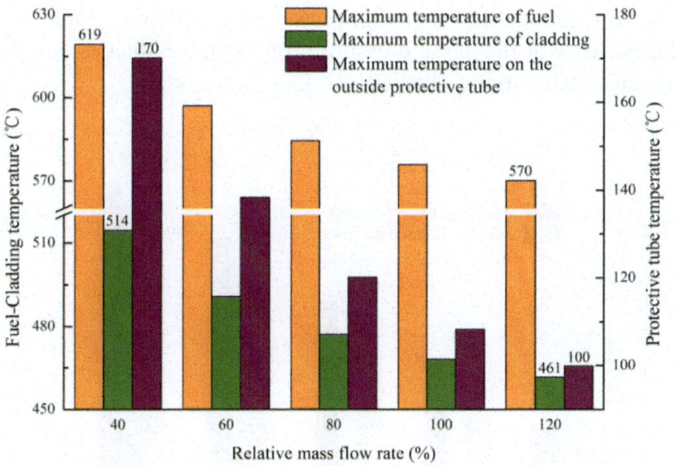

Fig. 10. Temperature of fuel test rod at different mass flow rates

a reverse pressure drop of about 10 kPa, and it does not change with mass flow rate. With the decrease of mass flow rate, the proportion of pressure drop in resistance section significantly increased, this is due to the length of resistance section is short and the influence of gravity is small.

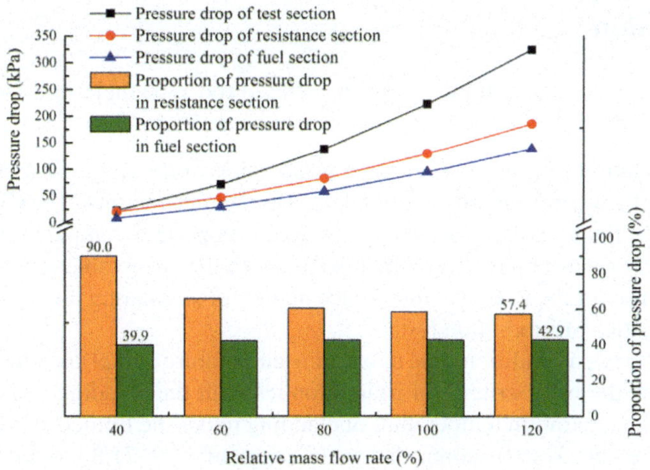

Fig. 11. Pressure drop of coolant at different mass flow rates and its proportion

For the flow channel of fuel section, the velocity, mass flow rate, and shunt ratio of outside narrow channel at different mass flow rates are shown in Fig. 12. It can be seen from the figure that with the decrease of mass flow rate, the velocity of annular channel and narrow channel are reducing, but the amplitude of reduction is different. From the perspective of the shunt ratio, the shunt ratio of narrow channel is decreasing. This is because the narrow channel has a greater resistance coefficient, which resulting in the velocity reduces faster at the same pressure drop, so the reduction amplitude of mass flow rate is greater than annular channel.

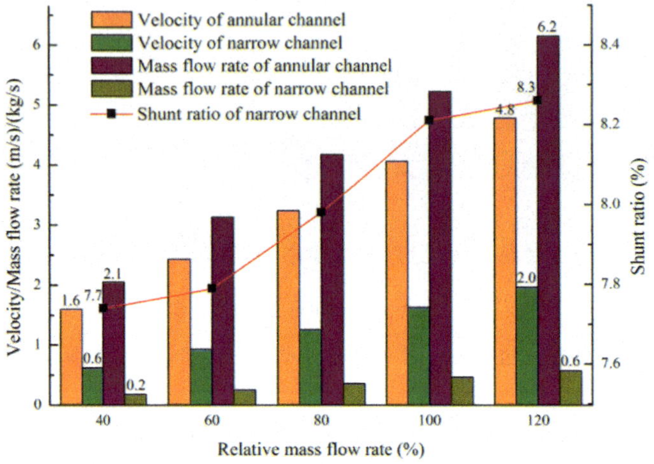

Fig. 12. Flow of different flow channels at different mass flow rates

4 Conclusion

In this study, the thermal hydraulic characteristics of rod-type fuel irradiation test section were numerically simulated. The main conclusions are as follows.

(1) The calculation results of seven type mesh were verified, the Case 5 mesh with a quantity of 12.7 million was finally selected as the mesh setting. The CFD simulation results and the hydraulic test data of pressure drop were compared, the maximum relative error was about 12%, which was within the acceptable range, and verified the reliability of the RNG k-ε turbulence model for simulating the thermal hydraulic characteristics of irradiation test.

(2) Under the design conditions, the maximum temperature of fuel and outside protective tube meets the basic criteria of irradiation test, but the cladding does not. In order to make the maximum temperature of cladding under the limited scope, the average LHR of fuel needs to be controlled in the range of 47.9–71.8 kW/m. For non-rated flow conditions, when the average LHR of fuel is 60 kW/m, the temperature of fuel test rod at different mass flow rates can meet the basic criteria of irradiation test and leave a safety margin.

(3) The pressure drop of test section is mainly concentrated in the fuel section and resistance section, and the pressure drop of inlet section and outlet section is negligible. Due to the length of resistance section is short and the influence of gravity is small, the proportions of pressure drop in resistance section significantly increased with the decrease of mass flow rate.

(4) The shunt ratio of outside narrow channel is less than proportion of flow area, it can be considered that the flow channel design is reasonable. With the decrease of mass flow rate, due to the larger resistance coefficient of narrow channel, the shunt ratio of narrow channel decreases.

In summary, the irradiation test is reasonably designed to achieve the expected irradiation goals.

References

1. Yang, W., Zhao, G., Zhang, L., et al.:Irradiation testing capabilities and irradiation testing technology of the HFETR. Nuclear Science and Engineering 38(6), 986–994 (2018).
2. Jin, S., Yang, W., Lei, J., et al.: Simulation study on flow resistance characteristic of rod fuel irradiation device test section. In: 29th International Conference on Nuclear Engineering, China (2022).
3. Jin, S., Zhang, L., Yang, W., et al.: Numerical simulation research on the flow and heat transfer characteristics of rod fuel irradiation device test section. Journal of Thermal Science and Technology 22(2), (2023).
4. Xing, C., Jesse C., J, Jones, W. F., et al.: Thermal hydraulic analysis of the AFIP-7 irradiation test in the Advanced Test Reactor--Model correlation and performance evaluation. Nuclear Engineering and Design 395, (2022): 111857.
5. Liao, H., Wang, Y., Li, Y., et al.: 3D fluid-solid coupling simulation for plate-type nuclear fuel assemblies under the irradiation condition. Progress in Nuclear Energy 126, (2020): 103428.
6. Zhang, L., Yang, W., Zhao, W., et al.: Design and analysis of irradiation test scheme for fast reactor fuel slug with HFTER. Nuclear Power Engineering 43(3), 101–106 (2022).
7. Yakhot, V., Orszag, S. A.: Renormalization group analysis of turbulence. I. Basic theory. Journal of scientific computing 1(1), 3–51 (1986).
8. Yang, W., Zhang, L., Si, J., et al.: Design and Experimental Study of Irradiation Temperature in Fuel Specimen. Nuclear Power Engineering 39(6), 43–48 (2018).

Modeling and Analyzing of Fuel with Missing Pellet Surface (MPS) Defect Based on Multiphysics Method

Rong Liu(✉), Xiaoyang Yuan, and Shengyu Liu

School of Electric Power, South China University of Technology, Guangzhou, China
rliu290147@hotmail.com

Abstract. Maintaining the integrity of cladding is essential to avoiding the release of fission gas product in light water reactors. The fuel operating with a missing pellet surface (MPS) due to manufacturing defects is considered as an important cause of cladding failure. Therefore, it is of significance to establish analysis code to investigate and understand the failure mechanism in that circumstance. Based on COMSOL Multiphysics finite-element platform and CAMPUS code, this paper presents a fuel performance code with 2D cross-section geometry of a fuel pellet with cladding. First, the code is verified by predicting the performance of fuel without MPS defect compared with CAMPUS code. Then, it is used to analyze the fuel performance as a missing pellet surface (MPS) defect is considered. Finally, the behaviors of fuel with different depths of MPS defect are compared, and the behaviors of systems of UO_2-Zircaloy and U_3Si_2-Zircaloy with MPS in the same condition are also investigated.

Keywords: Missing pellet surface · Multiphysics modeling · Fuel performance

1 Introduction

The cladding plays an important role in safe operation of light water reactor. Therefore, it is necessary to research the mechanisms of cladding failure. Pellet-cladding interaction (PCI) is considered one of the most important causes of cladding failure [1]. Classical PCI is related to stress-corrosion cracking mechanism, which is driven by the localized strains in the vicinity of a pellet crack in the presence of a chemical agent [2]. Non-classical PCI failure is associated with the existence of a missing pellet surface (MPS) defect [3]. The defect of missing pellet surface (MPS) was found to be an essential cause of cladding failure as early as last century. It has been evidently observed that cladding fractures adjacent to MPS defects occur in both BWR fuel [4] and PWR fuel [5]. Resulting from manufacturing or pellet mishandling, missing pellet surface(MPS) occurs as a local geometric defect, which will lead to uneven cladding stress distribution and sufficient local stress increase that may cause cladding failure in some situations. Therefore, it is necessary to model for fuel with MPS defect to understand the thermo-mechanical behaviors of cladding and pellet to predict and prevent the related failures.

J. Liu and Y. Jiao (Eds.): WRFPM 2023, SPPHY 299, pp. 152–162, 2024.
https://doi.org/10.1007/978-981-99-7157-2_16

State-of-the-art of modelling for nuclear fuels and structural materials is documented in a report by the Nuclear Science Committee (NSC) of the Nuclear Energy Agency (NEA) [6]. Various fuel performance codes have been developed by scholars from the world. 1.5D or 2D axisymmetric models for simulation are employed by most of the codes, such as FRAPCON, FRAPTRAN and FALCON [7–9]. There are also some codes developed to analyze fuel performance with 3D geometries such as ALCYONE and TOUTATIS [10, 11]. The BISIN code developed by INL on the MOOSE framework has a multidimensional coupling analysis capability for 2D and 3D fuel performance simulation [12]. Khvostov et al. analyze the impact of missing pellet surface (MPS) on the failure-related characteristics of the cladding based on FALCON code [13]. A procedure for coupling 2D full-length fuel rod models to detailed 3D models is presented to analyze fuel rod with MPS defect by Spencer et al. [14]. In our previous work of CAMPUS code [15], a 2D axisymmetric geometry is applied to analyze nuclear fuel performance. CAMPUS is a fuel performance flexible code developed on COMSOL Multiphysics platform that integrates LWR fuel performance models from BISON [12], MATPRO [16], and FAST code [17], and has been verified by being compared with BISON and FRAPCON. However, axisymmetric geometry is not applicable for the fuel with MPS defect and it is unable to capture the local effects of MPS defect. To investigate the problem, a fuel performance code with a 2D cross-section geometric representation based on CAMPUS code using COMSOL software is developed. Similar to CAMPUS code, heat generation and conduction, thermal expansion, densification, elastic strain, fission product swelling, grain growth, fission gas production and release, gap heat transfer, mechanical contact, fuel thermal and irradiation creep, cladding thermal and irradiation creep, gap/plenum pressure with plenum volume are all taken into account in this code. This code can be verified compared with CAMPUS code in the same condition without MPS defect, then it is used to analyze the performance of fuel with MPS defect.

2 Modeling Details

2.1 Material Properties

In our previous work, the material properties of UO_2 pellets and Zircaloy cladding have already been introduced [15]. And the material properties of U_3Si_2 fuel were also introduced or updated in our previous work [18, 19]. Therefore, in this paper, the material properties will not be presented.

2.2 Modeling Geometry and Mesh

In this work, a 2D planar geometry in Cartesian coordinate system is used to represent UO_2 fuel pellet with MPS defect and Zr-4 cladding (see Fig. 1). The figure shows the representation of defective pellet and the corresponding geometry of 2D finite-element model. In order to understand the thermal and mechanical performance of fuel and cladding, three special positions are marked on the cladding, namely A, B and C, which respectively stand for the positions at 0°, 90°and 180° from defect center.

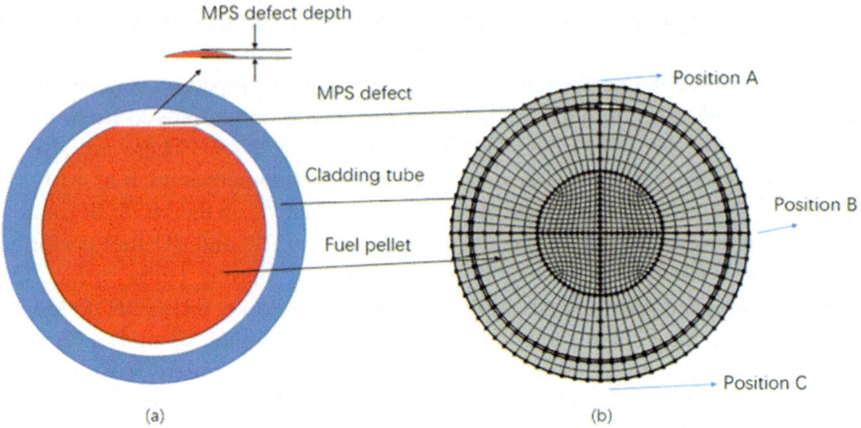

Fig. 1. (a) The geometric representation of defective pellet (b) The corresponding geometry and mesh division of 2D finite-element model.

2.3 Fuel Input Parameters

The input parameters of fuel used in this work are presented in the Table 1 below.

Table 1. Parameters summary.

Parameters	Value
Fuel pellet radius (mm)	4.09575
Gap width (μm)	82.55
Thickness of cladding (μm)	571.5
Fuel enrichment	4.9%
Initial fuel density	95% theoretical
Fast neutron flux (n/m^{2}s)	9.5×10^{17}
Coolant pressure (MPa)	15.5
Coolant temperature (K)	530
Coolant convection coefficient (W/m^2 K)	7500
Linear average power (W/cm)	200

2.4 Modeling Approach

The calculation of this model is the same as CAMPUS code except deformation mechanics. Since the geometry is different from 2D axisymmetric geometry of CAMPUS code, 2D Plane strain and generalized plane strain in the COMSOL structural mechanics module are employed to capture the mechanical behaviors of fuel pellet and cladding in 2D

CSM. A non-linear backward-difference formulation (BDF) in COMSOL platform is utilized to calculate the time-derivatives in the model. A system of linear equations generated from the combinations of the weak-form equation definitions and the finite-element mesh is solved by a direct solver called Multifrontal Massively Parallel sparse direct Solver (MUMPS).

3 Results and Discussions

In this section, the correctness of the code in this work can be verified by being compared with CAMPUS code in the same condition. The correctness of CAMPUS code has already been verified under normal operating condition in our previous work [15]. Then, a change of the geometry corresponding to fuel with MPS defect is considered. The analysis of the thermal and mechanical behaviors of fuel pellet with MPS defect will be presented. Also, the comparisons of fuel with different depths of MPS defect and the combinations of UO_2-Zircaloy and U_3Si_2-Zircaloy, are discussed.

3.1 The Verification of This Code

The evolutions of temperature are calculated by CAMPUS code and the code in this work to confirm the accuracy of this code. UO_2-Zircaloy system is used to contrast. According to the Fig. 2, CAMPUS code and the code in this work are found to obtain very similar results of temperature trends as the fuel experiences densification, fission gas release and gap closure in the same condition, therefore the code developed in this paper is reasonable for predicting fuel performance.

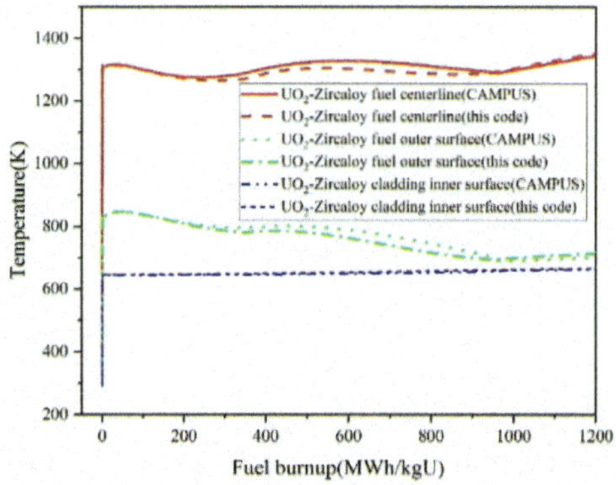

Fig. 2. Verification of temperature of UO_2-Zircaloy.

3.2 The Analysis of Fuel with MPS

In this section, the thermal and mechanical performance of fuel with MPS will be introduced. The UO_2-Zircaloy of fuel and cladding system with 0.1 mm deep defect is presented.

During the power history, since the abnormal gap heat transfer performance adjacent to the defect, the temperature of cladding at position A is lower than other positions (see Fig. 3). Therefore, the cladding thermal stress adjacent to the defect is lower than other positions. Under the coolant pressure and less thermal stress, a greater bending moment and radial contraction will occur at the cladding adjacent to the defect. The evolutions of cladding inner surface displacements are presented in Fig. 4, the cladding is found to experience an asymmetric deformation and deforms into an approximate ellipse during early stage of power history. Consequently, the cladding stress concentration has been obviously observed at cladding adjacent defect center. The von Mises stress distributions at some points of time are shown in Fig. 5.

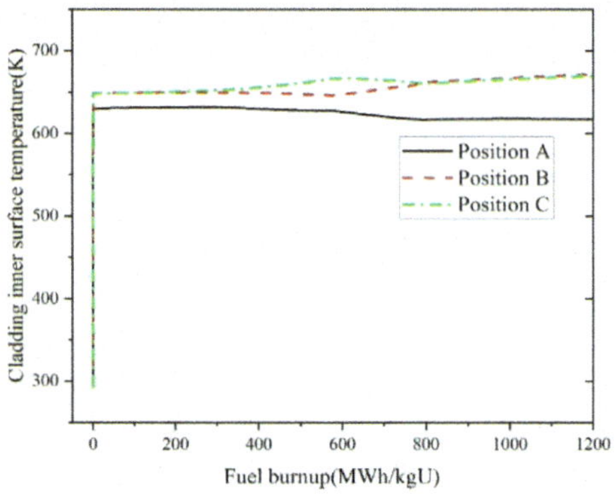

Fig. 3. The evolutions of cladding inner surface temperature.

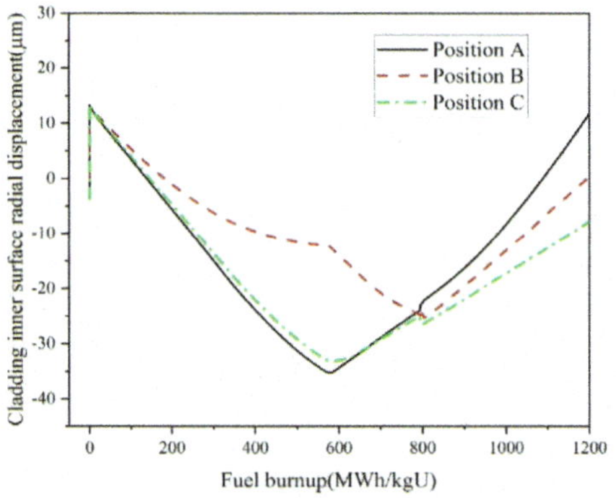

Fig. 4. The evolutions of cladding inner surface displacements at position A, B and C.

It is observed that, the peak stress could occur at the inner or outer surface of cladding adjacent to defect. The evolutions of von Mises stress of cladding inner and outer surface at position A are presented in Fig. 6, it could be found the cladding outer surface stress continues to increase while the inner surface stress starts to decrease after a maximum

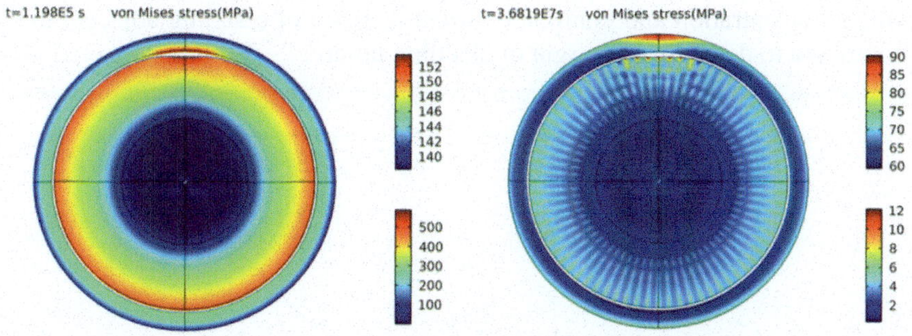

Fig. 5. The evolutions of von Mises stress distributions when t = 1.198E5s and t = 3.687E7s, the first legend is for cladding, the second is for fuel.

as burnup increases. And there are some irregularities and sudden changes due to the gap closure at some key positions in the curves.

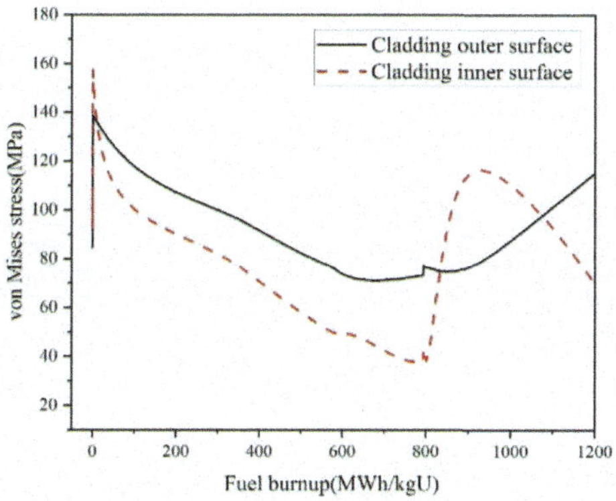

Fig. 6. The evolutions of von Mises stress of cladding adjacent to the defect center.

The von Mises stress of cladding outer surface at position A is always greater than that at position B and position C (see Fig. 7). But as burnup increases, the cladding von Mises stress at position A continues to increase while the trends of von Mises stress of cladding at position B or position C tend to be flat. This situation, in which the risk of cladding failure may increase significantly, is mainly ascribed to gap closure. The evolutions of gap size at the above three positions are presented in Fig. 8. As is shown in Fig. 8, the trends of gap size are different at three positions. The gap at position C closes firstly, and the gap at position A still remains unclosed as burnup getting larger. Since the unclosed gap, the temperature distribution of the fuel and cladding is also more uneven. The temperature of fuel region around defect center is obviously higher than other regions while the opposite happens on the cladding due to the lager difference of heat transfer performance. For the lack of proper support, the stress of cladding adjacent to the defect will continue to increase with the presence of a continuous bending moment.

The stress concentration and continuous stress increase of cladding adjacent to defect will contribute to decrease of margin of cladding failure.

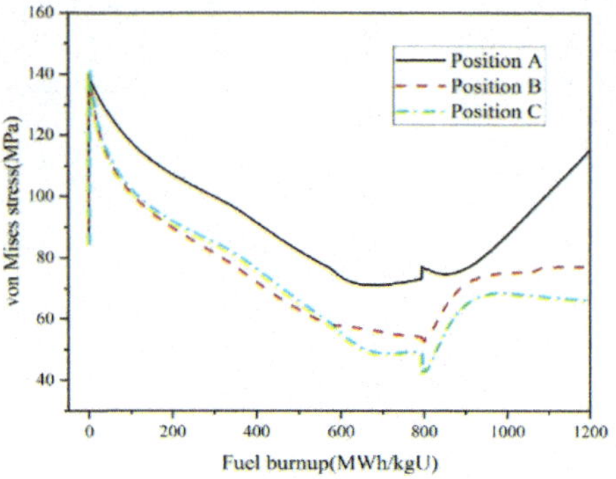

Fig. 7. The evolutions of von Mises stress of cladding outer surface at three positions.

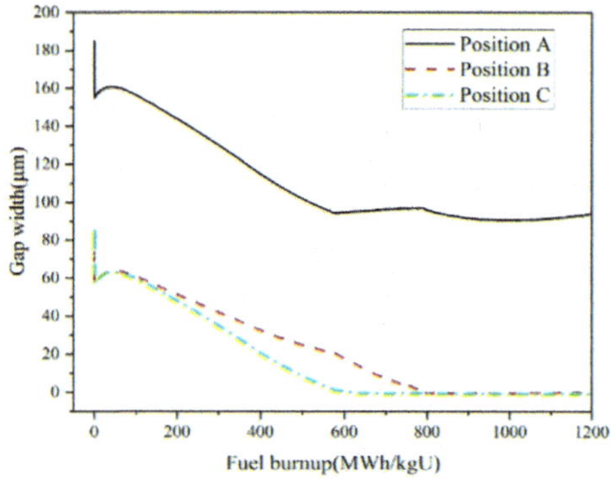

Fig. 8. The evolutions of gap size at position A, B and C.

3.3 Comparisons of Different Depths of MPS

In this section, fuels of UO_2-Zircaloy system with no defect, with a 0.1 mm deep defect and a 0.2 mm deep defect of MPS are modeled to demonstrate the effects of different defect depths. The temperature of centerline and cladding inner surface adjacent to defect center is presented in Fig. 9. The centerline temperature will rise with MPS, and the deeper the defect is, the higher the centerline temperature will be. Though the opposite happens at cladding inner surface due to deterioration of gap heat transfer performance around the defect.

As the depth of MPS increases, the stress concentration of cladding also becomes severer. It can be found in Fig. 10 that, the von Mises stress at cladding outer surface adjacent to defect center is obviously greater when the depth of MPS changes from 0.1

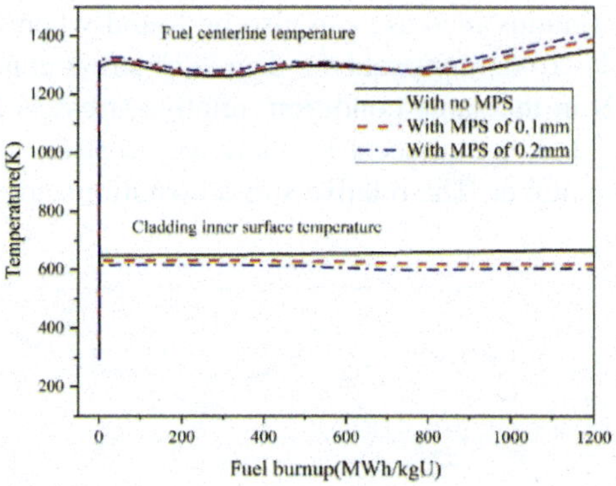

Fig. 9. The evolutions of temperature of fuel with MPS of different depths.

mm into 0.2 mm. Compared with fuel with 0.1 mm deep defect, the radial displacements of cladding is larger when the defect depth is 0.2 mm, and that indicates the cladding experiences a more serious asymmetric deformation. Therefore, the greater bending moment and stress concentration occur at cladding adjacent to defect. Also, fuel with deeper MPS defect starts to contacts with cladding and the gap starts to close earlier. It could be concluded that, the probability of cladding failure will increase when the MPS defect is deeper.

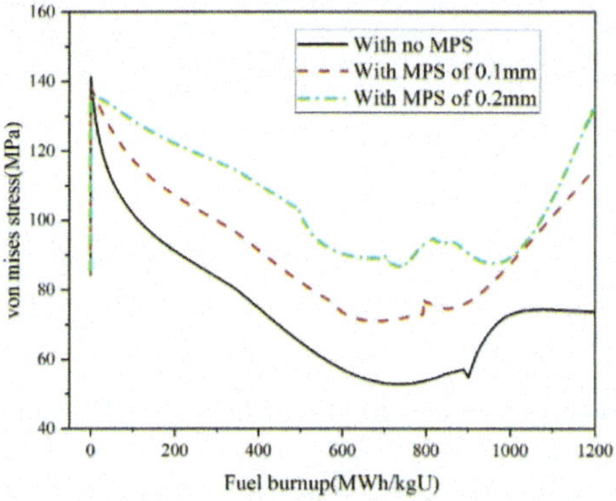

Fig. 10. The evolutions of von Mises stress of fuel with MPS of different depths at cladding outer surface adjacent to defect center.

3.4 Comparisons of Different Fuels with MPS

Based on our previous work [18], UO_2-Zircaloy and U_3Si_2-Zircaloy systems are modeled to compare the performance of UO_2 and U_3Si_2 fuels with MPS of the same depth in this section. Under the same power history and size of fuel and cladding, similar behaviors are observed in UO_2-Zircaloy and U_3Si_2-Zircaloy systems with MPS defect. Cladding stress concentration and continuous increase can also be found when fuel of U_3Si_2 operates with MPS (see Fig. 11). To demonstrate the degree of stress concentration of UO_2 and U_3Si_2 fuels with MPS in the same condition, relative stress is defined as the ratio of concentrated stress of cladding adjacent to the defect to stress of a certain position on cladding far away from defect. The relative stress evolutions are presented in Fig. 12.

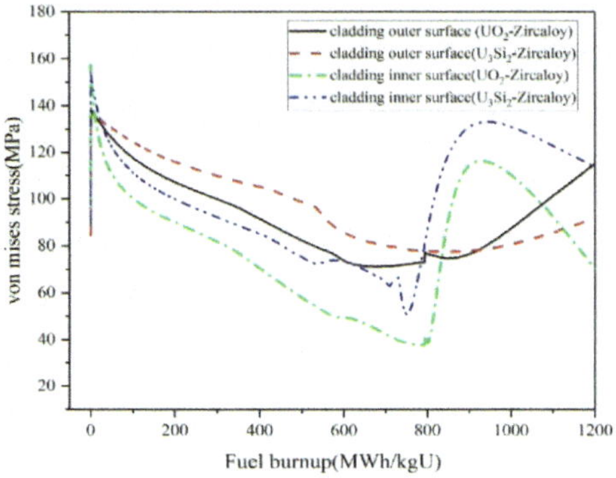

Fig. 11. The stress comparisons at cladding outer and inner surface adjacent to defect center.

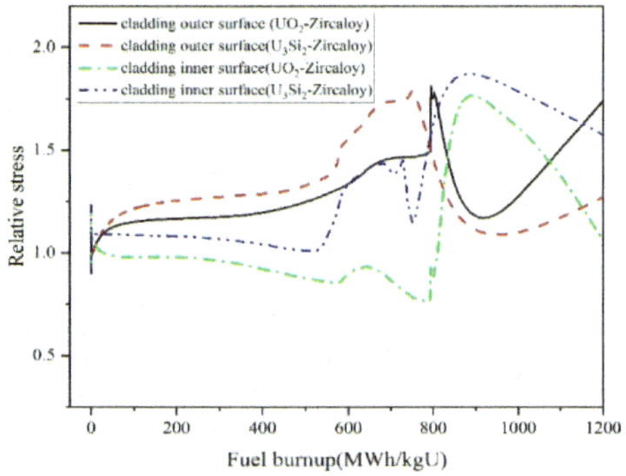

Fig. 12. The cladding relative stress evolutions of UO_2-Zircaloy and U_3Si_2-Zircaloy systems.

It is found that cladding inner surface relative stress of U_3Si_2-Zircaloy system is always larger than that of UO_2-Zircaloy system, while cladding outer surface relative stress of UO_2-Zircaloy is smaller at first, and then becomes lager as the burnup increases. The higher thermal diffusivity and lower heat capacity of U_3Si_2 fuel than that of UO_2 fuel, which lead to the lower cladding temperature and gas pressure of U_3Si_2-Zircaloy system, mainly contribute to it. The coolant pressure and creep will restrain cladding expansion and play a leading role in cladding deformation during early stage of power history. Since the lower gas pressure and cladding temperature of U_3Si_2-Zircaloy system, there is less thermal strain to counteract the strains produced by coolant pressure, which will lead to greater cladding deformation. The evolutions of radial displacements are presented in Fig. 13.

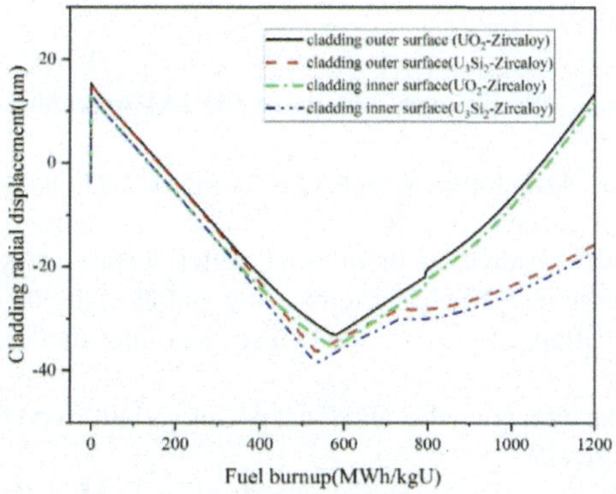

Fig. 13. The evolutions of radial displacements at cladding outer and inner surface adjacent to the defect of UO_2-Zircaloy and U_3Si_2-Zircaloy systems.

As can be seen from the above figures, the greater deformation of cladding adjacent to the defect of U_3Si_2-Zircaloy system will cause a greater degree of stress concentration when the cladding is contracting. As the burnup increases, the cladding starts to expand and the thermal stress and gas pressure mainly contribute to it, so that the rate of continuous stress increase and degree of stress concentration of U_3Si_2-Zircaloy at cladding outer surface tend to be smaller due to lower gas pressure and cladding temperature. From the above discussions, as accident tolerant fuel, U3Si_2 fuel shows no overall superiority with MPS defect compared with UO_2 fuel, and the only superiority of U_3Si_2 fuel with MPS is the lower rate of continuous stress increase and degree of stress concentration at cladding outer surface adjacent to the defect as burnup reaches a certain depth.

4 Conclusions

The fuel with missing pellet surface (MPS) defect is modeled to investigate its performance in this work. The fuel with different depths of defect and different fuels with defect in the same condition are also investigated. The main conclusions of this paper include: (1) With MPS defect, temperature of fuel centerline will rise and fuel region around the defect center is obviously higher than other regions, while the temperature of cladding is lower at that location. Therefore, the cladding deforms asymmetrically and stress concentration occur at the cladding adjacent to defect; (2) The deeper MPS defect will enlarge the temperature difference and the degree of cladding deformation and stress concentration; (3) Unlike the superiority showed in normal operating condition without defect, U_3Si_2 fuel is found to behave the severer cladding stress concentration compared with UO_2 fuel on the whole. But some limitations should not be ignored in 2D geometric representation applied in this code, such as capturing the effects of out-of-plane mechanical and thermal boundary conditions or other irregular defects. Developing 3D model for more accurate simulation to capture the effects of those local defects is necessary.

Acknowledgements. The financial supports from the Natural Science Foundation of Guangdong Province (Grant No. 2020A1515010648) and Universities Characteristic Innovation project of Guangdong Province (Grant No. 2022KTSCX006).

References

1. Dangouleme, D., et al. "Preliminary results of the IAEA review on fuel failures in water cooled reactors." (2009).
2. Capps, Nathan Allen. "Development and Application of 3-D Fuel Performance Modeling to Assess Missing Pellet Surface Influence on Pellet Clad Interaction and Clad Failure." (2016).
3. Capps, Nathan, et al. "Evaluation of missing pellet surface geometry on cladding stress distribution and magnitude." Nuclear Engineering and Design 305 (2016): 51–63.
4. Groeschel, F., et al. Failure root cause of a PCI suspect liner fuel rod. No. IAEA-TECDOC--1345. 2003.
5. Aleshin, Y., et al. "The effect of pellet and local power variations on PCI margin." Proceedings of Top Fuel (2010): 26–29.
6. Bartel, T. J., et al. State-of-the-Art Report on Multi-scale Modelling of Nuclear Fuels. No. NEA-NSC-R--2015–5. Organisation for Economic Co-Operation and Development, 2015.
7. Berna, G. A., et al. FRAPCON-3: A computer code for the calculation of steady-state, thermal-mechanical behavior of oxide fuel rods for high burnup. No. NUREG/CR-6534-Vol. 2; PNNL-11513-Vol. 2. US Nuclear Regulatory Commission (NRC), Washington, DC (United States). Div. of Systems Technology; Pacific Northwest National Lab.(PNNL), Richland, WA (United States); Idaho National Lab.(INL), Idaho Falls, ID (United States), 1997.
8. Geelhood, K. J., et al. "FRAPTRAN 1.4: a computer code for the transient analysis of oxide fuel rods." US Nuclear Regulatory Commission, Office of Nuclear Regulatory Research, NUREG/CR-7023 1 (2011).
9. Rashid, Y., R. Dunham, and R. Montgomery. "Fuel analysis and licensing code: FALCON MOD01." EPRI Report 1011308 (2004).
10. Thouvenin, G., et al. Multidimensional modeling of a ramp test with the PWR fuel performance code ALCYONE. American Nuclear Society, 555 North Kensington Avenue, La Grange Park, IL 60526 (United States), 2007.
11. Bentejac, Fabrice, and Nicolas Hourdequin. "TOUTATIS: an application of the CAST3M finite element code for PCI three-dimensional modeling." Proc. Pellet-clad Interaction in Water Reactor Fuels (2005): 9–11.
12. Hales, J. D., et al. BISON theory manual the equations behind nuclear fuel analysis. No. INL/EXT-13–29930. Idaho National Lab.(INL), Idaho Falls, ID (United States), 2016.
13. Khvostov, G., W. Lyon, and M. A. Zimmermann. "Application of the FALCON code to PCI induced cladding failure and the effects of missing pellet surface." Annals of Nuclear Energy 62 (2013): 398-412.
14. Spencer, B. W., et al. "3D modeling of missing pellet surface defects in BWR fuel." Nuclear Engineering and Design 307 (2016): 155-171.
15. Liu, Rong, et al. "Multiphysics coupled modeling of light water reactor fuel performance." Progress in Nuclear Energy 91 (2016): 38–48.
16. Allison, C. M., et al. "SCDAP/RELAP5/MOD3. 1 code manual, volume IV: MATPRO–A library of materials properties for light-water-reactor accident analysis." DT Hagrman, NUREG/CR-6150, EGG-2720 4 (1993): 4–234.
17. Prudil, Andrew Albert. FAST: a fuel and sheath modeling tool for CANDU reactor fuel. 2013.
18. Liu, Rong, Wenzhong Zho Cai. "Multiphysics modeling of accident tolerant fuel-cladding U3Si2-FeCrAl performance in a light water reactor." Nucle u, and Jiejinar Engineering and Design 330 (2018): 106–116.
19. Liu, Shengyu, Rong Liu, and Chengjie Qiu. "Transient fuel behavior analysis of U3Si2 fuel and FeCrAl cladding based on multiphysics method." Progress in Nuclear Energy 159 (2023): 104648.

AFA 3G Operating Experience

G. Bolsée and G. Gentet(✉)

Framatome SAS, Fuel, 2 Rue Professeur Jean Bernard, 69007 Lyon, France
guy.gentet@framatome.com

Abstract. The AFA technology was developed in the 80's and since then has been continuously improved by Framatome to provide utilities with high performing products for more demanding plant operating conditions. Thermal hydraulic performance enhancements and transition to new materials for cladding, guide tubes and grids accomplished through the AFA 2G and 3G generations, accompanied the utility needs. The burn-up increase and the achievement of more demanding conditions was supported by the development of the $M5_{\mathbf{Framatome}}$ alloy which allows the AFA 3G fuel assembly to achieve a licensed Fuel Assembly (FA) burn-up of 62 GWd/t. $M5_{\mathbf{Framatome}}$ alloy brings to the fuel assembly a high resistance against corrosion and hydriding allowing burn-up increase. Moreover, the use of $M5_{\mathbf{Framatome}}$ alloy provides the AFA 3G fuel assembly with a higher maneuverability allowing Extended Low Power Operating conditions and allows that LOCA margins would not be hindered by the presence of highly irradiated fuel assemblies for high burn-ups. The AFA 3G grid design results from improvements of the original AFA design and enables the AFA 3G fuel assemblies to support optimized fuel management and plant power up-rates. The fuel assembly performance can be further enhanced by addition of MSMGs (Mid Span Mixing Grids). The AFA 3G grid has also been optimized to facilitate the reshuffling of the cores. The evolution of the fuel managements mainly characterized by increased burn-ups and increased resident times required further improvement of the fuel assembly design with regards to the increased risk of Grid to Rod Fretting and Fuel Assembly bow in particular for 14 ft—4 loop plants. The AFA 3G 14ft design has been reinforced by the implementation of new structural material, the Framatome Q12 zirconium alloy, featuring enhanced creep resistance to preclude the risks of Fuel Assembly bow. This was done in a two-step approach with the implementation of Q12 MONOBLOC guide tubes followed more recently by the implementation of Q12 AFA 3G grids. The current AFA 3G fuel assembly design is the result of the integration of an extensive Operating Experience acquired by Framatome in many different types of reactors and more particularly in the 12ft and 14ft 17x17 plants. Framatome delivered more than 43000 AFA 3G 17x17 fuel assemblies providing an outstanding Operating Experience that will be highlighted in the paper.

Keywords: Fuel assembly · Operating experience · AFA 3G · $M5_{\mathbf{Framatome}}$ · Q12 · MONOBLOC · HTP · GAIA and TRAPPER are trademarks or registered trademarks of Framatome or its affiliates · In the USA or other countries

J. Liu and Y. Jiao (Eds.): WRFPM 2023, SPPHY 299, pp. 163–171, 2024.
https://doi.org/10.1007/978-981-99-7157-2_17

1 Introduction

Framatome delivered for the first time its AFA 3G technology in reloads in 1998 in 17×17 12ft reactors. The AFA 3G is the third generation of the AFA technology which uses bi-metallic grids, namely Alloy 718 springs mounted on zirconium alloy strips. Since the first introduction, the AFA 3G technology has been extended to 14ft reactors including EPR and adapted to fit with the evolution of utilities needs such as power uprates, extended cycle lengths, increased discharge burn-up. Continuous improvement of the AFA 3G technology has been key to support the evolution of the PWR market.

By end of 2022, more than 43000 AFA 3G fuel assemblies, have been delivered, among them 21250 being 14 ft fuel assemblies. The AFA 3G fuel assembly is presently licensed up to 62 GWd/t.

2 Customer Requests for Product Performance Evolution

2.1 Burn-Up Capacity and Longer Cycles

Over the past 25 years, the cycle length and the discharged burn-up significantly increased from just above 30 GWd/t up to values which are presently above 50 GWd/t for the 17× 17 plants. Extended operational cycles up to 24 months have emerged as an important strategic goal for many global utilities operating nuclear power plants which is translating in higher performance requirements for the fuel assemblies.

The AFA 3G technology accompanied this evolution through the implementation of new materials such as $M5_{\textbf{Framatome}}$ for cladding and structure and more recently Q12 structural material. The AFA 3G is licensed by the French Safety Authorities up to a burn-up of 62 GWd/t obtained in a fuel management corresponding to 18-month cycles with one fourth of the core replaced during each reshuffling.

2.2 Thermal-Hydraulic Performance

With the generalization of advanced fuel managements featuring in-out loading schemes, which better protect the vessel against the fast neutron degradation and improve the fuel cycle economy, high thermal-hydraulic performances are requested with and without power uprates which can reach 17% over the initial nominal power. From the beginning, AFA 3G technology took benefit from the development of an optimized spacer grid design combined with Mid Span Mixing Grids (MSMG) to enhance the thermal hydraulic performance of the fuel assembly by more than 15%. Today, more than 2950 (17x17 and 15x15) fuel assemblies with MSMG's have been delivered from Framatome European manufacturing facilities and more than 11000 manufactured in Asia by Framatome partners.

2.3 Maneuverability

A high level of maneuverability is required by the utilities which operate a high number of Nuclear Power Plants (NPP). This need is also increasing with the growing share of renewables in the electricity generation. Maneuverability encompasses the capability of a fuel to allow a fast return of the NPP to the grid, and the flexibility of the fuel to support power variations without excessive limitations on the power steps. It also covers the capacity of a fuel assembly to operate at low power over long periods, e.g. several weeks. Framatome advanced cladding $M5_{\textbf{Framatome}}$ and improved pellet geometry and manufacturing process preventing Missing Pellet Surface, are key AFA 3G attributes to enhance Pellet-Cladding Interaction robustness with direct benefit on Fuel maneuverability.

2.4 Reliability

The lack of Fuel rod reliability is a source of coolant contamination that can lead to plant shut down if its level goes beyond an acceptable threshold. The fuel assembly which is not leakproof cannot be reinserted and must be repaired. Aside from the potential Pellet-Cladding Interaction damages mentioned above, usual causes for reliability defects are either Grid-To-Rod-Fretting or debris fretting. The development of the AFA 3G fuel assemblies benefited from an optimum testing protocol to validate the robustness of the grid designs to prevent the Grid-To-Rod-Fretting phenomenon. Moreover, anti-debris nozzles have been developed to stop the debris and protect the fuel rods: the TRAPPER bottom nozzle is the standard anti-debris feature of the AFA 3G with a fully demonstrated efficiency.

The 2022 fuel rod failure rate (number of unloaded failed rods divided by the total number of unloaded rods) of AFA 3G assemblies delivered by Framatome is well below the rate of 1.0x10–5.

3 AFA 3G Technology Improvements

3.1 Q12 for High Dimensional Stability

In order to reinforce the dimensional stability of the fuel assemblies, Framatome developed the Q12 alloy. This ultra-low tin Zr1NbSnFe quaternary alloy is an evolutionary development of the current $M5_{\textbf{Framatome}}$ alloy and thus benefits from the extensive industrial experience already gained with the latter. The alloy Q12, derived from $M5_{\textbf{Framatome}}$ alloy, is obtained by adding low tin (0.5 wt%) and iron contents (0.1 wt%). This alloy, therefore, exhibits a niobium content of 1% and controlled oxygen (0.135 wt%) and sulfur contents. Tin, an element in solid solution in zirconium, improves resistance to irradiation creep. Its content is limited in order not to degrade the corrosion kinetics, especially at elevated burn-ups. Iron, in combination with niobium, is a key element for ensuring good corrosion performance, especially in high-duty conditions. For structural components, these modifications provide higher irradiation creep strength while maintaining suitable corrosion resistance. Q12 is processed in the same way as $M5_{\textbf{Framatome}}$. In particular, the low temperature process is applied, which involves accomplishing the heat treatments after the cold forming operations below a defined temperature limit in order to avoid β-Zr particles formation. Fully recrystallized microstructures with fine grains and uniformly distributed precipitates are obtained.

Q12 Free growth and creep performance

An irradiation experiment on Q12 (and $M5_{\mathbf{Framatome}}$ as a reference) was carried out in a commercial PWR. Tube specimens were placed inside the guide thimbles of PWR host assemblies at axial high-flux locations. The free growth and axial creep samples were designed to enable water flow inside and outside the test specimens, avoiding radial differential pressure and leading to double face corrosion of the tubes. The irradiation conditions were therefore representative of those experienced by guide thimbles, with a temperature range of 300–330 °C.

The experimental irradiation data after 8 cycles in D24 PWR show a low and stable Q12 alloy free growth within the applicable range of fluence for FA. Free growth breakaway, which is typical of recrystallized zirconium alloys, is observed but the increase in growth rate occurs beyond the fluence range for PWR fuel assembly. The irradiation growth behavior of the Q12 alloy appears very close to that of the $M5_{\mathbf{Framatome}}$ alloy irradiated in comparable conditions (Fig. 1).

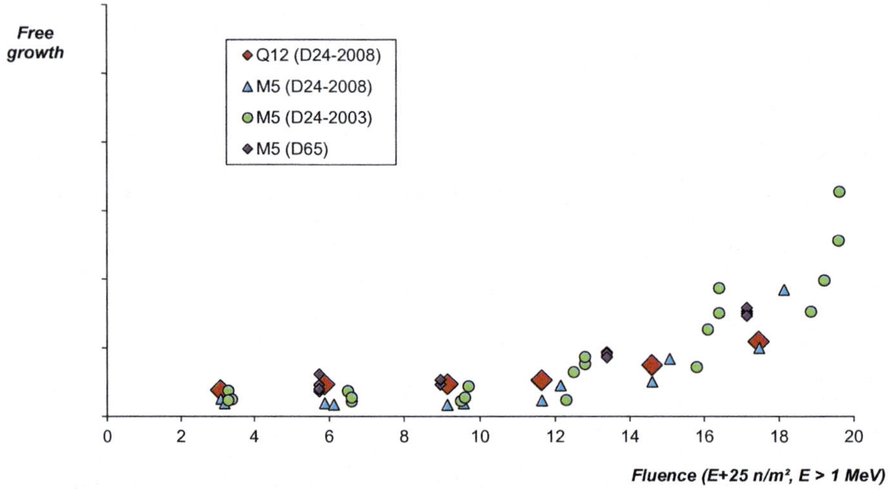

Fig. 1. Q12 free growth—comparison with M5

The evolution of the elongations due to creep (excluding the contribution of free growth) measured in D24 PWR on Q12 and $M5_{\mathbf{Framatome}}$ samples against the fluence is presented in Fig. 2. The results show that Q12 has a stronger creep resistance than M5, as expected due to the tin addition in Q12, with the secondary creep rate being reduced by a factor of ~2 under these irradiation conditions. For the expected PWR FA fluence of 12x10 + 25 n/m^2 at end of life, the Q12 creep strain is approximately one-half that of $M5_{\mathbf{Framatome}}$ at the tested stress level of -10MPa.

Fuel assembly axial growth for fuel assembly equipped with Q12 guide thimbles

The length increase of the fuel assemblies with Q12 guide thimbles has been measured up to 64.7 GWd/t on several assembly designs. Currently, the database is composed of 129 measurements including 47 measurements on 14ft AFA 3G fuel assemblies.

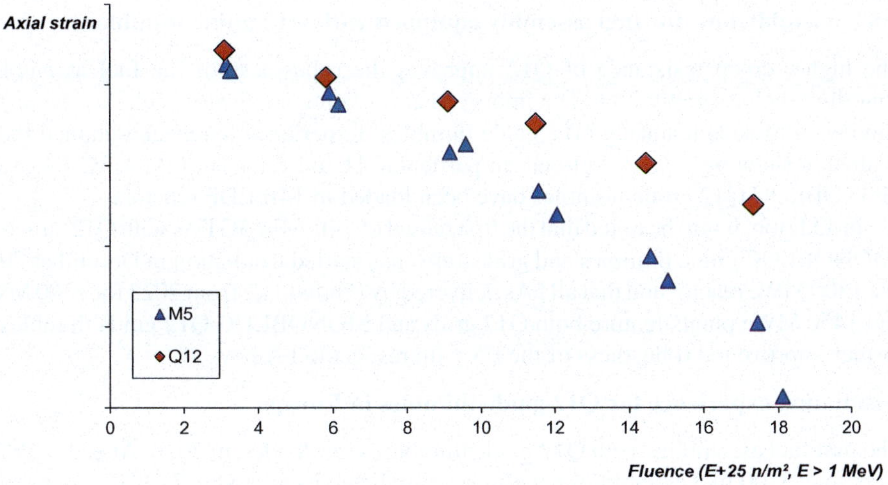

Fig. 2. Q12 irradiation axial creep: comparison with $M5_{\mathbf{Framatome}}$ (free growth elongations are subtracted)

These data show a predictable and stable axial growth (max 0.12%) till high burn-up (Fig. 3).

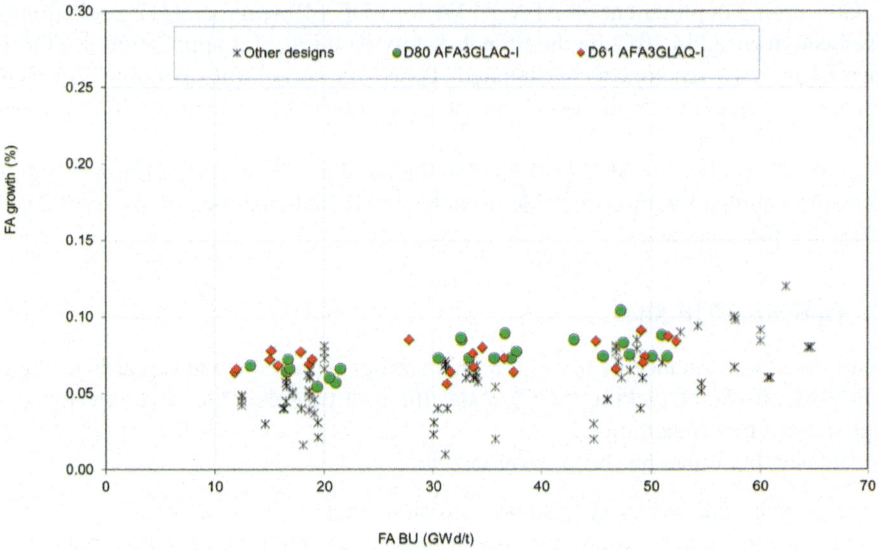

Fig. 3. Fuel Assembly (FA) growth of FA equipped with Q12 guide thimbles

Fuel assembly bow for fuel assembly equipped with Q12 guide thimbles

The higher creep resistance of Q12 improves the robustness of the fuel assemblies regarding fuel assembly bow. The improvement is being confirmed with the increasing number of reloads featuring Q12 guide thimbles. Experience feedback obtained today is already showing a positive trend, in particular 41 full reloads of AFA 3G FAs with MONOBLOC Q12 guide thimbles have been loaded in 14ft EDF reactors.

In addition, it is to be noted that the first reload of 14ft AFA 3G FAs with Q12 structure (MONOBLOC guide thimbles and grid straps) has started irradiation in December 2022 in a 1300 MWe reactor and that all FAs delivered by Framatome from 2023 for 1300 MWe and 1450 MWe plants feature both Q12 grids and MONOBLOC Q12 guide thimbles to further improve the robustness of the FA with respect to FA bow.

Irradiation experience for Q12 guide thimbles in Europe

The first fuel assemblies with Q12 guide thimbles were loaded in 2010. At end of 2022, more than 3100 fuel assemblies have been or are being irradiated in 24 PWRs in France, Germany, Sweden and Switzerland including 41 full reloads of 14ft AFA 3G FAs with Q12 guide thimbles in 15 PWRs and the first reload of 14ft AFA 3G FAs with Q12 structure (guide thimbles and grid straps) was loaded in a 1300 MWe reactor in 2022 (Table 1). A burn-up of 64.7 GWd/t has been achieved on Lead Test Assemblies and of 53 GWd/t so far on reloads.

The associated in reactor experience feedback confirms the good behavior observed on AFA 3G design with Q12 guide thimbles.

Concerning deployment on AFA 3G FA for EDF 14ft reactors, Q12 guide thimbles have been licensed in 2012 by the French Safety Authority for application in EDF 14ft reactors and two reactors were supplied. Based on successful operating experience, decision was made to deploy to all 14ft EDF reactors. More than 500 FAs are being delivered each year.

Since 2018, all 14 ft AFA 3G FAs manufactured and delivered in Europe by Framatome are equipped with Q12 guide thimbles (EPR included, beginning from the first core or the first reload).

3.2 Optimized 718 Alloy

To improve performance of alloy 718 grid springs in reactor with regards to the risk of Stress Corrosion Cracking (SCC), a specific heat treatment has been developed and implemented by Framatome.

The heat treatment has been developed to:

- reduce material sensitivity to stress corrosion cracking
- keep the mechanical properties close to those of the reference 718 material

This material is the new standard for AFA 3G, for 12ft and 14ft reactors including EPR: the first fuel assemblies with optimized grids springs were introduced in reactor in 2019 and by the end of March 2023, more than 4700 fuel assemblies have been delivered to 34 PWRs in France.

Table 1. Framatome operating experience of FAs equipped with Q12 guide thimbles in Europe (Status end of 2022)

Fuel array	Country	Number of reactors	First insertion	FA design	Number of FAs	Achieved or (projected at end of current cycle) burn-up (GWd/t)
15x15	Switzerland	1	2013	HTP	102	64.7
	Sweden	1	2012	AFA 3G	4	51.3 (54.1)
16x16	Germany	2	2012	HTP	8	52.5
17x17	France	15	2013	AFA 3G	2749	53
		1	2022	AFA 3G	62	(16.5)
		1	2018	GAIA	4	31.2
	Sweden	2	2012	GAIA	170	57.6
18x18	Germany	3	2010	HTP	32	62.4

3.3 $M5_{Framatome}$ Cladding Material

$M5_{Framatome}$ was developed by Framatome to meet burn-up extension and the use of high duty fuel management schemes for lower operating costs, higher flexibility and strong safety margins requested by today's nuclear market. Framatome has a huge in-pile experience feedback for M5Framatome at elevated burn-up and in high duty conditions.

An extensive irradiation program has been in progress for $M5_{Framatome}$ fuel rods since 1989 in several power reactors spread across the globe with a wide range of different reactor environments. In the meantime, the $M5_{Framatome}$ alloy has reached commercial maturity and it is now a standard cladding tube material.

Table 2 provides data regarding Fuel Assemblies featuring $M5_{Framatome}$ cladding and supplied by Framatome in Europe, Asia and Africa. Among these Fuel Assemblies, 2 869 achieved a burn-up above 52 GWd/t, the maximum fuel assembly burn-up being 68 GWd/t and the maximum fuel rod burn-up being 80 GWd/t. The burn-up distribution of these $M5_{Framatome}$ fuel assemblies is given on Fig. 4.

Numerous inspections and measurements on fuel rods confirmed the good operational behavior of the $M5_{Framatome}$ cladding material with respect to waterside corrosion and hydrogen pickup, circumferential creep and axial growth up to very high burn-up.

$M5_{Framatome}$ with High Iron Content

In parallel to the massive deployment of $M5_{Framatome}$ claddings, Framatome maintained an important R&D effort to continue to reinforce further the good performances of this advanced alloy anticipating more demanding needs in term of fuel management or power uprates. As an example, the influence of the iron content, beyond the concentration targeted in production has been thoroughly assessed by Framatome with regards to the

Table 2. Framatome operational experience with $M5_{Framatome}$ cladding material (Status 2022/12)

Fuel array	Country	Number of reactors	First insertion	Number of FAs	Number of rods	Max FA burn-up (GWd/t)	Max FR burn-up (GWd/t)
17X17-14ft	BELGIUM	1	2000	336	88 704	54	59
	FRANCE-1300	16	1996	4 538	1 197 952	63	69
	FRANCE-N4	4	2005	3 136	827 904	51	55
17X17-12ft	BELGIUM	2	2009	636	167 904	55	60
	FRANCE	34	1993	7 679	2 026 248	57	**80**
	SOUTH AFRICA	2	2002	852	224 928	58	63
	SPAIN	1	1999	4	1 056	46	51
	SWEDEN	2	1998	830	217 968	60	66
	U.K	1	2008	753	198 792	52	56
18X18	GERMANY	3	1993	1 633	489 132	64	70
16X16	GERMANY	7	1993	2 637	620 496	63	69
15X15	BELGIUM	1	1998	916	186 080	54	59
	GERMANY	1	2004	240	49 200	59	65
	NETHERLANDS	1	2004	480	98 400	59	63
	SWEDEN	1	2000	324	66 096	61	65
	SWITZERLAND	1	2005	9	475	68	72
14X14	BELGIUM	2	1993	693	123 725	55	60
Total		83		26 410	6 774 266		

(*) 43 reactors in China and 1 reactor in Brazil are currently supplied with $M5_{\mathbf{Framatome}}$ cladding fuel assemblies manufactured under license.

benefit for corrosion in high duty conditions. Extended out of pile and in-pile characterizations have been carried out with $M5_{\mathbf{Framatome}}$ material featuring a higher iron content [1] which demonstrated that the main $M5_{\mathbf{Framatome}}$ properties are not impacted or even improved by the iron content up to 1000 ppm:

- improvement on corrosion resistance in very demanding conditions
- no impact on creep and mechanical properties
- no impact on hydrogen uptake and growth in reactor
- no impact on accidental behavior (LOCA and RIA) and licensing.

Equivalent or even improved global in-reactor performance and behavior in accident conditions and no significant change of the microstructure up in $M5_{Framatome}$ are observed which nowadays supports the increase of the maximum iron content in the $M5_{Framatome}$ without any impact on the licensing file. Delivery of $M5_{Framatome}$ with high iron content started in 2021. By the end of March 2023, more than 2700 AFA 3G Fuel Assemblies

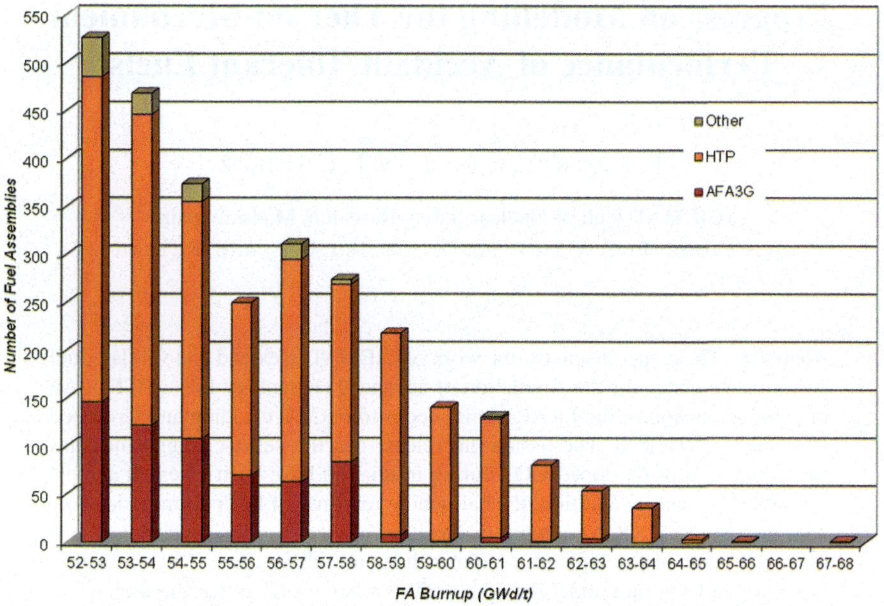

Fig. 4. Framatome $M5_{\mathbf{Framatome}}$ high burn-up experience

with fuel rod cladding featuring high iron content have been delivered to EDF reactors in France. $M5_{\mathrm{Framatome}}$ with high iron content is the standard cladding material to be delivered by Framatome in Europe, USA, South Africa, and Brazil.

4 Conclusion

The AFA 3G fuel assembly technology demonstrated a very high level of performance for 12ft and 14ft bundles with an irradiation experience of more than 43000 FAs acquired over more than 30 years. Throughout these years, Framatome has regularly improved the AFA 3G technology either to supply margin against emergent issues revealed by the introduction of new and more demanding reactor operation conditions, or to allow optimal use of the fuel assemblies regarding fuel cycle economy, power plant uprate and maneuverability.

Reference

1. D. Kaczorowski, J. P. Mardon, P. Barberis, P.-B. Hoffmann and J. Stevens, « Impact of Iron in M5 », In: Zirconium in the Nuclear Industry: 17th International Symposium, 2014, vol. STP 1543, p. 159-183.

Progress on Modelling the Thermo-Mechanical Performance of Accident-Tolerant Fuels

P. Aragón(✉), F. Feria, and L. E. Herranz

CIEMAT, Unit of Nuclear Safety Research, Madrid, Spain
{pau.aragon,francisco.feria,luisen.herranz}@ciemat.es

Abstract. This paper discusses the work of CIEMAT to extend and validate fuel performance codes for the simulation of advanced technology fuels (ATFs). An in-house extension of the FRAP-family codes to FeCrAl cladding and Cr-doped UO_2 fuel is described. The fission gas release and the pellet-cladding mechanical interaction of Cr-doped UO_2 during irradiation have been assessed against experimental data. In addition, the ballooning response of FeCrAl under loss-of-coolant accident (LOCA) conditions has been evaluated, taking Zircaloy-4 as a reference material to emphasize major behavioural differences. The simulation of Cr-doped UO_2 fuel under reactor conditions has revealed that the fission gas release is highly sensitive to the pre-processing of input data. Moreover, accounting for the enhanced viscoplasticity of Cr-doped UO_2 is essential in accurately describing pellet-cladding mechanical interaction during power ramps; otherwise, the cladding deformation will be overestimated. As for the LOCA scenario, the superior mechanical strength of FeCrAl entails a delay in the prediction of the burst time, as well as a reduced balloon size. This work has been carried out under the IAEA/CRP ATF-TS framework.

Keywords: Fuel performance modelling · Accident tolerant fuel · Cr-doped UO_2 · FeCrAl · FRAPCON · FRAPTRAN

1 Introduction

The analysis of in-pile behaviour of advanced technology fuels (ATFs) under normal and off-normal conditions requires the development and validation of fuel performance codes. Research institutions worldwide are actively engaged in investigating their fundamental properties and performance to eventually encapsulate the resulting understanding in fuel performance codes. CIEMAT is contributing to the technology readiness of ATFs by identifying and addressing the key modelling needs to accurately predict their thermo-mechanical response.

In this context, the ongoing project of the IAEA known as Testing and Simulation for Advanced Technology and Accident Tolerant Fuels (ATF-TS) aims to promote the development and deployment of ATF concepts for use in existing and future nuclear power reactors. Among the key specific objectives of the ATF-TS project are to benchmark fuel performance codes against data from experimental programmes relevant to

J. Liu and Y. Jiao (Eds.): WRFPM 2023, SPPHY 299, pp. 172–182, 2024.
https://doi.org/10.1007/978-981-99-7157-2_18

advanced fuel and cladding concepts, and to develop a methodology for evaluating the safety of ATF performance during LOCA events with a view to nuclear power plant applications.

The Unit of Nuclear Safety Research of CIEMAT has been actively participating in the ATF-TS project. The objective of this paper is to summarize the main outcomes of the modelling activities conducted in relation to Cr-doped UO_2 pellets and FeCrAl cladding in Work Tasks 2.1 (Fuel rod modelling and simulation) and 3.1 (Validation of fuel performance codes for LOCA analysis), respectively.

2 Cr-Doped UO_2 Pellets

2.1 Background on Cr-Doped UO_2 Properties

The addition of relatively low amounts of Cr_2O_3 to the UO_2 matrix (<1 wt%) is beneficial for fuel performance due to two main reasons: (i) improved fission gas retention and (ii) enhanced viscoplasticity [1–3]. The primary effect of Cr-doping is the promotion of grain growth during sintering, resulting in a larger grain size in Cr-doped UO_2 than in conventional fuel. This feature leads to a longer diffusion length, which is associated with reduced fission gas release since the intra-granular fission gas has to travel further to reach the grain boundaries. Moreover, Cr-doping enhances the thermal creep rate of the pellet and lowers the subsequent stresses imposed in the cladding during power transients, making the latter less susceptible to failure by degrading mechanisms like stress-corrosion cracking (SCC).

While other oxide dopants, such as TiO_2 and Nb_2O_5, can also be used to achieve a large grain microstructure, studies have shown that these dopants may have prohibitive effects on other properties [3]. For instance, the intra-granular fission gas diffusivity in doped UO_2 has been reported to be significantly higher than in undoped UO_2 due to changes in the oxygen potential [4], leading to increased fission gas release. Depending on the type of dopant and its concentration, the effect of the increased diffusion length may eventually be offset by this effect. Therefore, achieving desirable fission gas release properties for Cr-doped fuel requires balancing the positive effect of increased grain size against the negative influence arising from the enhanced diffusivity of the fission gas.

In addition to the above-mentioned properties, Cr-doped pellets also show greater densification during sintering than standard UO_2, resulting in a higher mass of fissile material in the pellets and a reduced porosity volume fraction. This translates into reduced in-pile densification and thus higher dimensional stability when in the reactor.

2.2 FRAPCON-4.0 Extension to Cr-Doped UO_2 Simulation

This section aims to discuss the extension of FRAPCON-4.0 to simulate Cr-doped UO_2 pellets. FRAPCON-4.0 is a 1.5-D fuel performance code developed by Pacific Northwest National Laboratory (PNNL) to perform steady-state calculations of Light Water Reactor (LWR) fuel rods, including slow power ramps [5]. By default, FRAPCON-4.0 is limited to modelling UO_2 and MOX fuel based on material models and correlations reported in the MATPRO library [6].

Although the material properties and behavioural aspects of Cr-doped and undoped UO_2 pellets are roughly similar [1], there are differences in their fission gas release and viscoplastic behaviour, as mentioned above. This paper focuses on the former phenomenon, with code modifications related to modelling the viscoplasticity of the pellet left for future work.

A fission gas diffusivity model developed by Cooper et al. [4] for Cr-doped UO_2 has been implemented into FRAPCON-4.0. The model is applied as a correction to the diffusivity correlation proposed by Turnbull for undoped UO_2 [7] and acts to enhance fission gas release. Figure 1 shows the dependence of diffusivity on temperature as described in the Forsberg-Massih model [8], which is the recommended option by PNNL, as well as in the models from Turnbull and Cooper.

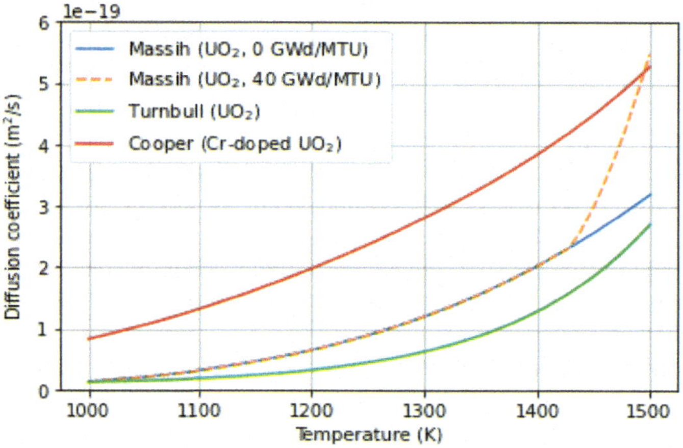

Fig. 1. Fission gas diffusion coefficient as a function of temperature.

The effect of increased diffusion length is implicit in the dependence of the fission gas release models included in FRAPCON-4.0 on the grain size diameter. A minor modification to the code consists of adding an option to specify the grain size diameter through the input file. Note that, in the default version of FRAPCON-4.0, the grain size diameter is hard-coded to 10 μm.

Finally, the correlation used to calculate swelling due to the build-up of solid fission products, which relates the fractional volume change of the pellet as a linear function of burnup, has been changed from 0.062 to 0.068% per MWd/kgU based on the post-irradiation examination report from the Halden IFA-716.1 rod [9]. This is roughly a 10% increase relative to the solid swelling rate used in FRAPCON-4.0 [6]. Other material models and correlations included in FRAPCON-4.0, such as melting temperature, density, thermal conductivity, specific heat capacity, emissivity, thermal expansion, and densification have not been modified. However, it is possible to account for the reduced in-pile densification of Cr-doped UO_2 pellets through the input file.

2.3 Validation Under Reactor and Fast Power Ramp Conditions

The predictive capabilities of the FRAPCON-4.0 extension to simulate Cr-doped UO_2 have been assessed under reactor and power ramp conditions using experimental data from Halden (IFA-716 rod 1) and Studsvik (SCIP-II rod Aa2), respectively. The main outcomes of the simulations are summarized in this section.

Specific objectives of IFA-716.1 were to study the effect of dopant concentration (diffusion coefficient) and grain size (diffusion length) on the fission gas release of Cr-doped UO_2, as well as its densification and swelling behaviour. The rod 716.1 was manufactured by AREVA and irradiated in Halden reactor conditions at about 235 °C D_2O coolant temperature and 34 bar pressure during 842 full power days. It was equipped with a fuel thermocouple attached to its top, a fuel stack elongation detector, and a pressure transducer that was reported faulty around 625 days (27.8 MWd/kgU). Table 1 provides additional information on the fuel rod specifications.

Table 1. Fuel rod specifications in IFA-716.1 and SCIP-II Aa2.

Parameter	IFA-716.1	SCIP-II Aa2
Fuel stack length [mm]	417.5	480
Pellet diameter [mm]	9.13	8.67
Pellet grain size [μm]	70	49
Pellet density [g/cm^3]	10.50	10.70
Cladding material	Zircaloy-4	Zircaloy-2/L
Cladding outer diameter [mm]	10.75	10.05
Cladding thickness [μm]	725	605
Fill gas pressure (@ 20 °C) [bar]	10	13.8 (*)
Free volume [cm^3]	5.8	23.8 (*)

(*)Expert judgement

The in-pile results from the 716.1 rod after final unloading showed almost no densification (0.1–0.2%). Therefore, an expected density increase of 0.15% has been fed into the code through the input file.

To simulate IFA-716.1, the first step is to pre-process the Halden data (e.g., power history) for FRAPCON-4.0. While the user can use any condensation algorithm, the code developers suggest avoiding time steps shorter than 0.1 days because the fission gas release models are based on steady-state and slow power ramp data and thus do not reflect the release rates expected for rapid power changes [5]. Consequently, inter-granular resolution is neglected for time steps shorter than 0.1 days. However, the maximum linear heat rate (LHR) evolution from IFA-716.1 shows many power ramps that elapse a very short period of time (<0.1 days). In this light, two condensation algorithms have been used. The first algorithm (1) respects the recommendation from the developers and yields time steps longer than 0.1 days. The second algorithm (2) includes an additional condition: if the maximum LHR change between two consecutive times is larger than

0.1 kW/m/min, the data point is included in the condensed data set. Figure 2 shows the maximum LHR evolution obtained for each condensation algorithm.

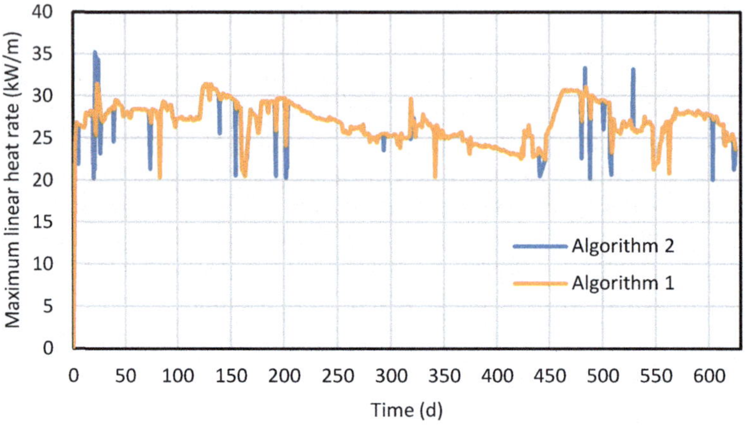

Fig. 2. Maximum linear heat rate evolution for the two condensation algorithms.

Figure 3 shows the evolution of the rod internal gas pressure. The Forsberg-Massih fission gas release model along with the new fission gas diffusivity model were used in the simulation. At the beginning-of-life, the simulation results are similar. After 320 days, however, there is a sudden power increase that causes a significant deviation. The condensation algorithm that captures power ramps (2) leads to a higher accumulation of fission gas along grain boundaries due to the disabled resolution in some time steps. Consequently, over 1% of the fission gas is released into the free volume of the rod at 320 days, leading to an increase in internal pressure. The same explanation applies to the deviation after the power uprate at around 450 days. Notably, the simulation results from Algorithm 2 closely match the experimental data, with a 6% fission gas release at 625 days (i.e., when the pressure transducer was reported to fail).

Figure 4 shows the evolution of fuel centreline temperature at the top of the test rod. Although both the simulation and experimental data follow a parallel trend after the first 30 days of irradiation, the simulation underpredicts the experimental data significantly. At the start of irradiation, the simulation shows cooling of the pellet on average, whereas the experimental data shows an increase in fuel centreline temperature. This suggests an overestimation of heat transfer across the gap. Neither the build-up of fission products in the pellet nor the amount of fission gas released into the gap can explain the observed differences at such an early irradiation time. The differences may be attributed to different relocation behaviours between Cr-doped and undoped UO_2 pellets, but the reasons behind this are still unknown. In this light, an ad-doc calibration of the relocation model is foreseen to assess its impact on the predicted internal gas pressure. If, as postulated, Cr-doped UO_2 pellets undergo a much smaller relocation, the fuel centreline temperature will increase with respect to the prediction shown in Fig. 4 and, consequently, internal pressure estimates will deviate from the ones reported in Fig. 3, overestimating the measured values.

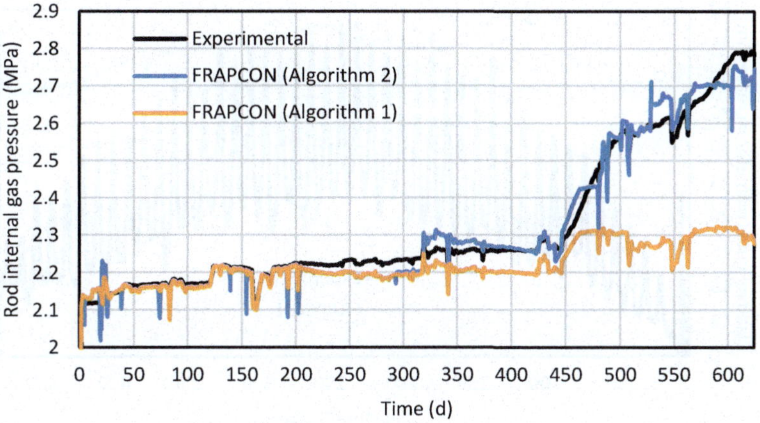

Fig. 3. Rod internal gas pressure evolution.

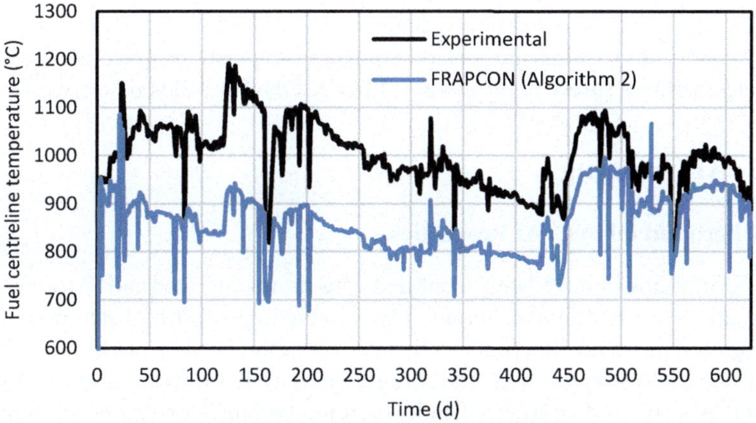

Fig. 4. Fuel centreline temperature evolution.

The SCIP-II Aa2 test aimed to investigate the pellet-cladding mechanical interaction (PCMI) that occurs during fast power ramp conditions in Cr-doped UO_2 fuel. The test utilized a 480 mm segment of a standard BWR fuel rod manufactured by AREVA and irradiated in Oskarshamn 2 to an average burnup of 28 MWd/kgU. Table 1 provides additional specifications for the test rodlet. The Aa2 rodlet was ramped to 45 kW/m with a 12-h hold time in the Studsvik's R2 reactor. Given that fission gas release is not relevant in this scenario (measured experimentally at 1.7%), the previous discussion on the sampling algorithm does not apply.

Post-irradiation examinations were conducted after the ramp test to measure the change in cladding outer diameter along the axial direction. The simulations carried out using the extension of FRAPCON-4–0 to Cr-doped UO_2 indicate that the rigid pellet approach leads to an overprediction of the cladding outer diameter after the power ramp (see Fig. 5). This was expected due to the enhanced viscoplasticity of the Cr-doped UO_2

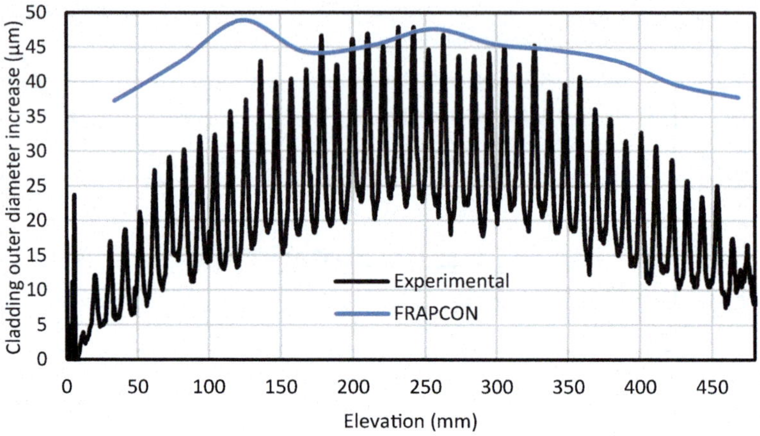

Fig. 5. Cladding outer diameter increase after the power ramp test.

pellets [3]. Therefore, the accurate simulation of PCMI with Cr-doped UO_2 pellets under the above conditions requires the consideration of viscoplasticity as a critical component. Nevertheless, from a safety point of view, FRAPCON-4.0 yields conservative results.

3 FeCrAl Cladding

3.1 Background on FeCrAl Properties

Ferritic FeCrAl alloys have been identified as a promising alternative to conventional Zr-based alloys as cladding materials due to their improved high-temperature steam oxidation resistance and superior mechanical strength [10, 11]. Its higher modulus of elasticity and yield strength make it more resistant to deformation and breakage under stress. This is particularly important under high-temperature conditions, in which several experimental findings suggest that FeCrAl undergoes limited or no ballooning due to creep [12, 13]. On the other hand, FeCrAl cladding must be thinner to compensate for its higher neutron absorption cross-section. Overall, FeCrAl has the potential to increase the so-called fuel coping time.

3.2 FRAPTRAN Extension to FeCrAl Cladding Simulation

The fuel performance code FRAPTRAN-2.0 has been extended to FeCrAl cladding simulation to assess its behaviour under a LOCA scenario. Developed by PNNL, FRAPTRAN-2.0 calculates the transient thermo-mechanical response of a single fuel rod to Design Basis Accidents (DBAs) at burnup levels up to 62 MWd/kgU [14]. The default version of the code is limited to modelling Zr-based claddings, and (burnup-dependent) slowly varying phenomena such as densification, swelling, creep, and irradiation growth are not calculated. Nevertheless, the state of the fuel rod at the onset of any transient may be read from an initialization file generated by FRAPCON-4.0, which contains information on the effect of burnup.

The FeCrAl correlations implemented in FRAPCON-4.0 (for base irradiation) and FRAPTRAN-2.0 (for LOCA transient) are described in [15, 16]. Specifically for the latter, the study involved developing an instantaneous plasticity model from BISON stress-strain curves [17], deriving an overstress failure criterion, and implementing a high-temperature creep correlation sourced from the open literature.

3.3 Verification Under LOCA Conditions

This section presents a comparative analysis of the thermo-mechanical response of FeCrAl and Zircaloy-4 cladding during the heat-up phase of a LOCA scenario. For that purpose, the Halden LOCA test IFA-650.10 [18] has been simulated using the extension of FRAPTRAN-2.0 mentioned above. The test used a 440 mm segment cut from a standard PWR fuel rod irradiated in Gravelines 5 to an average burnup of 61 MWd/kgU. Detailed information on its power history and axial power distribution can be found in [19]. Table 2 summarizes the fuel rod (radial) geometry required for the base irradiation simulation, for which the extension of FRAPCON-4.0 described in [15] has been used. The FeCrAl simulation used the same specifications as those used for Zircaloy-4 in the open literature, except for the radial geometry of the fuel rod. This is to (i) exclude any external thermal-hydraulic effects (i.e., identical outer cladding diameter) and (ii) avoid affecting the PCMI (i.e., identical gap dimensions). As a consequence, a larger pellet diameter and a thinner cladding result in the FeCrAl simulation. Further information on the fuel rod radial geometry can be found in [16].

Table 2. Fuel rod (radial) geometry of the IFA-650.10 mother rod.

	FeCrAl	Zircaloy-4
Pellet diameter [mm]	8.57	8.19
Gap thickness [μm]	84	
Cladding outer diameter [mm]	9.5	
Cladding thickness [μm]	382 [20]	570

Figure 6 shows the evolution of the internal gas pressure in the rodlet. Cladding failure is evident by the sudden drop down to the test rig pressure at 217.5 s (Zircaloy-4) and 263 s (FeCrAl). The burst time of FeCrAl is delayed by about 21% compared to Zircaloy. During the first 175 s of the transient, the behaviour of Zircaloy-4 and FeCrAl is almost identical. However, at around 180 s, the Zircaloy-4 simulation reaches a maximum, while the same does not occur in the case of FeCrAl until 225 s within the transient. This, along with the different behaviour observed thereafter, can be explained by the superior mechanical strength of FeCrAl and its effect on the free volume of the rodlet.

The onset of plastic deformation occurs around 165 s and 210 s for Zircaloy-4 and FeCrAl, respectively, triggering a significant increase in the free volume of the rodlet. This increase eventually dominates over the effect of the rising temperature, resulting

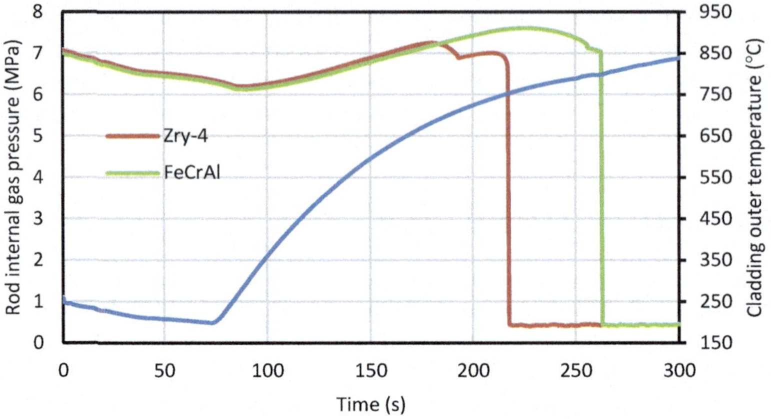

Fig. 6. Rod internal gas pressure evolution. The cladding outer temperature at the ballooning node fed into the code (identical for both cladding materials) is also shown.

in the internal pressure reduction. The delayed onset of plastic deformation consistently reflects the higher yield strength of FeCrAl compared to Zircaloy-4.

As shown in Fig. 7, permanent hoop strains at the burst node in both cladding materials start increasing as soon as the yield strength is exceeded. However, FeCrAl deforms permanently at a lower rate than Zircaloy-4, as it has a higher resistance to high-temperature creep, and therefore takes longer to reach the instability strain (5%). It is important to note that the discontinuities at 193 s and 256 s seen for Zircaloy-4 and FeCrAl, respectively, are due to the transition to the FRAPTRAN-2.0 large deformation model (BALON2) at the instability strain.

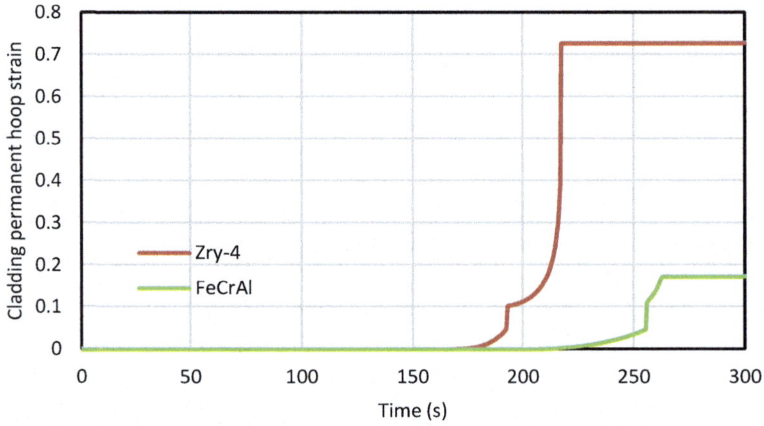

Fig. 7. Cladding permanent hoop strain evolution.

4 Conclusions

This paper summarizes the main insights gained from CIEMAT's modelling activities under the IAEA/CRP ATF-TS framework to extend and validate the FRAP-family codes for Cr-doped UO_2 pellets and FeCrAl cladding.

- With regard to Cr-doped fuels, comparisons to IFA-716.1 suggest that radial relocation should be notably attenuated relative to UO_2. In addition, attention should be paid to the input data condensation algorithm, so as not to miss key features in the fuel rod power history. Under power ramp conditions, viscoplasticity should be taken into account, the rigid-pellet approximation being no longer applicable.
- As for FeCrAl cladding, comparisons set with Zircaloy-4 cladding based on IFA-650.10 confirm its improved performance under LOCA conditions, delaying failure time and leading to significantly less deformation.

Further investigations are currently ongoing to find out optimal approaches to simulate additional fuels and scenarios.

References

1. AREVA, "Incorporation of Chromia-Doped Fuel Properties in AREVA Approved Methods," 2016.
2. Y. Che, G. Pastore, J. Hales, and K. Shirvan, "Modeling of Cr2O3-doped UO2 as a near-term accident tolerant fuel for LWRs using the BISON code," *Nuclear Engineering and Design*, vol. 337, pp. 271–278, Oct. 2018, https://doi.org/10.1016/j.nucengdes.2018.07.015.
3. National Nuclear Laboratory, "Provision of Information on Chromium Doped Fuel for Use in Light Water Reactors - Issue 1," 2019.
4. M. W. Cooper *et al.*, "Fission gas diffusion and release for Cr2O3-doped UO2: From the atomic to the engineering scale," *Journal of Nuclear Materials*, vol. 545, Mar. 2021, https://doi.org/10.1016/j.jnucmat.2020.152590.
5. K. Geelhood, W. Luscher, P. Raynaud, and I. Porter, "FRAPCON-4.0: A computer code for the calculation of steady-state, thermal-mechanical behavior of oxide fuel rods for high burnup," 2015.
6. W. Lusher, K. Geelhood, and I. Porter, "Material property correlations: Comparisons between FRAPCON-4.0, FRAPTRAN-2.0, and MATPRO," 2015.
7. J. A. Turnbull, C. A. Friskney, J. R. Findlay, F. A. Johnson, and A. J. Walter, "The diffusion coefficients of gaseous and volatile species during the irradiation of uranium dioxide," *Journal of Nuclear Materials*, vol. 107, pp. 168–169, 1982.
8. K. Forsberg and A. R. Massih, "Diffusion theory of fission gas migration in irradiated nuclear fuel UO2," *Journal of Nuclear Materials*, vol. 135, pp. 140–148, 1985.
9. K. Fuglesang, "In-pile results from the fission gas release mechanisms study in IFA-716 after final unloading," Halden, Norway, 2016.
10. R. E. Stachowski, R. B. Rebak, W. P. Gassmann, and J. Williams, "Progress of GE development of accident tolerant fuel FeCrAl cladding," in *Top Fuel 2016*, Boise, 2016.
11. K. A. Terrani, S. J. Zinkle, and L. L. Snead, "Advanced oxidation-resistant iron-based alloys for LWR fuel cladding," *Journal of Nuclear Materials*, vol. 448, no. 1–3, pp. 420–435, 2014, doi: https://doi.org/10.1016/j.jnucmat.2013.06.041.
12. C. P. Massey, K. A. Terrani, S. N. Dryepondt, and B. A. Pint, "Cladding burst behavior of Fe-based alloys under LOCA," *Journal of Nuclear Materials*, vol. 470, pp. 128–138, Mar. 2016, https://doi.org/10.1016/j.jnucmat.2015.12.018.

13. S. B. Bell, K. A. Kane, C. P. Massey, L. A. Baldesberger, D. Lutz, and B. A. Pint, "Strength and rupture geometry of un-irradiated C26M FeCrAl under LOCA burst testing conditions," *Journal of Nuclear Materials*, vol. 557, Dec. 2021, https://doi.org/10.1016/j.jnucmat.2021.153242.
14. K.J. Geelhood, W.G. Luscher, J.M. Cuta, and I.A. Porter, "FRAPTRAN-2.0: A Computer Code for the Transient Analysis of Oxide Fuel Rods," 2016.
15. P. Aragón, F. Feria, and L. E. Herranz, "Modelling FeCrAl cladding thermo-mechanical performance. Part I: Steady-state conditions," *Progress in Nuclear Energy*, vol. 153, Nov. 2022, https://doi.org/10.1016/j.pnucene.2022.104417.
16. P. Aragón, F. Feria, and L. E. Herranz, "Modelling Fecral Cladding Thermo-Mechanical Performance. Part II: Comparative Analysis with Zircaloy Under Loca Conditions," *Progress in Nuclear Energy*, vol. 163, 2023, https://doi.org/10.1016/j.pnucene.2023.104838.
17. Idaho National Laboratory, "FeCrAl power-law hardening stress," 2021. https://mooseframework.inl.gov/bison/source/materials/tensor_mechanics/FeCrAlPowerLawHardeningStressUpdate.html (accessed Oct. 19, 2021).
18. P. Bennett, "LOCA testing at Halden, the tenth experiment IFA-650.10," Halden, Norway, 2010.
19. I. Vallejo and L. E. Herranz, "HALDEN LOCA Tests: Simulations of IFA-650.3, .5 and .10 with FRAP Serie Codes. Preliminary Results and Comparisons to Experimental Data," in *Enlarged Halden Programme Meeting (EHPGM)*, 2014.
20. IAEA, "Analysis of Options and Experimental Examination of Fuels for Water Cooled Reactors with Increased Accident Tolerance (ACTOF)," 2020.

Additive Manufacturing Process Design For Thimble Plug Assembly

Guo-peng Qin[1(✉)], Yu-Shan Huang[2], Xin Tong[2], and Li-ying Zhang[1]

[1] CNNC Jianzhong Nuclear Fuel Co., Ltd., Yibin, China
qad366@163.com

[2] Guangzhou Shinengine AM Technology Co.Ltd., Guangzhou, China
huangys@shinengine.cn, tongxin_gz@husun.com.cn

Abstract. The thimble plug assembly is a kind of associated core components in pressurized water reactor (PWR) nuclear power plants, and the existing mature process is manufactured by machining parts and assembly welding. However, the traditional subtractive manufacturing process requires a variety of profiles of different specifications to be customized for processing different parts, and the raw material preparation cycle is long, which is not conducive to the rapid manufacturing and research and development of prototype products. In order to adapt to the development trend of advanced nuclear fuel manufacturing technology at home and abroad and the demand for localization of raw materials, CNNC Jianzhong Nuclear Fuel Co., Ltd. (CJNF) has carried out a series of alternative studies on additive manufacturing process for the thimble plug assembly. This paper introduces the latest research progress of CJNF on additive manufacturing of thimble plug assembly from the perspectives of additive manufacturing structure design, 3D model and selective laser melting (SLM) process design, and molding size optimization design, and welding process. The research results are of great significance to improve the manufacturing technology level of domestically associated core components.

Keywords: Thimble plug assembly · Additive manufacturing · Selective laser melting · Nuclear grade stainless steel · Structural design · Process design

1 Introduction

The thimble plug assembly is a kind of associated core components in pressurized water reactor (PWR) nuclear power plants, and its main function is to limit the flow of core coolant in the nuclear fuel guide pipe. The shape of the typical thimble plug assembly is shown in Fig. 1, and the main structure consists of three parts——compression system, support plate with a flow hole, and thimble plug. The product material mark is austenitic stainless steel and nickel-based superalloy. The traditional products need to customize 8 kinds of raw material processing parts first, and then connect them to form parts though 3 kinds of welding structures. Due to the long preparation time of raw materials and the complicated process, the entire manufacturing cycle of the thimble plug assembly

J. Liu and Y. Jiao (Eds.): WRFPM 2023, SPPHY 299, pp. 183–189, 2024.
https://doi.org/10.1007/978-981-99-7157-2_19

is very long, which is not conducive to the development of new product prototypes. In order to meet the needs of rapid prototyping and manufacturing of a new generation fuel, but also comply with the development trend of international advanced technology, CNNC Jianzhong Nuclear Fuel Co., Ltd. (hereinafter referred to as CJNF) has carried out a series of alternative research work on additive manufacturing technology.

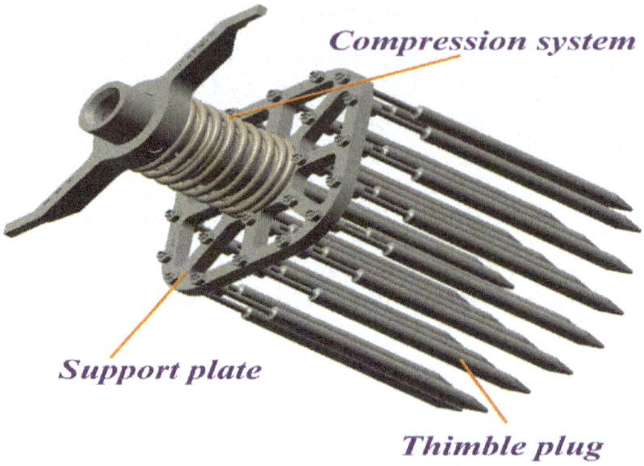

Fig. 1. Outline diagram of the thimble plug assembly

1.1 Study Conditions

Additive technology route: Selective Laser Melting (SLM) [1].

Raw material: 022Cr19Ni10 Low cobalt metal powder (15 μm ~ 53μm).

Equipment type: Double laser molding equipment.

Assembly method: Additive and subtractive mixed TIG welding.

1.2 Research Results

The preliminary research of CJNF adopts alternative solution, in order to minimize the material specification required, reduce the number of special processes, and shorten the time of product manufacturing cycle. The research direction is as follows.

1.3 Additive Structure Design

Due to the presence of buffer springs and welded connection structures in the thimble plug assembly, it is difficult to perform additive manufacturing molding as a whole. Considering the economy of manufacturing, the best solution is the additive manufacturing of complex profile parts, and the guide pins, pins, nuts of small parts with low manufacturing difficulty, and the elastic functional parts (Buffer springs) are manufactured from subtractive materials. This scheme uses the most economical additive + subtractive combination manufacturing method, under the premise of ensuring that the external dimension of the thimble plug assembly product remains unchanged, excluding the remaining process uncertainties, focusing on the mechanical, corrosion resistance and irradiation creep resistance of the additive manufacturing material itself. The overall manufacturing route obtained by CJNF is: using additive manufacturing methods to Integral print (welded structural parts of the original 2 parts), thimble plugs, compression rods; Using traditional technology to manufacture pressing system springs and supporting small parts; The entire thimble plug assembly is then assembled and welded, as shown in Fig. 2.

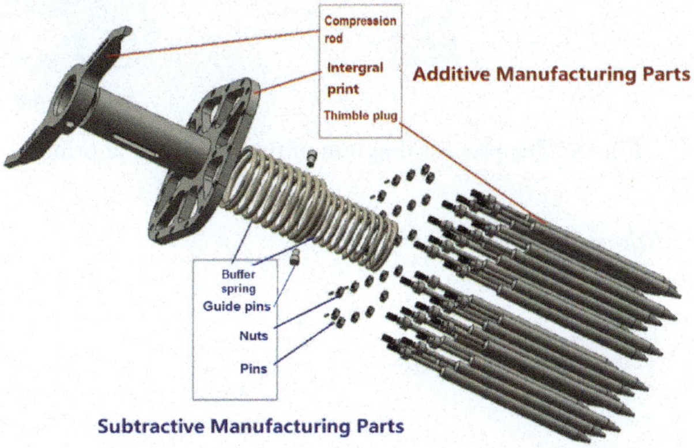

Fig.2. Illustration of preliminary AM alternatives for thimble plug assembly

1.4 Additive Manufacturing Parts Design

According to the product structure design, the two separate parts and the corresponding welded structure are eliminated. The guide cylinder and the support plate are additive manufactured as one part ----Integral print. As shown in the Fig. 3, the traditional method uses board and bar materials to make the guide cylinder and support plate, and then weld it together. The additive method uses powder to form integral print directly without welding.

Similarly, the additive method can also print the compression rod and thimble plug parts directly without custom profiles, as shown in Fig. 4. This design directly reduces the 4 most difficult raw material profiles and 1most demanding structural weld, optimizes the time-consuming production preparation process, and significantly shortens the entire production cycle.

1.5 SLM Process Design

In order to improve the manufacturing economy and efficiency, CJNF adopts a variety of parts hybrid layout design in the printing process. At the same time, considering that the mixed arrangement may cause the printing wind field disorder [2] and affect the stability of the scraper [3], the multi-part mixed arrangement optimization experiment is carried out, as shown in the left of Fig. 5. According to the shape of the part, the typesetting

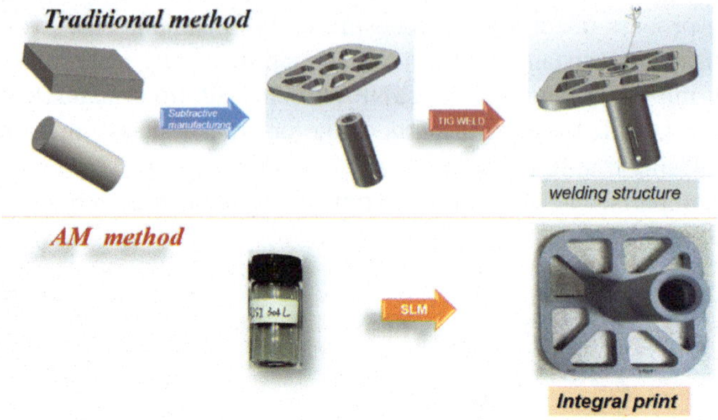

Fig. 3. The process diagram of the AM integral print

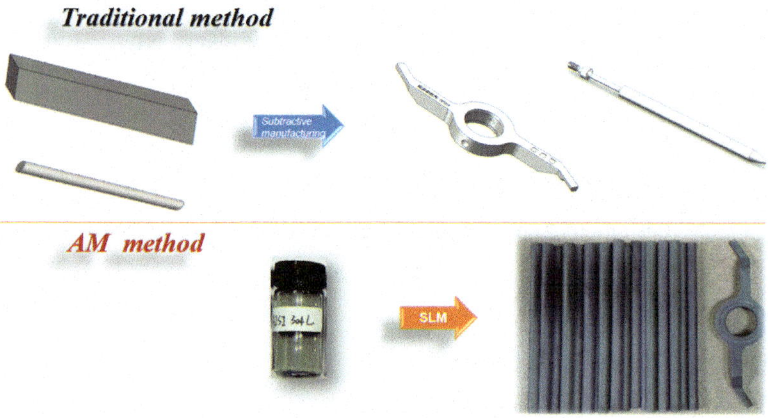

Fig.4. The process diagram of AM compression rod and thimble plug

position is optimized to ensure that the entire printing substrate has a good slag blowing wind field environment and a small amount of powder spreading scraper loss. At the same time, in order to ensure that the quality of the printing process is controllable and does not affect the printing quality of the parts, a variety of furnace damage inspection specimens are designed to ensure the stable and controllable manufacturing process of additive manufacturing parts. The video screenshot to the right of Fig. 5 shows the SLM molding process of the parts of thimble plug assembly.

1.6 Molding Size Optimization

The SLM printing process is characterized by thermal build-up [4] and material shrinkage, so internal stress concentrations are inevitable. Even heat treatment and mechanical vibration cannot completely eliminate internal stress. Especially for the structure of thimble plug plate barrel parts, there are typical characteristics of both cylinder and plate shape, and it is difficult to take into account the shrinkage deformation rate in all

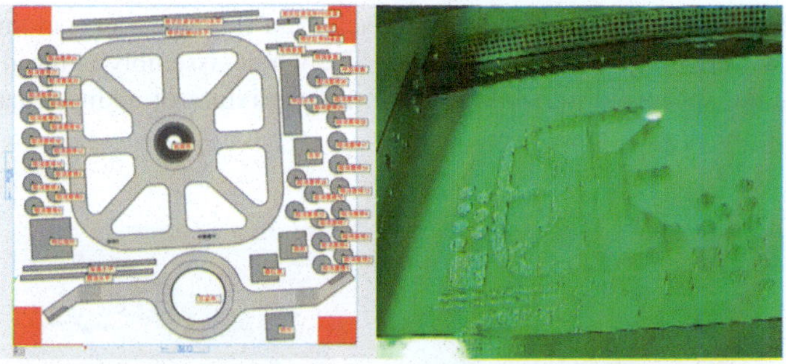

Fig.5. Schematic diagram of SLM process design

directions [5]. Moreover, the deformation of one structure will also cause the shape and position tolerance of another structure to change, such as the warping deformation of the plate structure will affect the verticality and axial channeling of the center cylinder; Another example is that the shrinkage deformation of the diameter of the central cylinder will affect the flatness and four-side parallelism of the plate structure. This requires process optimization of the molded dimensions in advance. There are generally two optimization methods for CJNF, one is to modify the printing model in advance according to the forming formation to reserve the deformation margin, and later use mechanical processing to remove the margin to ensure the product size; Second, after collecting deformation data for multiple times, the deformation trend is analyzed by simulation calculation, and then the inverse deformation structure is designed for the printing model to offset the deformation of the printing molding. By optimizing the molding size, CJNF has basically succeeded in achieving the full control of the size of the SLM molding process. Figure 6 shows a comparison between the design model and the actual print.

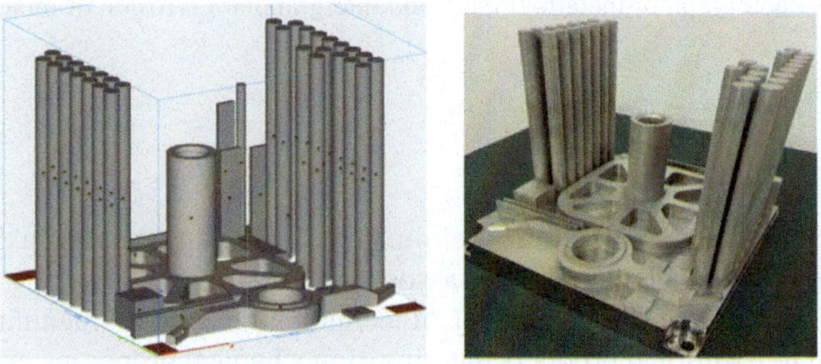

Fig. 6. Comparison of design model with actual products

1.7 Welding Process Design

After all the parts to be added and subtracted are machined, the compression rod, inner spring, outer spring, guide pin are assembled and TIG welded to form a pressing system, as shown in the left of Fig. 7. The compression system is then assembled with a thimble plug, and the anti-loosening connection is achieved by TIG spot welding of the bar bundle nut, as shown on the right of Fig. 7. The whole assembly is through guide pin welding and nut spot welding, and finally forms a mixture of additive and subtractive material [6] thimble plug assembly.

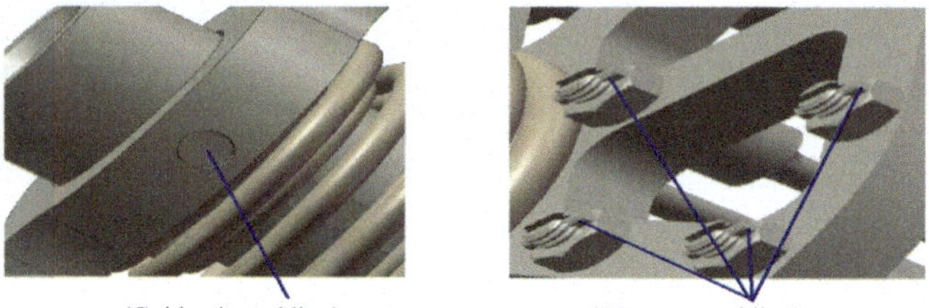

(Guide pin welding) (Nut spot welding)

Fig. 7. Schematic diagram of the welding process design of thimble plug assembly

The metallographic photograph and intergranular corrosion photograph of welds with additive and subtractive materials is shown in Fig. 8.

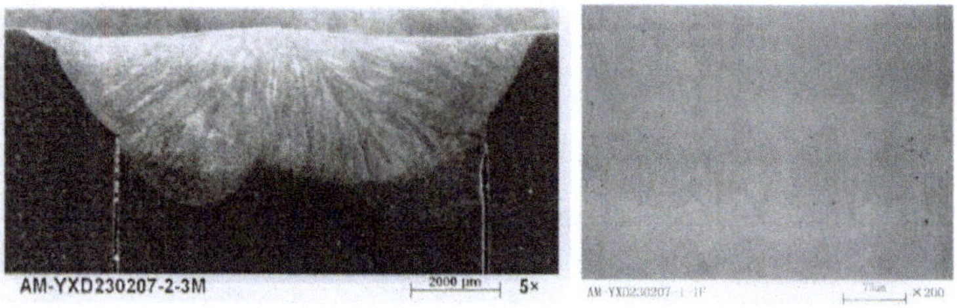

Fig. 8. Photograph of metallographic and intergranular corrosion of mixed welds

1.8 Conclusions

CJNF has successfully completed the preliminary application research on the alternatives additive manufacturing technology of associated core components. At present, three thimble plug assemblies produced by SLM technology have been manufactured, which are equivalent to traditional products in terms of mechanical properties, phase structure, chemical composition and external dimensions. After comparative analysis, compared with the traditional process, the additive manufacturing solution reduces the number of raw material specifications by 50%, reduce the manufacturing time by 50%, and the manufacturing economy is greatly improved. This achievement is of great significance for improving the technology level and design ability of domestically fuel.

References

1. XU Yang. A review of metal 3D printing technology[J]. China Metal Bulletin, 2019(2): 104-105. DOI: CNKI:SUN:JSTB.0.2019-02-066
2. ZHANG Lihao, QIAN Bo, ZHANG Chaorui, MAO Jian, FAN Hongri. Summary of development trend of metal additive manufacturing technology[J]. Materials Science and Technology, 2022, 30(1): 42–52. https://doi.org/10.11951/j.issn.1005-0299.20210111
3. CARLOS D B, GU D D, SCHMIDT M, et al. On the selection and design of powder materials for laser additive manufacturing[J]. Materials & Design, 2021, 204: 109653. https://doi.org/10.1016/j.matdes.2021.109653
4. KVRNSTEINER P, WILMS M B, WEISHEIT A, et al. High-strength damascus steel by additive manufacturing[J]. Nature, 2020, 582(7813): 515-519. https://doi.org/10.1038/s41586-020-2409-3
5. YUAN Hong, HE Gening, LI Lei, et al. Development and application of 3D printing technology in nuclear power[J]. Science &Technology Vision, 2020(17): 118–122. https://doi.org/10.19694/j.cnki.issn2095-2457.2020.17.40
6. DONG Yiwei, ZHAO Qi, LI Xiaolin. Key technology and development of increase and decrease composite processing[J]. Metal Working(Metal Cutting), 2016(13): 7–12.

Research and Application of Radioactive Control Methods on the Primary Circuit of PWR Fuel Cladding with Loss of Air Tightness

Zhang Xianggui(✉), Yang Hongye, Ma Fei, Wu Wenqi, and Chen Guangjun

Jiangsu Nuclear Power Corporation, Jiangsu, China
{zhangxg01,yanghy02,mafei01,wuwq,chengj02}@cnnp.com.cn

Abstract. During the operation of a pressurized water reactor nuclear power plant, a large number of radionuclides are generated by the fuel fission and contained within the fuel rod. When the fuel cladding suffers from a loss of air tightness, some inert gases and iodine in the gaseous form are easily released into the primary coolant, causing an abnormal increase in radioactivity in the primary circuit and associated systems, which may result in an increase in the radioactivity of the effluent that cannot be discharged. At the same time, the radiation dose of the plant in the nuclear island control area increased. In this paper, a method to control radioactivity in the primary loop following loss of fuel cladding air tightness is studied, and the operation mode during shutdown state transition is optimized, which has been applied in a domestic nuclear power plant. The application of the research results shows that when a domestic nuclear power plant experienced a loss of fuel cladding air tightness, I-131 in the primary coolant was controlled below 3.0E4Bq/L and Xe-133 below 1.0E6Bq/L before the last main pump was shut down and the reactor was uncovered. The qualified emission of radioactive effluents and the collective radiation dose of personnel did not increase significantly. The research results of this method have achieved good application effect, and have reference significance for similar types of reactors.

Keywords: Loss of fuel cladding air tightness · Radioactivity control · Shutdown state transition · Operation mode optimization

1 Introduction

The nuclear fuel cladding is a sealed shell of nuclear fuel, which acts as the safety barrier of nuclear power plants, containing fission products and preventing their leakage [1]. Once the fuel cladding is broken, the fission products will be released into the coolant of the primary circuit, causing an abnormal increase in the radioactivity of the primary circuit and the connected systems. Some radioactive fission products will be released into the working environment or the external environment, causing an increase of the collective radiation dose of the personnel, and even causing adverse effects on public health. Therefore, in the event of fuel cladding damage, it is necessary for nuclear power plants to effectively control primary circuit radioactivity.

J. Liu and Y. Jiao (Eds.): WRFPM 2023, SPPHY 299, pp. 190–199, 2024.
https://doi.org/10.1007/978-981-99-7157-2_20

Based on the actual situation of nuclear fuel cladding damage in a domestic nuclear power plant, this paper studies the operation mode in the shutdown state transition process under the loss of fuel cladding air tightness, reduces the primary coolant radiation level, and controls the collective radiation exposure of personnel during overhaul to be equivalent to that under normal conditions. At the same time, in response to the impact on the effluent, we will study the optimization of three waste treatment and ventilation operation to ensure that all radioactive effluent is discharged in a qualified manner and avoid any impact on the environment.

2 Abnormal Fuel Cladding Damage in a Domestic Nuclear Power Plant

In 2022, a nuclear power plant in China discovered an abnormal increase in the activity levels of fuel integrity characteristic nuclides such as radioactive iodine and inert gases in the coolant of the primary circuit. The I-131 sampling detection activity was approximately 2.21E4 ~ 3.09E4Bq/L (previously undetected), while other inert gases such as Xe-133, X-135, Kr-87, Kr-88, Kr-85m, etc. also increased by about one order of magnitude. The changes in the radiochemical analysis data of the primary circuit for 5 days after discovering abnormalities are as follows (Table 1).

Table 1. Primary circuit release data before and after anomalies

Time	Unit status	I-131 (Bq/L)	I-133 (Bq/L)	Total γ (Bq/L)	Kr-85m (Bq/L)	Kr-87 (Bq/L)	Kr-88 (Bq/L)	Xe-133 (Bq/L)	Xe-135 (Bq/L)
One month before the anomaly	Full power operation	Undetected	7.12E + 03	3.96E + 06	1.92E + 03	3.96E + 03	5.63E + 03	4.36E + 03	1.98E + 04
One day before the anomaly	Full power operation	Undetected	5.86E + 04	5.06E + 06	3.71E + 04	6.20E + 04	9.05E + 04	1.91E + 05	1.71E + 05
Discovering the anomaly	Full power operation	2.67E + 04	7.29E + 04	5.45E + 06	4.58E + 04	6.88E + 04	1.06E + 05	2.96E + 05	2.83E + 05
One day after the anomaly	Full power operation	2.32E + 04	7.44E + 04	5.83E + 06	3.61E + 04	5.81E + 04	8.47E + 04	2.52E + 05	2.50E + 05
Two days after the anomaly	Full power operation	2.26E + 04	6.61E + 04	5.77E + 06	4.63E + 04	7.82E + 04	1.17E + 05	3.33E + 05	3.17E + 05
Three days after the anomaly	Full power operation	3.54E + 03	2.51E + 04	4.78E + 06	3.50E + 04	4.05E + 04	8.50E + 04	2.99E + 05	2.33E + 05
Four days after the anomaly	Full power operation	7.42E + 03	2.53E + 04	5.49E + 06	4.90E + 04	5.17E + 04	1.06E + 05	4.25E + 05	3.04E + 05
Five days after the anomaly	Full power operation	2.22E + 03	1.87E + 04	5.11E + 06	4.06E + 04	3.81E + 04	8.86E + 04	3.88E + 05	2.10E + 05

It can be seen that I-131 was detected for the first time in the radiochemical sampling analysis of the primary circuit, and the specific activity of the inert gas Xe-133 increased by more than 60 times. The total iodine and total γ activity also increased significantly. After the step increase in the activity levels of the radioactive nuclides mentioned above, the overall trend shows a stable trend, with significant decreases in I-131 and total iodine activity.

During the steady-state power operation of the unit, some fuel fission products (especially iodine and cesium) will precipitate and adhere to the inner surface of the fuel cladding. If the fuel rod cladding is damaged or the air tightness is lost, some inert gases and iodine present in gaseous form are easily released into the primary coolant. Any ratio of short-lived to long-lived isotope specific activity can be used as an indicator of fuel integrity. Once these ratios undergo significant changes, it indicates the possibility of suspected fuel damage or loss of air tightness in the core. Usually, it is necessary to combine three indicators to judge [2–6]. Currently, the ratio commonly used internationally includes:

(1) A step jump of at least 0.25 higher in the ratio of I-131/I-133 indicates possible fuel damage or loss of air tightness. The ratio of I-131/I-133 was calculated according to the primary radiochemical sampling analysis data, as shown in Table 2 below. It can be seen that the I-131/I-133 ratio has a step jump of 3.09 higher than the judged fuel integrity anomaly value of 0.25, indicating the possibility of suspected loss of fuel air tightness in the core.

Table 2. Ratio of I-131/I-133 in the primary circuit

Time	Unit status	I-131(Bq/L)	I-133(Bq/L)	I-131/I-133
One month before the anomaly	Full power operation	Undetected	7.12E + 03	None
Discovering the anomaly	Full power operation	2.67E + 04	7.29E + 04	0.366255

(2) A step jump of at least 1 higher in the ratio of Xe-133/Xe-138 indicates possible fuel damage or loss of air tightness. The ratio of Xe-133/Xe-138 was calculated according to the primary radiochemical sampling analysis data, as shown in Table 3 below. It can be seen that the Xe-133/Xe-138 ratio has a step jump of 0.366 higher than the judged fuel integrity anomaly value of 1, indicating the possibility of suspected loss of fuel air tightness in the core.
(3) A step jump of at least 0.5 higher in the ratio of Kr-85m/Kr-87 indicates possible fuel damage or loss of air tightness. The ratio of Kr-85m/Kr-87 in the last half year was calculated according to the primary radiochemical sampling analysis data, as shown in Table 4 below. It can be seen that Kr-85m/Kr-87 ratio has no obvious step.

The activity of inert gas (Xe, Kr), iodine and other fuel fission products increased significantly. The method for judging the abnormal fuel integrity mentioned above indicates that there is a loss of fuel assembly air tightness in the core.

Table 3. Ratio of Xe-133/Xe-138 in the primary circuit

Time	Unit status	Xe-133(Bq/L)	Xe-138(Bq/L)	Xe-133/Xe-138
One month before the anomaly	Full power operation	4.36E + 03	Undetected	None
Discovering the anomaly	Full power operation	2.96E + 05	9.57E + 04	3.092999

Table 4. Ratio of Kr-85m/Kr-87 in the primary circuit

Time	Unit status	Kr-85m(Bq/L)	Kr-88(Bq/L)	Kr-85m/Kr-87
One month before the anomaly	Full power operation	1.92E + 03	5.63E + 03	4.85E-01
Discovering the anomaly	Full power operation	4.58E + 04	1.06E + 05	6.66E-01

3 Primary Circuit Radioactive Control During Unit State Transition

3.1 Control Power to Prevent the Expansion of Fuel Cladding Rupture

When the fuel cladding abnormality was discovered in the nuclear power plant, the fuel assembly with the longest running time inside the reactor experienced 108 power surges, approaching the upper limit value of 150 specified in the fuel assembly specification. According to EPRI's research reports such as "Fuel Reliability Guidelines: Pellet Cladding Interaction" and "IAEA Nuclear Power Series–Review of Fuel Damage in Water Cooled Reactors", as well as other research reports and relevant practical experience [6–10], the cumulative effect of short-term and periodic fuel cladding stress and strain caused by reactor power changes will reduce the strength of the fuel cladding, The design must prevent fatigue failure of the fuel cladding due to cyclic changes in stress and strain. The interaction between fuel pellets and fuel cladding (PCI effect) is an important cause of fuel cladding damage. The fuel pellets form an hourglass structure due to uneven thermal expansion and contact with the fuel cladding. The fuel cladding generates a ring ridge on the corresponding shoulder of the pellets, leading to local stress concentration around the ring ridge, which can lead to local fatigue damage of the fuel cladding. At the same time, due to the different coefficient of thermal expansion, when the reactor power is rapidly and significantly increased, the fuel pellet is more swollen than the fuel cladding, forcing the fuel cladding to be subject to tensile stress, which may lead to fuel cladding damage under excessive tensile stress.

To prevent the increase of fuel cladding damage due to sudden power changes, causing an increase in primary circuit radioactivity and radiation dose in the nuclear island control area building, and to avoid excessive radioactive nuclide processing load during overhaul, peak regulation should be avoided before overhaul. The rate of power increase and decrease should not exceed 5%/h, and the PCI effect on the fuel cladding should be stabilized for at least 3 h after power reduction to eliminate the impact. During the power reduction period, the primary radioactive sampling analysis frequency should be strengthened, and sampling should be conducted every 2h until the unit reaches the cold state.

3.2 Primary Circuit Water Quality Control

3.2.1 Overall System Process and Control Objectives

To limit the level of radioactivity in the primary circuit, it is necessary to separate radioactive substances. Usually, methods such as precipitation, solvent extraction, ion exchange, distillation, electrochemical separation, isotope exchange, etc. are used. Due to the limited conditions of the primary circuit, the introduction of impurity ions may cause the deviation of the water quality of the primary coolant. Therefore, adsorption of resin bed or membrane separation and water exchange are mainly used to reduce the radioactivity level in the primary circuit. Taking a domestic nuclear power plant as an example, the following system is used to reduce the radioactivity of the primary circuit. The relationship between the various systems is shown in Fig. 1.

(1) The volume control system completes the charging and discharging of primary coolant, and can introduce primary coolant into primary purification system for purification, or discharge primary coolant into coolant storage system;
(2) Coolant storage system mainly completes the function of receiving and storing the coolant of primary system, and can purify the coolant of primary system;
(3) The boron-containing water storage system is used to store the volume of boric acid necessary for the safe operation of the reactor plant;
(4) Fuel pool and boron containing water tank purification system can purify and remove mechanical and dissolved impurities in boron containing water storage tank.

The main contribution of radioactivity in primary coolant comes from inert gases such as Xe, Kr, halogen I and other metal ions K, Na, Cs, etc., but the inert gases are usually reduced by residual decay or purge discharge. Therefore, only halogen and other metal ions are considered in the purification process by using resin bed and water exchange method. According to the results of primary coolant release analysis before overhaul (Table 5), It can be seen that the main radioactivity at this time comes from iodine isotopes, Na-24, K-42 and Cs-138, among which the longest half-life of I-131 is 8.02 days, and its half-life in human body is 7.6 days, and I-131 is a class 1 carcinogen, which has a great influence on human thyroid. Therefore, I-131 is selected as the control target.

3.2.2 Control Measures and Effects

(1) When the unit enters the stage of boron injection at the minimum monitored power level, the backup primary purification system resin bed is put into operation, and the purification flow rate is increased as far as possible to reduce the primary radioactivity. After the main pump stops completely, the resin bed of the primary purification system has no running power, and the purification is withdrawn;

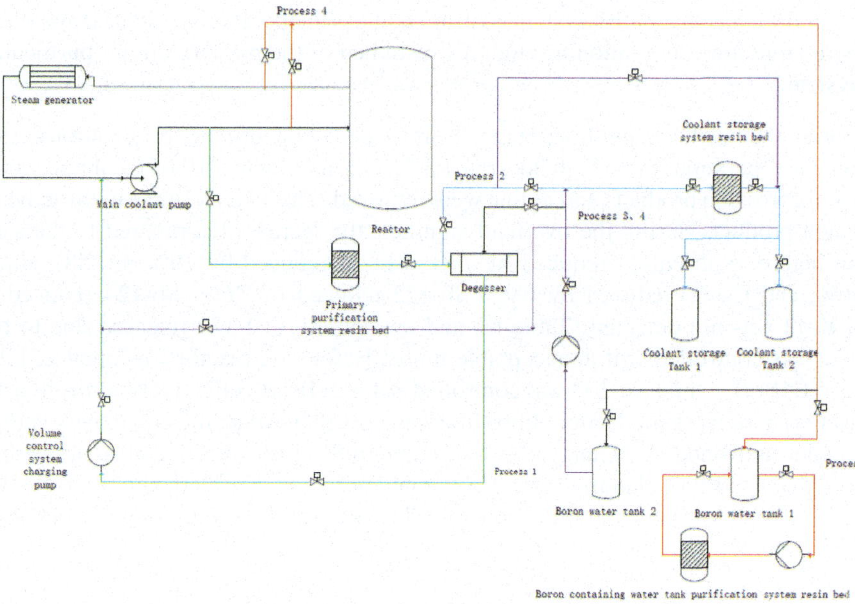

Fig. 1. Overall process of the system

Table 5. Primary coolant nuclides and their half-lives before overhaul

Radionuclide	I-131	I-132	I-133	I-134	I-135	Na-24	K-42	Cs-138
Half-life	8.02 days	2.3 h	20.8 h	52.6 min	6.61 h	15 h	12.36 h	32.2 min

(2) During the primary level reduction, the drainage shall be purified by the resin bed of the coolant storage system and then discharged into the coolant storage tank;
(3) Due to operation control needs, during the stay of a certain primary level platform, the primary water is changed, then it is drained to the coolant storage tank, and then the boric acid with low radioactivity is taken from the boron water tank by the pump, purified by the resin bed of the coolant storage system, and pumped to the primary circuit;
(4) During the whole maintenance and cooling period, the primary coolant is directly discharged to boron tank 1 through the pipeline, and the boric acid with low radioactivity is taken from boron tank 2 by the pump. After purification by the resin bed of the coolant storage system, the primary water is replenishment. At the same time, the primary water drainage received by boron tank 1 is purified by the resin bed of the purification system containing boron tank.
(5) In the ascending stage of the unit, the coolant in the resin bed degasser of the coolant storage system is used for circulating purification;
(6) During the low water level after refueling, the primary drainage passes through the resin bed of the coolant storage system;

(7) After the capacity control system is restored, the primary circuit is cleaned and rinsed with large flow circulation through the resin bed of the primary circuit purification system.

During the overhaul shutdown stage, the resin bed of the primary purification system was put into operation with a high flow rate for 31.2h, and a total of 10.65 Ci of radioactive activated corrosion products and iodine were removed. The activity of the main activated corrosion products during the overhaul significantly decreased compared to the same period before, with Na-24 reduced by 73%, Mn-54 reduced by 79%, Co-58 reduced by 56%, and Co-60 reduced by 68%. Sb-122 reduced by 76%, Sb-124 reduced by 68%, K-42 was not detected (Table 6), and the activity of I-131 released due to fuel damage was controlled within acceptable limits. Before the reactor was opened, I-131 was 1.37E4Bq/L, while Xe-133 was controlled at 4.32E4Bq/L, effectively controlling the exposure risk of workers. Under almost the same overhaul time, the collective overhaul dose under fuel cladding damage was 197.54man·mSv. The collective dose of overhaul without damaged fuel cladding was 228.90man·mSv, which effectively controlled the exposure risk of workers and reduced by 13.7% compared with that without damaged fuel cladding.

Table 6. Comparison of primary coolant nuclides during the same period after optimization implementation

Fuel cladding	Unit status	Na24(Bq/L)	K42(Bq/L)	Mn54(Bq/L)	Co58(Bq/L)	Co60(Bq/L)	Sb122(Bq/L)	Sb124(Bq/L)
Damaged	Cold state	5.48E + 02	None	4.37E + 03	3.85E + 04	2.07E + 03	3.11E + 04	2.46E + 04
Undamaged	Cold state	2.06E + 03	None	2.10E + 04	8.73E + 04	6.50E + 03	1.32E + 05	7.69E + 04
Reduction		73%	None	79%	56%	68%	76%	68%

4 Three Wastes and Ventilation Control During the Overhaul

Nuclear power plant has its particularity in that it stores a large amount of radioactive materials. It can cause harm to human body through both internal and external irradiation. Therefore, nuclear power plants strictly regulate and control radioactive substances. The Dampierre nuclear power plant in France experienced internal exposure caused by air pollution. During a refueling overhaul at the power plant, radioactive iodine in the coolant exceeded the standard when the primary circuit was opened, resulting in more than 70 people with internal I-131 contamination, of which more than 10 consumed more than 1% of the annual control limit. According to Regulations on Environmental Radiation Protection of Nuclear Power Plants [11], nuclear power plants must control the annual total emission of radioactive effluents per reactor. For a 3000MW thermal power reactor, the control values of its gaseous and liquid effluents are shown in Tables 7 and 8.

According to the sampling results of primary coolant, the damage of fuel cladding will lead to the increase of radionuclide in primary coolant, which will lead to the increase

Table 7. Control values of airborne radioactive effluents

Single unit	Light water reactor	Heavy water reactor
Inert gas	6×10^{14}Bq/a	
Iodine	2×10^{10}Bq/a	
Particle (half-life $\geq$ 8d)	5×10^{10}Bq/a	
Carbon-14	7×10^{11}Bq/a	1.6×10^{12}Bq/a
Tritium	1.5×10^{13}Bq/a	4.5×10^{14}Bq/a

Table 8. Control value of liquid radioactive effluents

Single unit	Light water reactor	Heavy water reactor
Tritium	7.5×10^{13}Bq/a	3.5×10^{14}Bq/a
Carbon-14	1.5×10^{11}Bq/a	2×10^{11}Bq/a(except tritium)
Other nuclides	5×10^{10}Bq/a	

of radionuclide emission in effluents. The measured results of primary coolant in cold state, except for halogen I-131 and inert gas Xe-133, all other radionuclide concentrations are lower than those without damage of fuel cladding in the same period. Therefore, halogen I-131 and inert gas Xe-133 are the most threatening nuclides. Relevant measures should be considered to reduce the concentration of nuclides in effluents.

(1) In the primary nitrogen purge stage, the primary purge exhaust is introduced into the radioactive gas purification system to remove the aerosol through glass fiber, and the inert gas is retained by the activated carbon bed to reduce the halogen content, and then discharged after treatment. Except for the inert gas, the radioactive activity of the chimney during the primary circuit purging stage is lower than that of the fuel cladding when the aerosols and iodine are undamaged, The volume activity of inert gas was also controlled within 2.0E6Bq/m3 (Table 9).

Table 9. Chimney radioactivity during primary purge stage

Monitoring objects	Aerosol volume activity (Bq/m3)	Iodine volume activity (Bq/m3)	Inert gas volume activity (Bq/m3)
No damage to fuel cladding	5.50E + 03	4.794	<3700
Damage to fuel cladding	1.13E + 03	<3.7	2.30E + 05

(2) Normally, during overhaul, the air pollution in the plant is reduced through ventilation and circulation purification before discharge. To reduce the amount of airborne

radioactive effluents, continuous purification can be carried out during the exhaust process. After purification and discharge, except for a slight increase in the activity of I-131 and Xe-133 in the ventilation, other nuclides are equivalent to those without damage to the fuel cladding (Table 10).

Table 10. Chimney radioactivity during maintenance

Monitoring objects	Inert gas (Bq/m3)				Halogen (Bq/m3)	
	Ar-41	Kr-88	Xe-133	Xe-135	I-131	I-133
Damage to fuel cladding	<6.79E + 01	<1.64E + 02	1.38E + 04	<3.36E + 01	4.17E-03	< 1.85E-03
No damage to fuel cladding	<9.30E + 01	<1.82E + 02	<2.01E + 02	<5.16E + 01	< 3.95E-04	< 2.26E-03
Monitoring objects	Aerosol (Bq/m3)				Tritium carbon (Bq/m3)	
	Cs-134	Cs-137	Co-60	Mn-54	H-3	C-14
Damage to fuel cladding	<2.84E-04	<2.59E-04	<3.00E-04	< 2.43E-04	2.44E + 02	2.56E + 02
No damage to fuel cladding	<2.36E-04	<2.97E-04	<2.42E-04	< 3.21E-04	1.63E + 02	7.81E + 01

(3) The radioactive substances in the radioactive solution in the floor drain will evaporate into the air, and the radioactive substances will be adsorbed on the surface of the floor drain. If there is a water film covering the radioactive substances, it can prevent the radioactive substances from directly entering the air. However, during the overhaul, it is impossible to ensure that every floor drain is covered by water film. Therefore, while ensuring that the drainage pipe is connected to the floor drain, the floor drain should be covered and sealed to prevent radioactive substances from spilling into the air.

(4) The radioactive liquid effluents during overhaul contain a large amount of I-131 and inert gas Xe-133. These nuclides cannot be fixed in the residual liquid by distillation, so I-131 is usually purified by ionic resin bed, while Xe-133 can only be pursed, discharged or residual decay, and the pollutants usually occur in concentration at the beginning of overhaul. The half-life of Xe-133 is 5.24 days, so it is necessary to reserve a water tank for radioactive waste liquid water at the initial stage of overhaul to be further disposed after Xe-133 decays. In the event of fuel cladding damage in a domestic nuclear power plant, the liquid effluent was discharged qualified throughout the entire overhaul period.

5 Summary

Based on the actual situation of the damaged nuclear fuel in a domestic nuclear power plant, the objective of the primary circuit radioactivity control is determined, and the operation mode in the process of shutdown state transition under the loss of the gas tightness of the fuel cladding is formulated, as well as the radioactivity control method of the primary circuit. At the same time, the influence of effluent is analyzed, and the operation mode of waste treatment and ventilation is optimized. This study has been well applied in a domestic nuclear power plant, effectively reducing the radiation level of the primary coolant. I-131 is controlled below 3.0E4Bq/L, Xe-133 below 1.0E6Bq/L, and other nuclides have decreased by more than 50% compared to the same period without fuel cladding damage, which better controlls the collective radiation dose of personnel during overhaul. The collective radiation dose of personnel is lower than that of previous overhaul. And it ensures the controllability of gaseous effluents during the primary circuit purging stage, equipment maintenance stage, and refueling stage. Except for inert gases and I-131, the measured volume activity of aerosols and other nuclides measured by chimney is equivalent to that without damage to the fuel cladding. Inert gases and I-131 are also controlled at a lower level. And all radioactive liquid effluents are also qualified to discharge. Although there are differences among different types of nuclear power plants, the operation of different reactor types can still be optimized based on this research method. Moreover, all optimized operating methods have achieved good results under the condition of fuel cladding damage, so they can still be used to reduce the impact on personnel and the environment even under non fuel cladding damage conditions.

References

1. YAN Changqi. Nuclear reactor engineering [M]. Harbin Engineering University Press.3rd edn. Harbin (2020).98–137.
2. FM-AC-4201.DZ Guidelines for the Supervision and Management of Nuclear Fuel Integrity
3. EPRI 3002010738 Fuel Surveillance and Inspection
4. EPRI 3002016031 Fuel Reliability Monitoring and Failure Evaluation Handbook
5. IAEA Nuclear Energy Series No.NF-T-2.5 Review of Fuel Failures in Water Cooled Reactors (2006–2015)
6. INTERNATIONAL ATOMIC ENERGY AGENCY. Fuel Failure in Normal Operation of Water Reactors.Experience,Mechanisms and Management (Proc.Tech. Comm.Mtg. Dimitrovgrad.1992). IAEA-TECDOC-709.IAEA.Vienna(1993).
7. INTERNATIONAL ATOMIC ENERGY AGENCY.Review of Failure in Water Cooled Reactors. Technical Reports Series No.388. IAEA.Vienna(1993).
8. INTERNATIONAL ATOMIC ENERGY AGENCY. Fuel Failure in Water Reactors.Causes and Mitigation (Proc.Tech. Mtg. Bratislava.2002). IAEA-TECDOC-1345.IAEA.Vienna(2003).
9. INTERNATIONAL ATOMIC ENERGY AGENCY. Structural Behaviour of Fuel Assemblies for Water Cooled Reactors (Proc.Tech. Mtg. Cadarache.France.2004). IAEA-TECDOC-1454.IAEA.Vienna(2005).
10. GARZAROLLI.F .VON JAN.R. STEHLE.H. The main causes of fuel element failure in water-cooled reactors.At.Euergy Rev.171 (1979) 31–128.
11. GB 6249–2011. Regulations for environmental radiation protection in nuclear power plants [S].4–5.

Sensitivity Analyses of Thermal Hydraulic Parameters in ATWS by Rods Failure-Loss of Offsite Power of the Third Generation Nuclear Power Plant

Mengying Liu(✉), Haode Xu, Qingyu Xie, and Peng Chen

Reactor Engineering and Safety Research Center, China Nuclear Power Technology Research Institute Co., Ltd, Shenzhen 518026, China
liumengying@cgnpc.com.cn

Abstract. The beyond basic design accidents of second and second plus generation nuclear plants are assessed in a method of best estimate assumption. The Nuclear Power Design Safety Regulations (HAF102–2004) are updated in 2016, the beyond basic design accidents are revised as Design Extend Conditions, which includes Design Extend Conditions without core melt (DEC-A) and Design Extend Conditions with core melt (DEC-B). It is pointed that conservative design should be considered appropriately in order to avoid the steep side effect. To meet the requirements of the new regulations, the ATWS by Rods Failure-loss of offsite power of the third-generation nuclear power plant is analyzed in this paper. Sensitivity analyses of thermal hydraulic parameters are calculated to study their influence on the accident consequences. It is proved that the steep side effect does not appear even the conservative is taken into account. However, the acceptance before is too strict to satisfy, the more proper acceptance should be used.

Keywords: The third-generation nuclear power plant · Design extend conditions · Anticipated Transient without Scram (ATWS) · Sensitivity analyses · Acceptance criteria · Steep side effect

1 Foreword

After absorbing the design, building and operation experience of Chinese nuclear power plant, The National Security Administration has revised the《Nuclear Power Plant Design safety rules》(HAF102–2004) in 2016. The concept of design extension conditions without core damage (DEC-A) and actual elimination is exported. Some requests are proposed for DEC-A. For example, uncertainty should be considered to avoid steep side effect. For DEC-A analysis, the IAEA has pointed out clearly that the best estimation method should be used. Determinism method, probability theory and engineering judgment should be combined to obtain a series of typical DEC-A conditions, some protecting measures should be put forward to improve the nuclear power plant safety, uncertainty should be considered to avoid steep side effect [1].

J. Liu and Y. Jiao (Eds.): WRFPM 2023, SPPHY 299, pp. 200–213, 2024.
https://doi.org/10.1007/978-981-99-7157-2_21

Anticipated without scram is a classical condition of DEC-A, the Second generation and second plus generation nuclear power plant used the best estimate values in their accidents analysis. This article studies the ATWS LOOP of a third-generation nuclear power plant, does some sensitivity analysis of the initial parameters and studies the range of influence. It proves that, even conservative, there is no steep side effect of the third-generation nuclear power plant [2].

2 Objects and Methods

2.1 Introduction

In the accident of anticipated without scram, the control rods cannot insert will lead the reactor lose the ability of controlling reactivity by control rods. Loss of Offsite Power (LOOP) will lead to loss of power for all auxiliary plant equipment, such as Reactor Coolant System (RCP) pumps, condensate and main feed water pumps, etc. After LOOP, RCP pumps coast down and coolant flow rate decreases. Although the reactor trip signal has been emitted, the control rods are still at high positions due to mechanical blockage. Therefore, the reactor trip is not realized. The coolant flow rate decreases and temperature rises may lead to a decrease in the DNBR margin. In the second and second plus generation nuclear power plant design, the acceptance criteria of the accident is that the minimum DNBR remains above the design limit. For this study object, deterministic DNBR design limit is 1.20 in the loss of flow conditions [3, 4].

2.2 Tools and Methodology

Firstly, the reactor physical code is used to calculate the physical parameters, e.g., the moderate temperature coefficients and doppler coefficients of the reactor. Secondly, the thermal hydraulic code is used to calculate the reactor accident process and simple DNBR trend. Finally, combining with the thermal hydraulic output parameters e.g., thermal power, pressure, temperature and coolant flow rate, the sub-channel code is used to calculate the minimum DNBR in detail. Figure 1 is the plant nasalization picture.

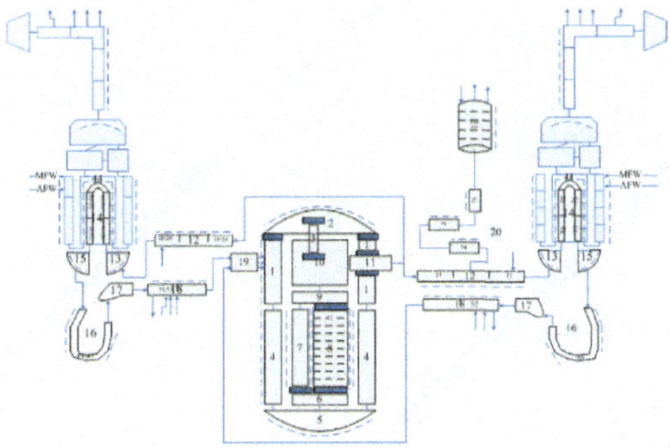

Fig. 1. The plant nodalization picture

2.3 Main Proposes

LOOP ATWS is an overheat accident, the increase temperature will lead the power decrease. It is supposed that all the control rods are out of the reactor. Since the lower temperature react is worse for this accident, a minimum moderator temperature coefficient in absolute value of a cycle is chosen. The initial state and uncertainty of nuclear power plant are in Table 1. The general accident scenario at basic condition (no uncertainty considered) is in Table 2.

As the pumps coast down at the beginning of the accident, the minimum DNBR appears early in the accident scenario. The DNBR value is mainly determined by the thermal power, temperature, pressure, the coolant flow rate. Sine the coolant flow rate is related to main pumps rotational inertia, other parameters are compared in figures.

Table 1. The initial state and uncertainty of the nuclear power plant

Parameter	Initial values	Uncertainty
Core power (%)	100	2
Pressurizer pressure (MPa abs)	15.5	0.25
Average temperature of reactor coolant (°C)	310	2.5
Water level of pressurizer (%)	53.1	7
VDA [ASDS] opening pressure (MPa abs)	8.6	0.15
Time for turbine trip(s)	2.2	2

Table 2. The general accident scenario

Events	Time(s)
Accident happens	0
Reactor trip signal	2.3
Turbine trips	4.5
Pressurizer safety valves opening	12.4
VDA opening	18.2
Reaching the minimum DNBR(1.25)	23.1

3 Sensitivity Analysis

3.1 Thermal Power

Considering the uncertainty of the core initial thermal power, the primary temperature, the thermal power, the primary pressure change curves are presented as Figs. 2, 3 and 4. The minimum DNBR is presented in Table 3.

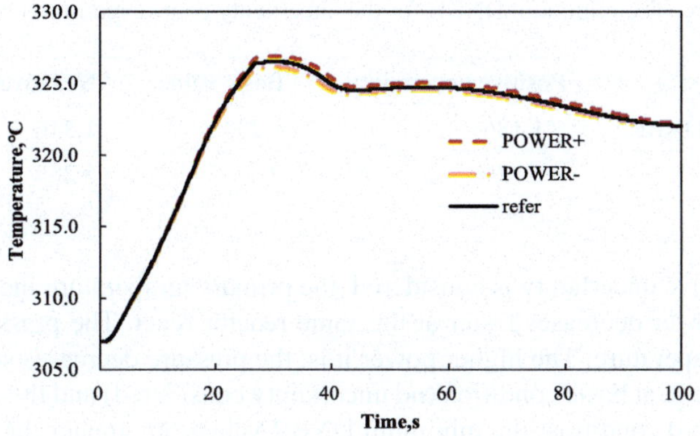

Fig. 2. The primary temperature for the uncertainty considered in initial power

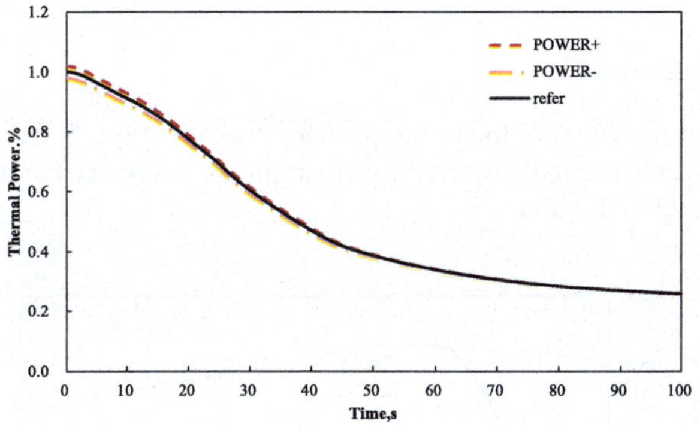

Fig. 3. The thermal power for the uncertainty considered in initial power

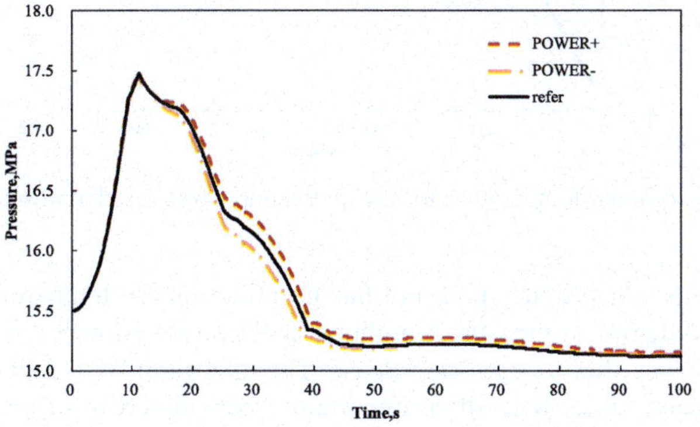

Fig. 4. The primary pressure for the uncertainty considered in initial power

Table 3. The minimum DNBR for the uncertainty considered in initial power

Parameter	Positive uncertainty	Basic value	Negative uncertainty
The minimum DNBR	1.186	1.250	1.316
Relative value	−5.12%	0	5.28%

If the positive uncertainty is considered, the primary temperature increased quicker, the thermal power decreases faster at the same reactor react. The pressure behaves as well as the temperature. The higher power it is, the pressure decreases slower.

From Table 3, at basic condition (no uncertainty considered) and the negative uncertainty considered condition, the minimum DNBR values are greater than the limit. But at the positive uncertainty considered condition, the minimum DNBR is less than the limit. If the acceptance criteria are the minimum DNBR remains above the design limit, it will not be met.

3.2 Primary Pressure

Considering the uncertainty of the initial primary pressure, Figs. 5, 6 and 7 show the primary temperature, the thermal power, the primary pressure change curves. The minimum DNBR is presented in Table 4.

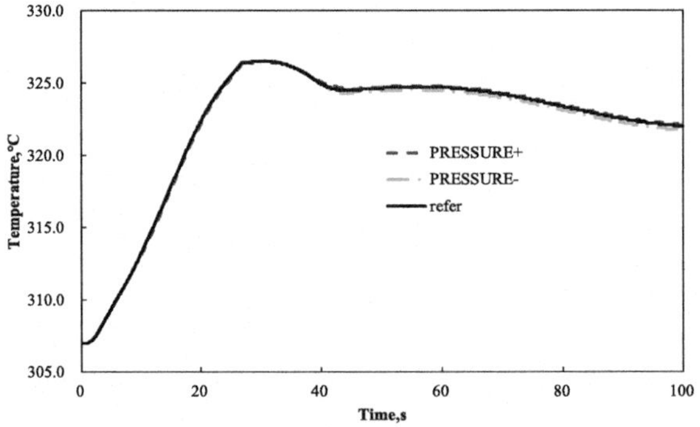

Fig. 5. The primary temperature for the uncertainty considered in primary pressure

The uncertainty of primary pressure has no effect on the temperature and thermal power. If the initial primary pressure is higher, it will lead the primary circuit release early, the pressure is lower after it begins to release. The minimum DNBR of these conditions are not in the same time. If positive uncertainty considered in primary pressure, the minimum DNBR value is greater the basic condition (no uncertainty considered). If negative uncertainty considered condition in primary pressure, the minimum DNBR value decrease, but still remains above the design limit.

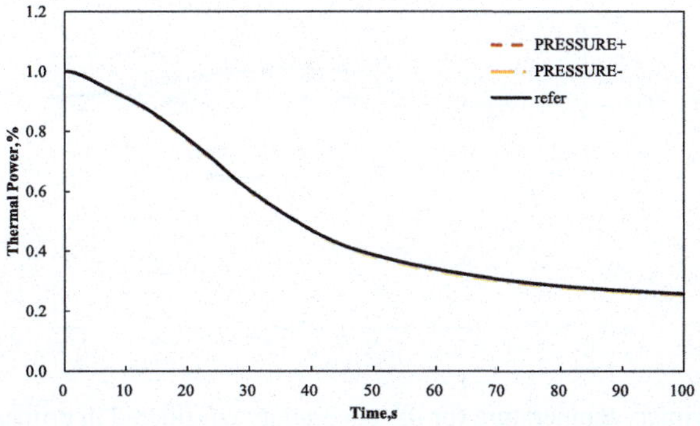

Fig. 6. The thermal power for the uncertainty considered in primary pressure

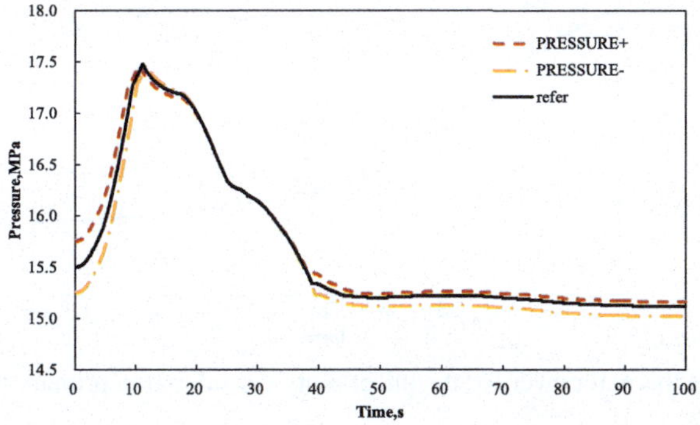

Fig. 7. The primary pressure for the uncertainty considered in primary pressure

Table 4. The minimum DNBR for the uncertainty considered in primary pressure

Parameter	Positive uncertainty	Basic value	Negative uncertainty
The minimum DNBR	1.255	1.250	1.243
Relative value	0.4%	0	−0.56%

3.3 Primary Temperature

Considering the uncertainty of the initial primary temperature, Figs. 8, 9 and 10 show the primary temperature, the thermal power, the primary pressure change curves. The minimum DNBR is presented in Table 5.

Even there is some difference of the primary temperatures, they have the same change trend since the thermal power at these conditions are all the same. Different temperatures have some effect on the primary pressure. The higher temperature is, the lower minimum DNBR value. All the minimum DNBR values remain above the design limit.

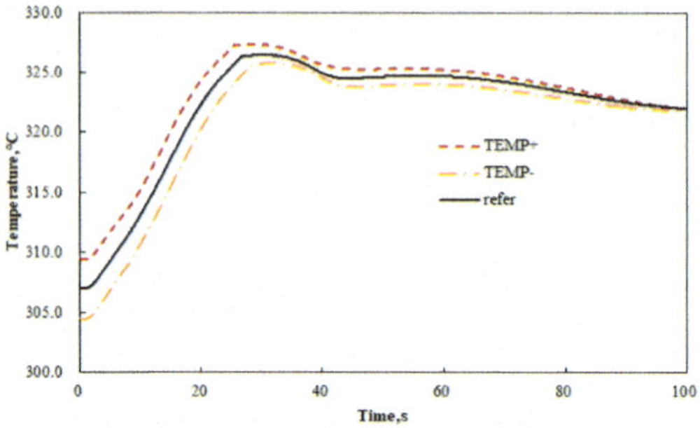

Fig. 8. The primary temperature for the uncertainty considered in primary temperature

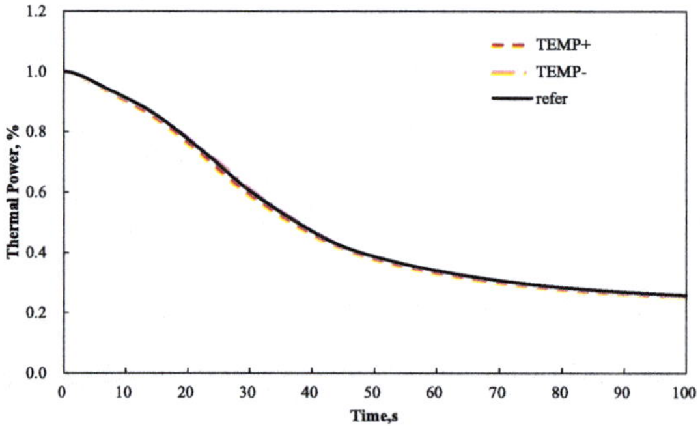

Fig. 9. The thermal power for the uncertainty considered in primary temperature

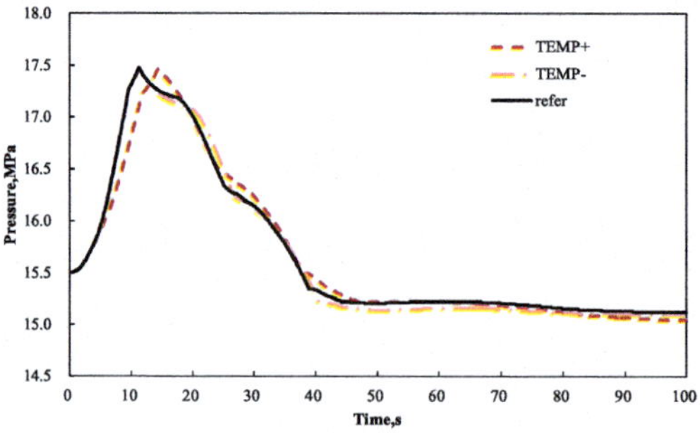

Fig. 10. The primary pressure for the uncertainty considered in primary temperature

3.4 Pressurizer Water Level

Considering the uncertainty of the pressurizer water level, Figs. 11, 12 and 13 show the primary temperature, the thermal power, the primary pressure change curves. The minimum DNBR is presented in Table 6.

If the positive uncertainty of pressurizer water level is considered, the primary pressure will reach the peak value earlier, the primary circuit will release earlier. The pressure is lower after it begins to release. After releasing, the pressure is higher than these two

Table 5. The minimum DNBR for the uncertainty considered in primary temperature

Parameter	Positive uncertainty	Basic value	Negative uncertainty
The minimum DNBR	1.239	1.250	1.268
Relative value	−0.88%	0	1.44%

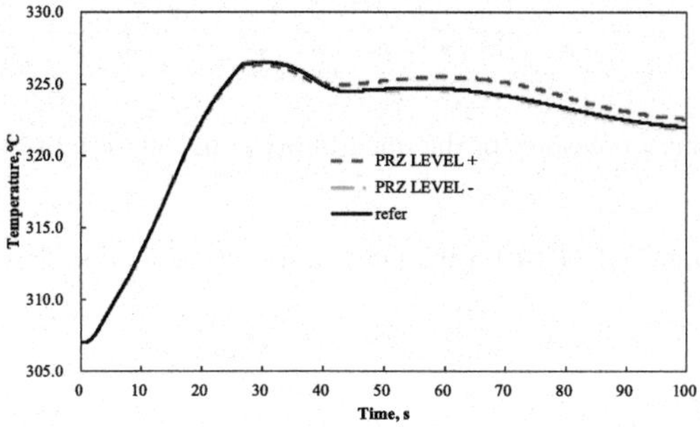

Fig. 11. The primary temperature for the uncertainty considered in pressurizer water level

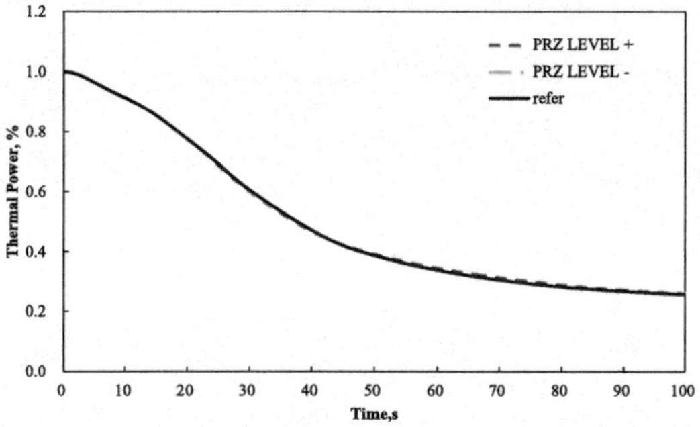

Fig. 12. The thermal power for the uncertainty considered in pressurizer water level

conditions, there is also some effect on the primary temperature. But it has little effect on the thermal power. For the minimum DNBR value time, the primary temperature is higher if negative uncertainty of pressurizer water level is considered, so the minimum DNBR of that condition is the smallest.

3.5 VDA Valve

Considering the uncertainty of open values of Atmospheric Steam Dump System (VDA), Figs. 14, 15 and 16 show the primary temperature, the thermal power, the primary pressure change curves. The minimum DNBR is presented in Table 7.

The open value of VDA has effect on the primary peak temperature. After loss of the main feed water, the secondary heat transfer ability becomes weak, the VDA opens.

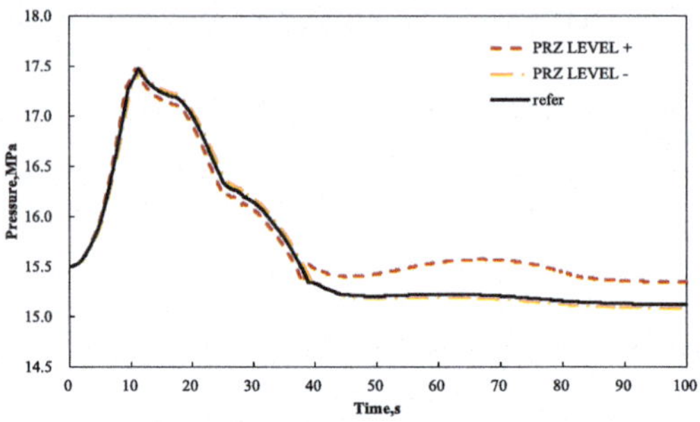

Fig. 13. The primary pressure for the uncertainty considered in pressurizer water level

Table 6. The minimum DNBR for the uncertainty considered in pressurizer water level

Parameter	Positive uncertainty	Basic value	Negative uncertainty
The minimum DNBR	1.252	1.250	1.249
Relative value	0.16%	0	−0.08%

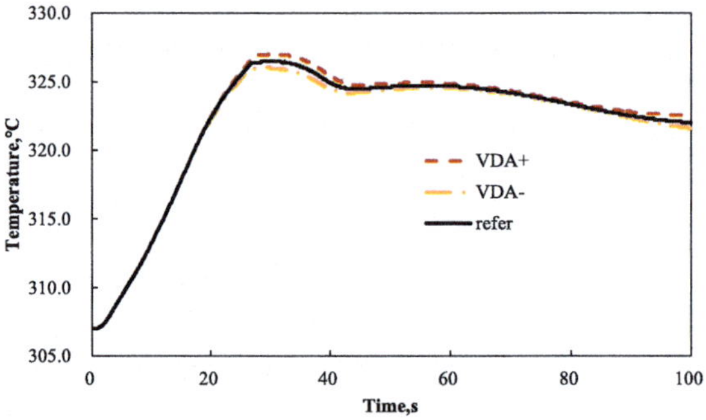

Fig. 14. The primary temperature for the uncertainty considered in VDA

If the positive uncertainty is considered, the VDA will open latterly. Thus, the heat will be brought out latterly, the primary peak temperature will be higher. If the negative uncertainty is considered, the VDA will open earlier, the heat will be brought out earlier, the primary peak temperature will be lower.

3.6 Turbine Trip

Considering the turbine trip time delay, Figs. 17, 18 and 19 show the primary temperature, the thermal power, the primary pressure change curves. The minimum DNBR is presented in Table 8.

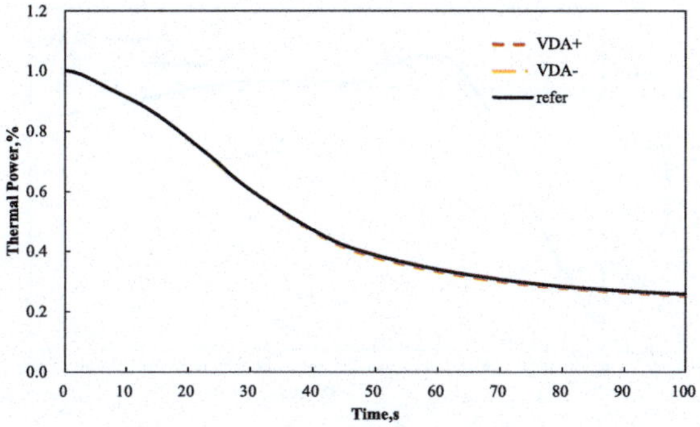

Fig. 15. The thermal power for the uncertainty considered in VDA

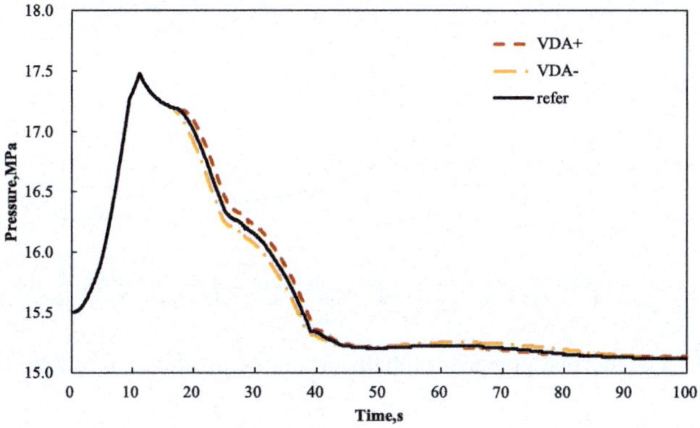

Fig. 16. The primary pressure for the uncertainty considered in VDA

Table 7. The minimum DNBR for the uncertainty considered in VDA

Parameter	Positive uncertainty	Basic value	Negative uncertainty
The minimum DNBR	1.239	1.250	1.256
Relative value	−0.88%	0	0.8%

If turbine trip time delay is considered, the secondary will lose the sink latterly, the primary temperature rises slowly, and the primary pressure reaches the peak value latterly. Meanwhile, the moderator feedback is less, the power is higher.

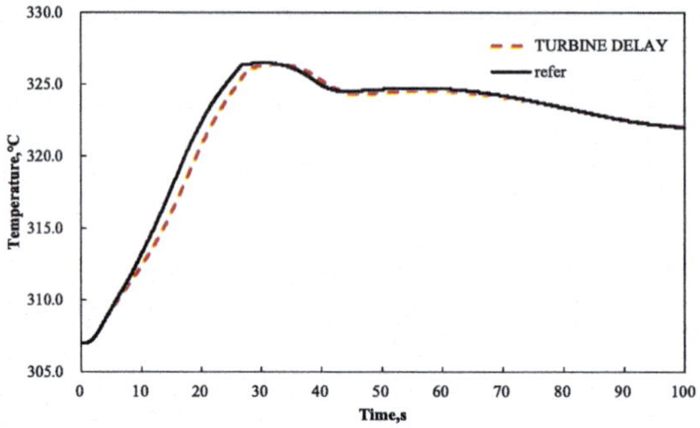

Fig. 17. The temperature for the uncertainty considered in turbine trip time

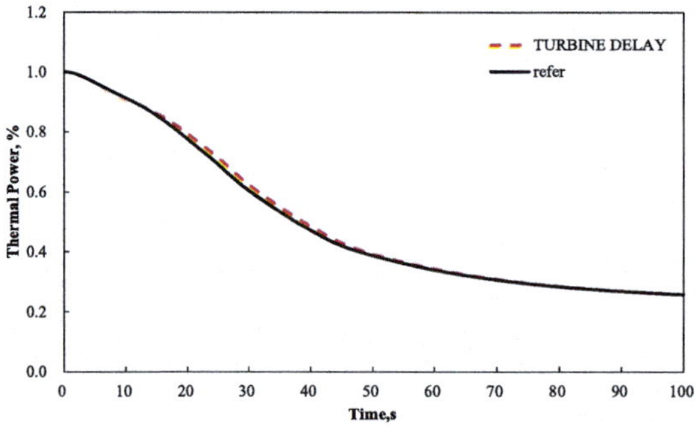

Fig. 18. The thermal power for the uncertainty considered in turbine trip time

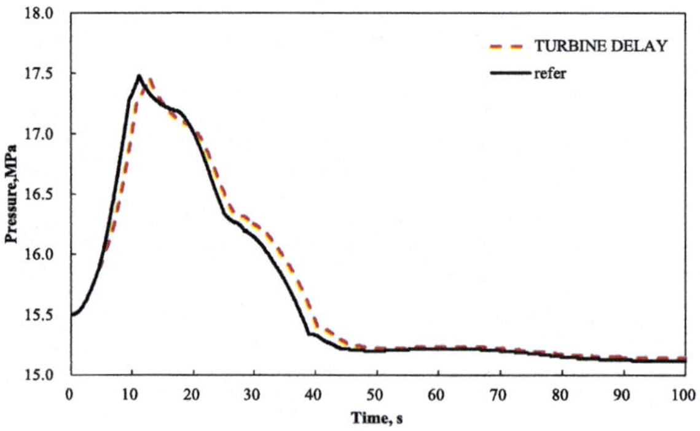

Fig. 19. The primary pressure for the uncertainty considered in turbine trip time

Table 8. The minimum DNBR for the uncertainty considered in Turbine trip time

Parameter	Turbine trip delay	Basic value
The minimum DNBR	1.201	1.250
Relative VALUE	−3.92%	0

4 Comprehensive Analysis

4.1 All Uncertainty Considered

Through the thermal hydraulic parameters' separate calculation, it can be sure that, the positive uncertainty of thermal power and primary temperature and open values of VDA are punishment for LOOP ATWS accident, the negative uncertainty of primary pressure, PZR water level and the turbine trip time delay are punishment. So, the main assumption of LOOP ATWS accident can be adopted as the Table .9.

Table 9. The all uncertainty considered

Parameter	Turbine trip delay	Basic value
Initial power	100%	102%
Pressure of PZR	15.5	15.25
Primary average temperature	310	312.5
Pressurizer water level	53.1	46.1
VDA opening pressure	8.6	8.75
Time for turbine trip	2.2	4.2

Figures 20, 21 and 22 show the primary temperature, the thermal power, the primary pressure change curves.

The minimum DNBR is presented in Table 10.

Form Table 10, it can be seen that if the uncertainty of all thermal hydraulic parameters are considered, the minimum DNBR is 1.120, it is less more than the limit value 1.20.

4.2 Acceptance Criteria

Results above proved that, the minimum DNBR will not be satisfied the limit if all uncertainty is considered for LOOP-ATWS accident. It's should pointed that the minimum DNBR remains above the design limit is a require for DBC-2, it is very strict for DEC-A, especially the uncertainty parameters are considered. If non-LOCA acceptance criteria of DBC-4 is used, (1) the amount of fuel rods experiencing DNB must remain lower than 10%; (2) the fuel pellet melting at the hot spot must not exceed 10% by

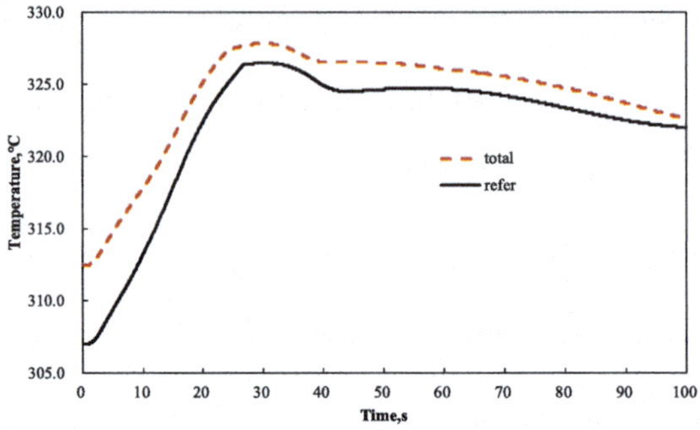

Fig. 20. The temperature for the all uncertainty considered

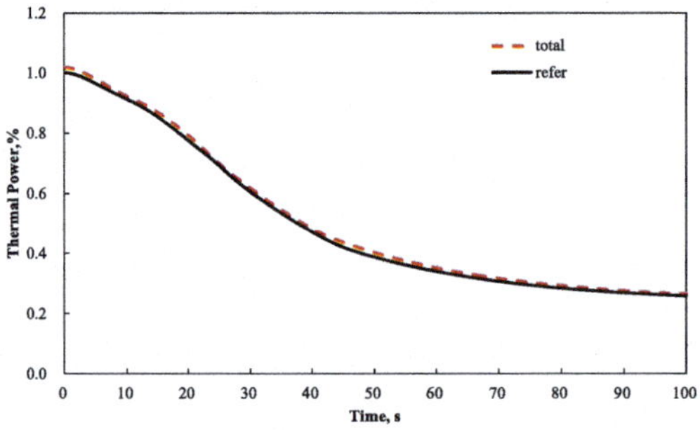

Fig. 21. The thermal power for the all uncertainty considered

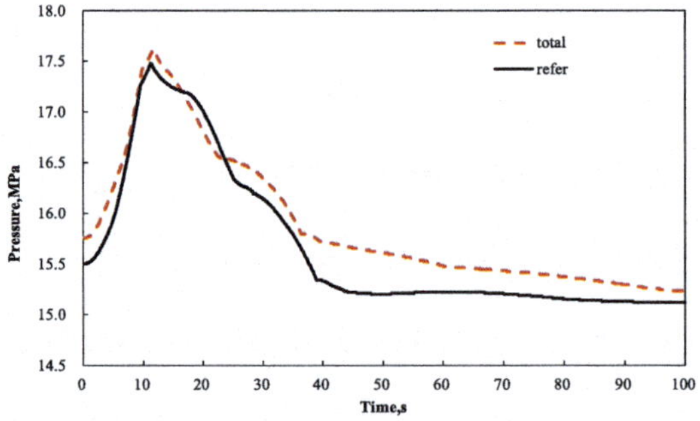

Fig. 22. The primary pressure for the all uncertainty considered

volume; (3) for cases not involving the rapid transient of oxidation of the cladding, the peak cladding temperature must remain lower than 1482 °C. Though more calculation, the amount of fuel rods experiencing DNB is 5.26%, lower than 10%. In the process of this accident, the core power is decreasing, the fuel pellet does not be melting, the peak cladding temperature remains lower than 1482 °C. Thus, the acceptance criteria are met.

Table 10. The minimum DNBR for the all uncertainty considered

Parameter	All uncertainty	Basic value
The minimum DNBR	1.120	1.250
Relative value	−10.4%	0

5 Conclusion

This paper did some uncertainty calculation of DEC-A LOOP ATWS of a third-generation nuclear power plant. The thermal hydraulic parameters effects alone and all are analyzed. Some conclusions are summarized:

(1) The thermal hydraulic parameters effects of the accident result are that: power>temperature>pressure>water level.
(2) The VDA opening pressure and turbine trip time affect the heat transfer of primary and secondary circuit, so they affect the result. The turbine trips as soon as earlier is better for this accident.
(3) If all the thermal hydraulic parameters uncertainty are considered, the DNBR acceptance criteria is too conservative to the accident. The acceptance of non-LOCA of DBC-4 should be used.
(4) It is proved that, even the uncertainty is considered, the acceptance is met, no steep side effect of the third-generation nuclear power plant happens.

References

1. Nuclear power plant design rules, HAF102–2016, 2016, 22–34.
2. Nuclear power plant deterministic analysis, Nuclear Safety guide rules. 2021.5.19.
3. Jiang Xiaohua, Li Guijie, Ningde 1,2 nuclear power plant ATWT transient analysis[R]. China Nuclear Power Technology Research Insititue, 2009.
4. HU Yisong, Li Kejia. Analyses for the Effect of Fuel Crud on DNBR[J]. Nuclear Science and Engineering, 2022, 44:994–997.

Confirmation of the Design Characteristics of the TVS-K Design After Operation in the PWR Reactor at Ringhals-3 NPP

A. Radostin[1(✉)], K. Lafchiev[2], and D. Jädernäs[2]

[1] TVEL, JSC, Moscow, Russian Federation
AlFRadostin@tvel.ru

[2] Studsvik Nuclear AB, Nyköping, Sweden

Abstract. In 2020, four TVS-K fuel assemblies were unloaded from Ringhals-3 NPP after completion of pilot operation. During four fuel cycles, the TVS-K reached a burnup of 54.3 MWday/kgU. To confirm the design characteristics and justification the reliability and performance of TVS-K materials, six fuel rods and the skeleton were delivered to the Studsvik Hot Cell Laboratories for post-irradiation examinations. This paper presents the results of the following studies:

- Examinations of the Skeleton and the Top Nozzle;
- Non-destructive examination of Guide Thimbles (GT), Instrumental Tube (IT) and Spacer grids (SG);
- SEM studies of SG including welds;
- Metallographic studies and determination of hydrogen content of GT and IT;
- TEM and EDX investigation of structural-phase state and element composition of GT and IT;
- Non-destructive examination of fuel rod claddings (E110opt);
- Metallographic studies and microhardness measurements of the top and bottom welds and fuel rod claddings and measurement of the gap;
- Fuel density measurement with fuel swelling assessment. SEM and EDX of fuel;
- Mechanical tests of fuel rod claddings and fuel rod springs relaxation;

The presented experimental data confirm the design characteristics and the high performance properties of the materials and TVS-K components and provide information on their corrosion resistance, strength, and dimensional stability under irradiation in the water-chemical regime of the PWR.

Keywords: TVS-K · Post-irradiation examinations

1 Introduction

The TVS-K 0AM4 was operated in Ringhals Unit 3, Sweden, for four reactor cycles. The TVS-K fuel assembly average discharged burnup was 54.3 MWd/kgU. After the completion of operation all fuel rods were extracted from this bundle. Six fuel rods and the skeleton were delivered to the Studsvik Hot Cell Laboratories for post-irradiation examinations.

J. Liu and Y. Jiao (Eds.): WRFPM 2023, SPPHY 299, pp. 214–223, 2024.
https://doi.org/10.1007/978-981-99-7157-2_22

The objective was to evaluate the condition of the materials and TVS-K components including fuel rods, grids and guide tubes after irradiation by various non-destructive and destructive examinations in order to support the licensing of the TVEL PWR fuel over the world.

This article summarizes the initial state skeleton examinations such as visual inspection, length measurements and bow measurements, skeleton stiffness test and some dimensional measurements of the top nozzle and top nozzle springs. The results of the study of fuel rods are also presented: visual inspection of fuel rods claddings, profilometry, diameter, ovality, rod bending measurements, oxide thickness measurements, microhardness of welds measurements, fuel-cladding gap determination, fuel density measurements, element content examinations by fuel pellet radius (Fig. 1).

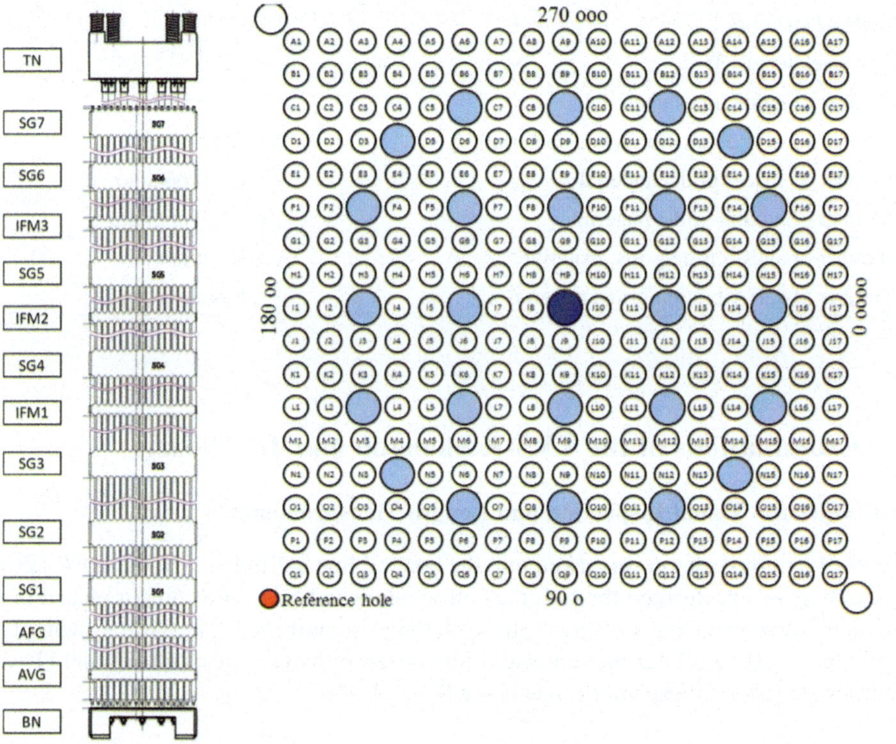

Fig. 1. Cross section and grids layout of TVS-K.

2 Design Parameters

The TVS-K LFA has been designed for operation in the Ringhals-3 NPP PWR reactor core. The TVS-K design has a square profile 17 × 17 fuel rod array with a 12.6 mm pitch and 9.5 mm fuel rod with 12 ft. Active fuel length. The load-bearing skeleton is formed by 24 guide tubes (GT), one instrumentation tube (IT), a top nozzle(TN) and a bottom nozzle(BN, seven spacer grids (SG), three intermediate mixing grids, one anti-fretting grid (AFG) and one anti-vibration grid (AVG) (Table 1).

Table 1. Main design values of TVS-K LFA

Items	Values
Fuel rod array	17 × 17
Fuel assembly pitch, mm	215.04
Number of fuel rods per assembly, cladding material	264, E110 opt alloy
Fuel rod pitch, mm	12.6
Fuel rod OD, mm	9.5
Fuel stack length, ft	12
Number of guide tubes, material	24, E635 alloy
Number of instrumentation tubes, material	1, E635 alloy
Number of spacer grids, material	7, E110 opt alloy
Number of mixing grids, material	3, E110 opt alloy
Number of anti-fretting grids, material	1, E110 opt alloy
Number of anti-vibration grids, material	1, Stainless steel

3 Examinations of the TVS-K Skeleton and Top Nozzle

3.1 Skeleton Visual Inspection and Length Measurements

The full-length skeleton was taken into the hot cell line at Studsvik Nuclear AB. Visual inspection was performed through the cell window and by a web camera placed in the hot cell. The designed condition of the skeleton was confirmed. The axial growth of the skeleton was 0.2%. After the removal of the nozzles, the lengths of all GT and IT were measured. The axial growth of the GTs and the IT was 0.4%.

3.2 Skeleton Stiffness Test

The skeleton was placed horizontally and supported only at the top and bottom nozzle.

The stiffness test was performed by loading on the 0/180 side. The test consisted of 5 loading cycles and totally 44 loading steps. The displacement on SG4 was recorded during each step, whereas the displacement of SG2 and SG6 were recorded only at the start and the end of each cycle. The lateral stiffness test was performed without the fuel rods. The skeleton stiffness for an amplitude bigger than 3 mm was 64 N/mm.

3.3 Skeleton Bow Measurements

The skeleton bow was measured by using an elastic cord, a measuring scale and a web camera. The cord was fastened in the middle of the skeleton. The measurements were performed before (with bottom and top nozzle) and after (without bottom and top nozzle) the stiffness test. At each spacer grid position the bow data was obtained by measuring the distance between the cord and the spacer edges.

The maximum measured bowing was 7–8 mm taking into account that the neighboring bundles had bigger values. The results are very similar and indicate that no significant deformation occurred during the stiffness test (Fig. 2).

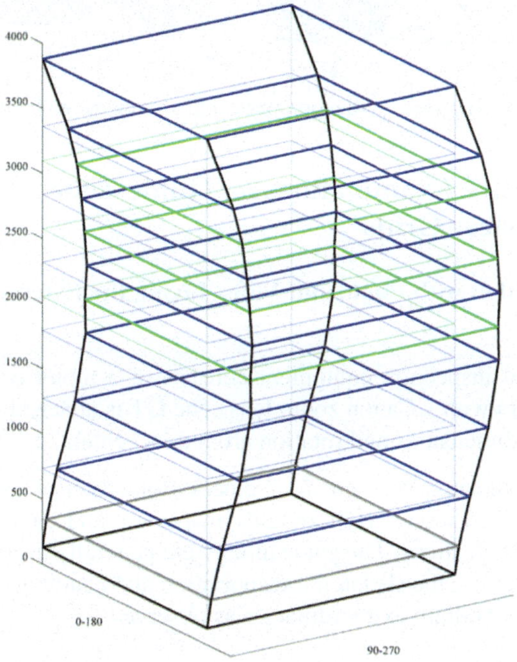

Fig. 2. Skeleton bow visualization.

3.4 Top Nozzle Examinations

Visual inspection and dimensional measurements of the top nozzle were performed after the top nozzle was removed from the skeleton (Fig. 3).

Imaging and video recording were performed with a web-camera.

The visual inspection showed that the top nozzle was in good condition and nothing extraordinary was found.

The height of the top nozzle, the diameter of the hold-down springs and the height of the locking plate springs were measured with a caliper.

The top nozzle was in good condition and nothing extraordinary was found.

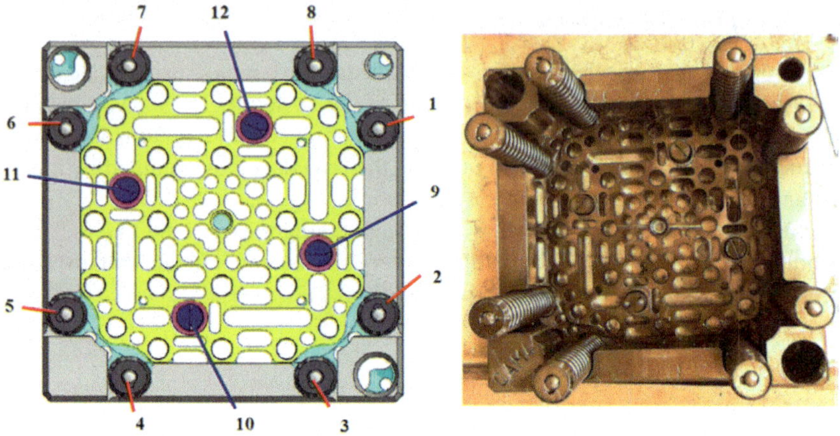

Fig. 3. Top view of the top nozzle and arrangement of springs.

4 Non-destructive Examinations on Guide Tube, Instrumentation Tube and Spacer Grids

4.1 Visual Inspection with Photo and Video Fixing the External State of the GT and IT

Method. The visual inspection equipment consists of a motor driven fixture and an Allied Vision camera with a Canon zoom lens. The GT is attached to a chuck allowing movement in three dimensions and rotation around its center.

Results. After the skeleton was cut, four pieces from each of the guide tubes were inspected visually. Each tube was thoroughly inspected, and images were recorded in eight circumferential positions. The guide tubes were generally in good condition. Some scratches and marks from the skeleton transportation and handling operations at the hot cell were present but nothing extraordinary was observed.

4.2 Length Measurements of the GT and IT

After the removal of the nozzles, the lengths of all GT and IT were measured from the tube bottom to the bottom of the AVG and from the tube top to the top of the SG7. The measurements were added to the average measured length between AVG bottom and SG7 top ends to receive the total length of each tube. The axial growth of the GTs and the IT was 0.4%.

4.3 Measurements of Oxide Film Thickness on the Outer Surface of the GT and IT by Eddy Current Method

Method. A standard oxide thickness measurement is performed every 0.5 mm in eight azimuthal locations along each tube piece.

Results. The average outer oxide thicknesses of GT and IT were very similar and the small differences in oxide thickness between them are within the method uncertainty. The average outer oxide thickness was around 12–15 μm at the bottom of the fuel assembly above the AFG grid and peaked at around 40–45 μm above the SG6.

5 SEM Studies of Spacer Grids

Method. The as cut sample is mounted in epoxy resin, ground, and polished.

The instrument was operated at a 20 kV electron acceleration voltage. A secondary electron detector was used to record the sample topography and a backscattered electron detector was used to visualize compositional variations in the sample.

Images acquired covering the whole radius of the cladding thickness are used for hydrogen content measurement. The basic assumptions when evaluating the amount of hydrogen with image analysis are:

The analyzed image is representative of the selected area and the area fraction of hydride equals the volume fraction of hydrides.

The amount of hydrogen in solid solution is negligible in comparison to the amount in the hydrides (at room temperature <20 parts per million (ppm)).

The hydrides are in δ-phase, ZrH1.62 corresponding to 62 atomic%hydrogen.

The determination of the hydride area fraction was performed with Image J image analysis software. Both the hydrogen content in each image and the weighted cross section mean value of hydrogen were calculated.

Results. In total thirteen SEM samples were prepared and examined, seven from the SG6 and six from the AFG. Inner straps, outer straps and welded joints between strap and spring and guide tube and strap were examined. When present in the sample the spring and the guide tube were examined too. The examinations included oxide and hydride imaging and hydrogen content measurements using SEM image analyses.

The average oxidation observed in the SEM was higher for SG6 than for the AFG. The oxide thickness on all examined samples was predominantly uniform. The average oxide thickness of the straps and springs was 8-9 μm in the SG6 samples and 2-3 μm in the AFG samples. The average oxide thickness in the GT was 40-42 μm at SG6 axial elevation and 17-19 μm at the AFG axial elevation.

The hydrogen content in the different spacer grid components was measured using three techniques: SEM image analyses, IGF (Inert Gas Fusion) and HVE (Hot Vacuum Extraction). There is a good agreement between the different techniques which were used. In consistence with the oxidation, the hydrogen content at the SG6 elevation was higher than at the AFG elevation.

6 Metallographic Studies of the GT and IT. Determination of Hydrogen Content in the GT and IT

6.1 Hydrogen Measurements

Method. The MALTE equipment in Studsvik built based on ASTM standard E146–83 was used.

The method is based on the principle that the amount of hydrogen released from the sample upon heating in vacuum is proportional to the increase in the pressure in a constant volume.

Hydrogen is extracted from the Zr samples by heating them in a vacuum system up to 1300 °C. The main part of the hydrogen content is determined from the volume and pressure of the gas.

Results. Fifteen samples, from different axial positions and were measured. The results are presented in Fig. 4. The hydrogen content of the three tubes is very similar for the respective axial positions. The trend of the results agrees very well with the trend of the oxide thickness measurements. The guide tube hydrogen content was around 80 ppm at the bottom of the fuel assembly between the AFG and the SG1. The maximum H-content measured was between SG6 and SG7 and was around 270–290 ppm.

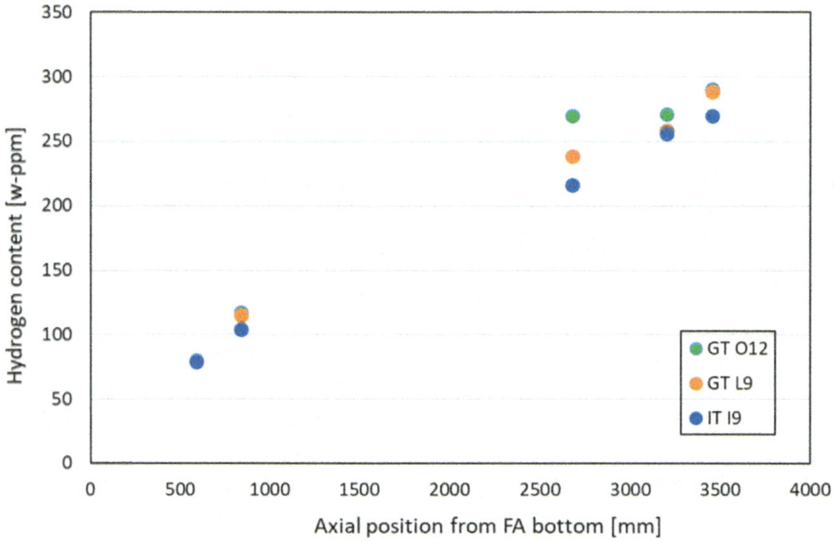

Fig. 4. Hydrogen content in the GT and IT

6.2 Guide Tube LOM

Method. The light optical microscopy (LOM) samples were cut from the GTs using a diamond cutting wheel and mounted in epoxy resin. After mounting, the surface to be examined was ground and polished.

LOM is performed in a Leica MEF 4 microscope and micrographs are recorded with a digital camera.

Results. Thirty LOM examinations were performed. The samples were cut at five different axial positions. At each position one cross section sample and one axial sample were examined in polished and etched condition for oxide and hydride imaging.

There was almost no difference in oxide thickness between the three GT. The outer and the inner oxide thickness was predominantly uniform. At the GT bottom the inner and outer oxide had similar thicknesses, whereas at the GT top the outer oxide thickness was slightly thicker than the inner, see Fig. 5.

The GT average oxide thickness was around 15–20 μm at the bottom of the fuel assembly above the AFG grid reaching a peak of around 40–41 μm above the SG6. The results are in very good agreement with the GT oxide thickness measurements performed by the eddy current technique.

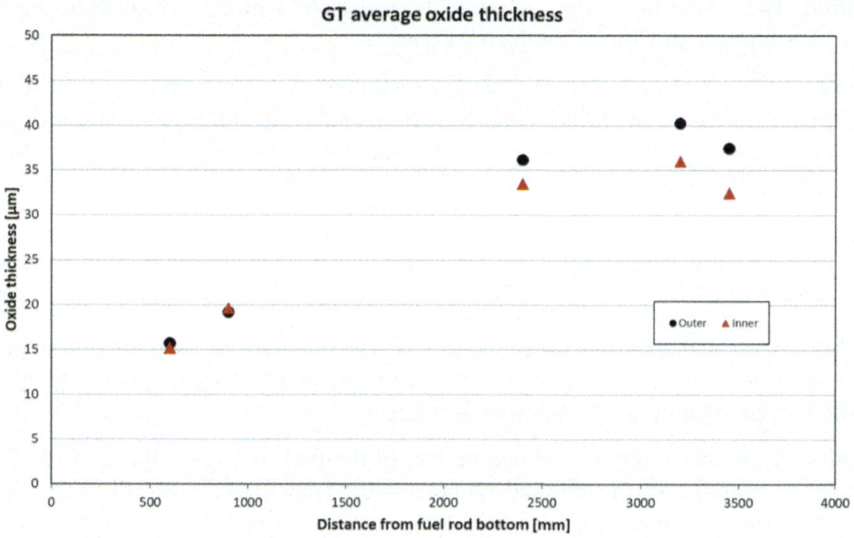

Fig. 5. GT average oxide thickness

6.3 Guide Tube TEM

Method. Four to five three-millimeter discs, 0.1–0.2 mm thick, were prepared from each position of interest on the tubes using conventional metallographic techniques and a disc puncher from Gatan. The discs were then electrolytically polished until electron transparent.

Secondary phase particles (SPP) were imaged in bright and dark field and the density and size as well as elemental composition were measured. The elemental composition was measured using EDS while EELS was used to measure the foil thickness for SSP number density measurements. Similar as for the SPPs above, the dislocation density was measured by measuring the amount of dislocation loops per unit foil volume.

Results. Initial TEM examinations of samples TEM1 and TEM2 from guide tube show that the samples contain two types of SPPs, one containing Zr-Nb and the other containing Zr-Nb-Fe. The Fe content in the SPPs in sample TEM2 was slightly higher. It is likely that the fast neutron fluence is lower in this region causing less dissolution of Fe from the SPPs. The SPP density is almost the same for the two samples but slightly lower for sample TEM2. The average SPP size is 100 and 109 nm for samples TEM1 and TEM2, respectively. Both samples have a- and c-loops.

7 Non-destructive Examinations of the Fuel Rod Claddings

7.1 Visual Inspection. Determination of Appearance Features. Determination of Tightness. Refinement of the Coordinates of the Cutting Samples

Method. The visual inspection equipment consists of a motor driven fixture and an Allied Vision camera with a Canon zoom lens.

Results. All six fuel rods were thoroughly inspected. All rods were in generally good condition. Scratches from the insertion/extraction of the rods were visible at some places, and some minor scratches and marks from handling of the rods were present. No SG fretting marks were observed.

7.2 Profilometry. Diameter Measurement. Ovality Measurement. Rod Bending Measurements

Method. A vertical profilometer rig was used. The diameter measurement system was calibrated prior to each measurement with a plug with two distinct diameters on different levels. The uncertainty of the diameter is $\pm 5\ \mu m$.

Results. Profilometry was performed on five of the fuel rods. The diameter of all five rods has decreased from the nominal diameter of 9.50 mm to an average diameter around 9.47–9.49 mm. The diameter measurement in the plenum has been influenced by the puncturing hole, but it is evident that the cladding at the plenum did not creep down and was close to its nominal value of 9.5 mm. The overall rod ovality was low for all measured rods. For all segments the pellet-pellet interfaces were visible as dips in the measurement signal.

7.3 Measurements of Oxide Thickness Against the Rod Height by Eddy Current Method

Method. The equipment has been calibrated with an unirradiated cladding tube. The linearity between the zero frequency and the free air frequency is checked by measuring on mylar films of different thicknesses on the cladding tube.

Results. A standard oxide thickness measurements were performed on all six rods. The eight azimuthal locations. The axial position of each measurement is adjusted and given from the fuel rod bottom end. The oxidation of all six fuel rods was low, around 5–6 μm at the fuel rod bottom and 12–13 μm at the fuel rod top.

For all measured segments the pellet-pellet interfaces were visible as dips in the measurement signal.

8 Density Measurements by of Fuel and Evaluation of Fuel Swelling. Element Content Examination (SEM). Element Content Examination by Pellet Radius

8.1 Hydrostatic Weighting. Density Distribution. Elemental Content of Cladding and Pellet. Determination of the Fuel-Cladding Interaction Width

Method. A cross sectional sample containing both pellet and cladding is measured. The measured value represents the total density for the pellet and cladding. To obtain the pellet density, the total density is corrected for the influence of the cladding volume and weight. The measurement method is based on the principle of Archimedes.

Results. The density was measured on nine samples from the selected fuel rods. The sample length was around 10 mm. No significant difference in fuel density was measured along the first fuel rod (10.15–10.2 g/cm^3). The fuel density measured on the fuel rod is slightly lower (10.10–10.16 g/cm^3) which could be explained with the higher burnup of second fuel rod.

8.2 Pellet Microstructure Examination (SEM). Radial Inter- and Intragranular Porosity Distribution (LOM)

Method. The light optical microscopy (LOM) samples were cut from the fuel rods using a diamond cutting wheel and mounted in epoxy resin. After mounting, the surface to be examined was ground and polished.

The fuel grain structure was revealed using an etchant preferentially attacking the grain boundaries of the UO_2. The etchant is a mixture of water, hydrogen peroxide and sulfuric acid. The cladding hydride etching was performed using an etchant containing water, hydrofluoric acid, nitric acid, and sulfuric acid.

Results. Ceramography was performed on in total 13 samples. The samples were cut from different axial positions with different local burnup (54.7, 57.6 and 54.5 MWd/kgU) but can be summarized in two groups based on the fuel microstructure observations. All these samples had a region with fully developed high burnup structure in the pellet periphery, which, depending on the sample, had a width between 20 and 70 μm. The second group are the samples with intermediate burnup on which no dark zone in the pellet center and no high burnup rim was present. The porosity and the grain size of these samples looked similar in all radial positions with both inter- and intragranular porosity and clearly visible grain boundaries. The pellet geometry was revealed by the axial samples from fuel rod O5. Nothing extraordinary was observed in the pellet chamfers and the pellet dishes.

9 Conclusion

The experimental data obtained during the examination confirm the design characteristics and the confirmed high performance properties of the materials and TVS-K components have allowed to successfully completing the licensing process for the first reload for PWR reactor.

Reactor Cores for Small-Sized Nuclear Power Plants (SNPP) and Floating Power Units (FPU)

M. Yu. Tuturkin(✉)

Afrikantov OKBM JSC, 15 Burnakovsky Proezd, Nizhny Novgorod 603074, Russia
tuturkinM@okbm.nnov.ru

Abstract. The Russian Federation, acting through ROSATOM, is rapidly developing small and medium-sized nuclear power industry on the basis of Small Modular Reactors (SMR). This segment of power industry is primarily aimed at supplying electric power to remote (including northern and island) territories, as well as areas with decentralized power supply. At present, the Russian Federation operates the floating power unit Akademik Lomonosov with KLT-40S reactor plants and 14–14 cores, and a series of multi-purpose nuclear-powered icebreakers (MNI) with RITM-200 reactor plants and 14–15-1 cores. The designs of the above-stated cores were taken as a basis for the development of the AS-14–15 core of the RITM-200N reactor plant for an SNPP and of the 14–17 core of the RITM-200S reactor plant for an upgraded floating power unit (UFPU), which was completed in 2022. Each type of the reactor cores has a number of specific features defined by the requirements for reliability (lifetime, stored energy, load following capability, etc.) and safety. The scope of work performed to justify the design performance capability and the lifetime characteristics includes: extensive calculations using verified and certified computational codes, comprehensive development work at the OKBM's test facilities, as well as years long irradiation tests carried out in the reactors of nuclear icebreakers and in the nuclear research reactor loop-type facilities with subsequent post-irradiation examinations. The basic design development of the reactor cores for the FPU with RITM-400M reactor plant (the Arctic version) and for the optimized floating power unit (OFPU) with RITM-200M reactor plant (the Tropical version) is currently underway. The outlined reactor cores have a reserve for a wide variety of characteristics and functional applications.

Keywords: Fuel assembly (FA) · Nuclear fuel · Reactor core · SNPP · UFPU · FPU · OFPU

1 Introduction

The issue of diversifying the uses of atomic energy has been raised since the birth of the peaceful atom. A small modular reactor (SMR) is another of many answers.

The SMRs should be able to provide electric power to a wide variety of consumers around the world. The more flexible power generation is provided by a power unit, the wider the scope of application for this type of reactors will be. Another advantage is the short construction time.

J. Liu and Y. Jiao (Eds.): WRFPM 2023, SPPHY 299, pp. 224–233, 2024.
https://doi.org/10.1007/978-981-99-7157-2_23

The Russian Federation, acting through ROSATOM, is rapidly developing small-sized nuclear power industry on the basis of SMRs. This segment of power industry is primarily aimed at supplying electric power to remote (including northern and island) territories, as well as areas with decentralized power supply.

This report provides an overview of reactor cores with fuel assembly configuration for small modular reactors developed based on more than 60 years of experience in designing and operating icebreaker cores.

Fig. 1. FPU *Akademik Lomonosov*

2 Implemented Projects

2.1 The 14–14 Core of KLT-40S Reactor Plant for the FPU Akademik Lomonosov

In 2020, the floating power unit Akademik Lomonosov with two KLT-40S reactor plants (Fig. 1) and 14–14 reactor cores having fuel assembly configuration was developed and put into commercial operation to provide electricity and heat to consumers in the Chukotka Autonomous District. This is the first-of-a-kind implemented project in the series of small-sized mobile transportable power units. The reactor unit designs include cores with fuel assembly configuration. Uranium dioxide dispersed in a silumin matrix is used as the fuel.

The vessel manufacturing and installation activities were performed in St. Petersburg [1].

At the beginning of 2023, the two 14–14 cores of the FPU Akademik Lomonosov produced a total of 2.5 TW-h during a 33,000-h period of operation.

The 14–14 fuel assembly is shown in Fig. 5b.

2.2 The 14–15-1 Reactor Core of RITM-200 Reactor Plant for Multipurpose Nuclear Icebreakers

In developing the solutions used in the FPU Akademik Lomonosov design, a 14-15-1 reactor core of the RITM-200 reactor plant was created with increased stored energy, lifetime, service life and capacity for a series of multipurpose nuclear icebreakers (Fig. 2) designed to escort ships along the Northern Sea Route and to explore and develop the Arctic.

Fig. 2. MNI *Arktika*

As of the beginning of 2023, three MNIs were put into operation: Arktika, Sibir, and Ural; two more MNIs are under construction: Yakutia and Chukotka.

The four 14-15-1 cores operating on the MNI Arktika and Sibir at the beginning of 2023 produced a total of 2.5 TW-h during a 40,300-h period of operation.

The 14-15-1 fuel assembly is shown in Fig. 5b.

3 Projects Under Construction

3.1 The AS-14–15 Reactor Core of RITM-200N Reactor Plant for SNP

Relevance. The accomplished activities related to creating the 14–14 reactor cores for the FPU Akademik Lomonosov and the engineering solutions taken for 14-15-1 reactor cores for multipurpose nuclear icebreakers aimed at reliable supply of energy to remote regions in the long term provided grounds for developing the AS-14-15 reactor core. In 2022, the development work was completed. The AS-14-15 reactor core of the RITM-200N reactor plant is designed for an SNPP (Fig. 3) aimed at power supply to the northern territory of the Republic of Sakha (Yakutia).

The objective of the first-of-a-kind SNPP is to supply power to Yakutia's power consumers, including mineral deposits.

Design Features. A fuel assembly with the core height increased was developed and applied to extend the fuel life.

The AS-14-15 fuel assembly is shown in Fig. 5a.

In 2028, it is scheduled to commission the pilot SNPP on the territory of Russia (Fig. 3).

Fig. 3. SNPP

3.2 The 14–17 Core of RITM-200S Reactor Plant for the UFPU

Relevance. In addition, to provide electric power to remote power consumers in the Far North and the Far East, including the Baisky GOK in the Chukotka Autonomous District, the basic design of the 14–17 core of the RITM-200S reactor plant intended for upgraded floating power units (UFPU) was completed in 2022. The main difference between the designs of the UFPU and the FPU Akademik Lomonosov is the upgraded reactor core with the reactor plant.

At the beginning of 2023, two UFPUs are under construction with 4 RITM-200S reactor plants, the hulls for which are being manufactured at the Chinese shipyard Wison (Nantong) Heavy Industry Co Ltd. Two more UFPUs are scheduled for construction in the nearest future. The issues related to replication of the UFPU as a sustainable source of power supply to oil and gas fields are being actively elaborated nowadays.

Design Features. The 14–17 core design of the RITM-200S reactor plant for the UFPU is based on the experience of creating and operating the 14–14 cores with fuel assembly configuration for FPU Akademik Lomonosov with KLT-40S reactor plants, 14-15-1 cores for MNI with RITM-200 reactor plants, the results of development work performed while developing the basic design of AS-14-15 core for an SNPP with RITM-200N reactor plant, as well as on the experience in developing, manufacturing, testing and operating the icebreaker cores.

The 14–17 fuel assembly is shown in Fig. 5a.

In 2027, it is scheduled to commission the UFPU.

Evaluation of the Economic and Environmental Effects. The coefficient of potential economic effect over the entire life cycle of the UFPU with a 14–17 core will be approximately 3.1 to the unit of cost.

With the intended use of UFPU as a source of electricity at the Baimsky GOK, the total potential economic effect in comparison with the power sources using diesel fuel is approximately 3 billion USD, and in comparison with the power sources using coal and LNG it is approximately 0.5 billion USD [2] (Fig. 4).

Fig. 4. UFPU

4 Projects Under Development

4.1 The Reactor Core of RITM-400M for the FPU of Arctic Version

The FPU of Arctic Version is designed within the project of GlobalTech LLC aimed at supplying power to the complex for production, storage and shipment of LNG and gas condensate [3].

The fundamental distinctive feature of the FPU with a RITM-400M reactor plant is the increase in the unit capacity of the core and of the power unit, respectively.

The development of the core basic design for the FPU with RITM-400M reactor plant is scheduled for 2023–2024.

4.2 The Reactor Core of RITM-200M for the OFPU of Tropical Version

The tropical version of the OFPU was elaborated based on the layout solutions of the UFPU design and is intended mainly for electric power supply to island territories in the southern latitudes of our planet. The service life and the interval between repairs of the main equipment has been increased due to minor structural and in-vessel modifications.

The basic design development of the OFPU RITM-200M core has been scheduled for 2024–2025.

5 Verification of the Reactor Cores

5.1 Scientific and Technical Basis for the Development

The core designs apply the materials, reference components and engineering principles that have been repeatedly proven and tested in the reactor cores of similar design, including the icebreaker ones.

A unique knowledge base has been accumulated by analyzing the results of many years of experience in operating various reactor cores, which allows us to guarantee the safety and efficiency of the applied design solutions (Fig. 5).

5.2 Design Computational Validation

The development of basic designs of the cores included neutronic, thermal-hydraulic and strength calculations performed via verified and certified computational codes using to the full advantage the experience gained in designing propulsion reactor cores.

The designs include reactivity control and emergency protection systems to ensure that the core is subcritical at any point in the fuel life.

The reactivity is compensated for by gadolinium-bearing burnable absorber rods in the fuel rod bundle and by movable control rods with dysprosium and boron content.

The most important results of the core development are the increase of the energy resource of the developed AS-14-15 core for an SNPP and 14–17 core for a UFPU up to 8.0 TW-h. The general view of the reactor core layout is shown in Fig. 6.

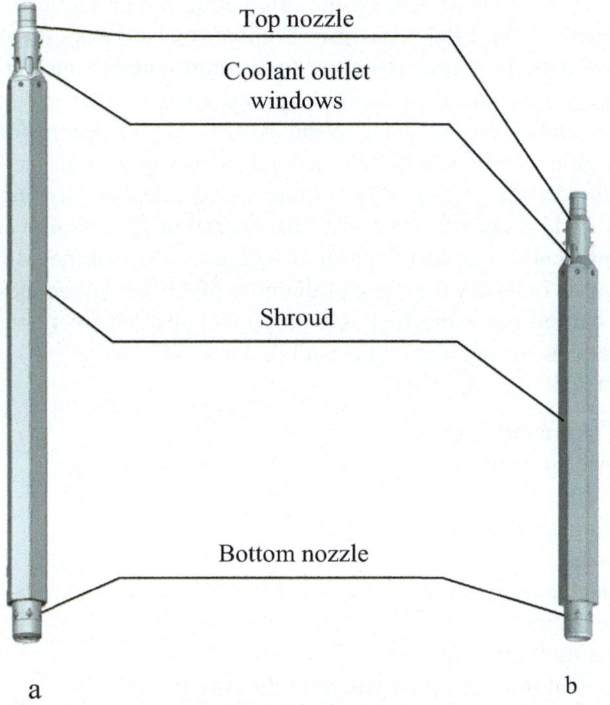

Fig. 5. Fuel assembly

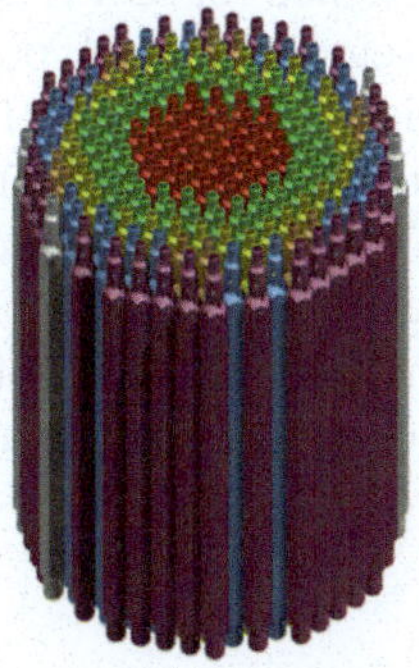

Fig. 6. General view of the reactor core layout

5.3 Design Experimental Validation

The scope of work performed to justify the design performance capability and the lifetime characteristics includes: comprehensive development work at the OKBM's test facilities, as well as years long irradiation tests carried out in the reactors of nuclear icebreakers and in the nuclear research reactor loop-type facilities with subsequent post-irradiation examinations.

Guided by the objective to develop the original design principles of the 14–14 fuel assembly designed for the FPU Akademik Lomonosov, and using the available experimental facilities, a 14-15-1 fuel assembly for a multi-purpose nuclear icebreaker was developed with more advanced structural components.

The fuel assemblies provided for in the AS-14-15 core design for SNPPs, in the 14–17 core design for UFPUs, and fuel assemblies provided for in the FPU design and OFPU design (Fig. 5a) are in many ways similar and identical to those incorporated in the 14–14 core and 14-15-1 core designs (Fig. 5b). Therefore, the research and development work (R&D) aimed at design justification of fuel assemblies of the 14–14 core and 14-15-1 core were used to justify the design reliability of the subsequent designs. The R&D activities were carried out using fuel assembly mockups, lead test assemblies, various dummies and models of fuel assemblies and fuel assembly units.

Those R&D activities included:

- lifetime and hydraulic testing,
- thermal-physical testing,
- hydraulic testing,
- thermal cycling testing,
- impact resistance testing,
- integrated mechanical testing,
- aerodynamic testing,
- tests using a simulated core,
- studies of coolant flow rate distribution at the core inlet,
- testing of burnable poison rod mockups and absorber rod mockups,
- analysis of coolant temperature distributions in RITM-200 reactor plant.

The overall size of the cores designed for SNPP, UFPU, FPU and OFPU is increased in height; therefore, the set of development work performed in 2021–2022, as part of the AS-14-15 core basic design development for RITM-200N reactor plant, included experimental studies aimed at justifying the reliability and performance of the fuel assemblies with extended length. The results of these studies were also used in design justification of the 14–17 fuel assemblies as they are identical in design.

The R&D work performed while developing the basic design of the AS-14-15 core with extended length included an extensive set of tests aimed at justifying the decisions made:

- lifetime and hydraulic testing,
- thermal-physical testing,
- hydraulic testing,
- thermal cycling testing,
- impact resistance testing,
- integrated mechanical testing.

The set of tests performed made it possible to substantiate the serviceability and reliability of the fuel assembly design, its component parts and units in compliance with the assigned longevity indicators and with regard to the operating conditions of the reactor cores designed for the SNPP, UFPU, FPU and OFPU. No inadmissible damages and destructions of the component parts were identified.

5.4 Unification of Component Parts

Component parts and materials provided for by the national and industry-specific standards were used to the maximum extent in the process of developing the core design.

The structure of the cores designed for the SNPP, UFPU, FPU and OFPU is unified to the maximum extent according to the structural components applied in the reference 14-15-1 core of RITM-200 reactor plant for the MNI. The AS-14-15 core uses the same fuel as the 14–14 core designed for the FPU Akademik Lomonosov. The 14–17 core uses the same fuel as the 14-15-1 core for the MNI.

By unifying the component parts of all the cores we increase the volume of series production of items, and consequently, reduce the production costs, the time of preproduction operations, the reject rates and the range of spare parts.

5.5 Fuel Justification

The dioxide fuel for the AS-14–15 core of SNPP. The base for the computational/experimental justification of the SNPP (UO_2) fuel was the development of this fuel for the 14–14 core of the FPU Akademik Lomonosov.

As part of the dispersion fuel justification intended for the FPU Akademik Lomonosov 14–14 core, a 14–14 loop fuel assembly with zirconium cladding of fuel rods was tested in the SSC RIAR MIR reactor with subsequent post-irradiation. The test results fully justify the fuel rod performance.

In 2012–2018, to validate the performance and predict the lifetime characteristics of fuel assemblies designed for cores of RITM-200 reactor plant and other advanced plants (including cores of RITM-200N and RITM-200M reactor plants), irradiation tests and post-irradiation examination of an experimental fuel assembly with fuel rods similar in design to those intended for the AS-14–15 core and a UO_2-based dioxide fuel composition were performed in the MIR reactor under conditions close to the operating ones. The inspection results justified the fuel rod performance under the AS-14–15 core operating conditions.

Intermetallic fuel for the 14–17 core of UFPU. The basis for the computational/experimental justification of the intermetallic fuel designed for the UFPU was the development of this fuel for the MNI 14-15-1 core.

The fuel was subjected to a complete cycle of tests, including tests in loop fuel assemblies of research reactors, mass tests and operation in icebreaker cores, and post-irradiation examination.

In order to substantiate the performance and predict the lifetime characteristics of fuel assemblies designed for the cores of RITM-200 reactor plant and other advanced plants (including the cores designed for the RITM-200S and RITM-400M reactor plants) and post-irradiation examinations of an experimental fuel assembly with fuel rods similar in design to those of the 14–17 core and with intermetallic fuel composition in a silumin matrix were performed in the MIR reactor.

The prototype fuel rods similar to 14–17 fuel rods in design, technology and applied materials were subjected to a complete cycle of tests which included tests in loop fuel assemblies of research reactors, mass tests and operation under conditions of icebreaker cores.

Fuel Rod Cladding. The unique Cr-Ni alloy is used as the fuel rod cladding material. This is due to its following advantages:

- high long-term strain capacity in the operating temperature range,
- resistance to radiation-induced embrittlement and retention of high short-term plastic properties under irradiation conditions in the operating temperature range,
- unique high corrosion resistance [4],
- proven performance under real operating conditions at nuclear icebreakers and in the MIR reactor loop-type facilities.

6 Conclusion

All of the cores mentioned in the paper are marine-type and have a number of specific features as compared to the cores of stationary power reactors, such as:

Table 1. Comparison of core parameters

Reactor core	14–14 FPU *Akademik Lomonosov* KLT-40S	14-15-1 MNI RITM-200	14–17 UFPU RITM-200S	AS-14-15 SNPP RITM-200N	14–19 OFPU [3] RITM-200M	14–18 FPU [3] RITM-400M
Reactor core type	FA configuration	FA configuration	FA configuration	FA configuration	FA configuration	FA configuration
Rated thermal power, MW	150	175	198	190	198	340
Reactor plant electric power, MW	35	36	58	55	58	95
Lifetime, h	21,000	75,000	75,000	75,000	75,000	75,000
EFPD, d	580	1070	1680	1750	2310	1650
Stored energy, TW·h	2.1	4.5	8	8	11	13.5
Fuel life, h	14,000	25,700	40,400	42,100	55,555	39,700
Reactor plant service life, years	40	40	40	60	40	40
Fuel type	Uranium dioxide	Intermetallic uranium	Intermetallic uranium	Uranium dioxide	Uranium dioxide	Intermetallic uranium
Fuel rod cladding material	Zr	Cr-Ni alloy	Cr-Ni alloy	Cr-Ni alloy	Cr-Ni alloy	Cr-Ni alloy

- greater load following capability;
- no partial refuelings (cores are completely loaded at the beginning of life and are completely unloaded during refueling operations);
- unique design solutions.

Table 1 provides the most important parameters of the operating cores and those under development.

The main development areas for reactor plants with reactor cores include:

- increase reactor capacity, lifetime and assigned service life of the main equipment,
- enhance economic efficiency,
- increase the reactor core stored energy to extend the refueling interval,
- create different options of power units with a wide range of parameters.

Each successive design continues the evolutionary development of advanced energy sources using reference technologies applied in nuclear shipbuilding.

Thus, these cores possess all the advantages of marine solutions (compactness, load following capability, resistance to external impacts, reference), meet the requirements of nuclear safety and allow providing high technical and economic performance; their operability and safety is justified by calculations, experiments.

and confirmed by successful operation of analogues and prototypes.

References

1. Rosatom; Rosenergoatom; Stations, Affiliates, and Projects, https://www.rosenergoatom.ru/development/innovatsionnye-razrabotki/razrabotka-proektov-aes-s-reaktorami-novogo-pokoleniya/plavuchie-atomnye-teploelektrostantsii-pates/.
2. A.E. Kontorovich, P.Yu. Sorokin, Collection of papers by laureates of the international contest of scientific and scientific-and-technical developments, innovations aimed at developing and exploring the Arctic and the Continental shelf, 2022, pp. 53–55. http://www.technodevelop.ru/arktika.
3. S.M. Brykalov, A.S. Balyberdin, D.A. Nyrkov, N.V. Sheshina, Ye.A. Guschina Research and Analysis Journal Arctic: Ecology and Economy. Selecting a Priority Floating Power Unit Type Based on Analysis of Technical and Economic Indices// Arctic: Ecology and Economics, volume 12, No. 4 (2022).
4. V.V. Novikov, V.I. Kuznetsov, V.A. Markelov, Ye.N. Mikheev, G.V. Kulakov: Description of Material Properties of the Fuel Element, Fuel Assembly and Absorber Rod of the Control and Protection System. JSC VNIINM, Moscow (2019).

Thermal-Hydraulic Characteristics of TVS-K Fuel Assembly

V. E. Lukyanov(✉)

Afrikantov OKBM, JSC Burnakovsky Proyezd, 15, Nizhny Novgorod 603074, Russia
lukyanov-ve@okbm.nnov.ru

Abstract. Description of the TVS-K fuel assembly designed for operation in PWR cores is presented. Results of experimental and computational justification of thermal-hydraulic characteristics of the TVS-K are given. It is shown that the design characteristics of the TVS-K provide the necessary thermal-hydraulic parameters and do not impose additional limitations on operation of homogeneous and mixed PWR cores.

Keywords: PWR · TVS-K fuel assembly · Hydraulic characteristics · Hydraulic compatibility · Critical heat flux

Nomenclature

CHF	Critical heat flux
DNBR	Departure from nucleate boiling ratio
EPRI	Electric Power Research Institute
IFM	Intermediate flow mixing
FA	Fuel assembly
MDNBR	Minimal departure from nucleate boiling ratio
NPP	Nuclear power plant
PWR	Pressurized water reactor
SG	Spacer grids
STDP	Statistical thermal design procedure

1 Introduction

Russian TVS-K 17 × 17 fuel assembly with 12 feet fuel stack has been developed for operation in PWR cores.

The TVS-K has six spacer grids with mixing vanes and three intermediate flow mixing grids with mixing vanes ensuring the high thermal-hydraulic margins.

The following investigations for justification of thermal-hydraulic characteristics of the TVS-K were carried out:

– hydraulic tests of fragments and mockups of the TVS-K including lifetime test of the full scale TVS-K mockup at PWR reactor core conditions;

J. Liu and Y. Jiao (Eds.): WRFPM 2023, SPPHY 299, pp. 234–242, 2024.
https://doi.org/10.1007/978-981-99-7157-2_24

- investigations of thermal-hydraulic characteristics and critical heat flux on TVS-K test sections.

Hydraulic characteristics and pressure loss coefficients of the TVS-K fuel assembly were determined based on hydraulic test results.

Critical heat flux tests were carried out on two water thermal test facilities. Representative CHF data were obtained for development and validation of the CHF correlation and for determination of the thermal diffusion coefficient. The CRK correlation for critical heat flux calculations was developed.

Fulfilment of thermal-hydraulic design criteria including the criterion for prevention of heat transfer crisis and the criterion for prevention of TVS-K lifting was confirmed.

Justification of TVS-K fuel assembly characteristics was carried out within the framework of the thermal-hydraulic design. Results are described in the technical reports.

The TVS-K lead fuel assemblies were successfully operated for four fuel cycles at Ringhals-3 NPP [1].

2 TVS-K Design Characteristics

The TVS-K fuel assembly consists of 264 fuel rods and U-Gd fuel rods. Fuel rods and U-Gd fuel rods are combined into a bundle 17 × 17 by spacer grids. Skeleton, consisted of 24 guide tubes, one instrumental tube and spacer grids, ensures the rigidity and geometrical stability of the TVS-K.

Installation of the TVS-K fuel assemblies in the core between upper and lower core support plates is provided by using the top and bottom nozzles.

The spacer grids have mixing vanes for coolant mixing and intensification of heat exchange.

In addition, three intermediate flow mixing grids with mixing vanes are installed in three upper spans between spacer grids for improvement of thermal characteristics of the TVS-K.

TVS-K design characteristics:

- 12 feet fuel stack;
- 6 spacer grids with mixing vanes;
- 3 IFM grids with mixing vanes;
- upper grid without mixing vanes;
- anti-fretting grid;
- anti-vibration grid;
- bottom nozzle with debris filter;
- top nozzle with holddown springs.

Figure 1 illustrates the arrangement of grids on TVS-K skeleton height selected from the condition of geometrical compatibility with reference FA.

TVS-K design characteristics provide:

- geometric stability of fuel bundle;
- hydraulic compatibility with the reference fuel;
- lifting prevention;

- departure from nucleate boiling prevention and reliable core cooling;
- prevention of rod bow and fretting of fuel rod cladding in the TVS-K lower span.

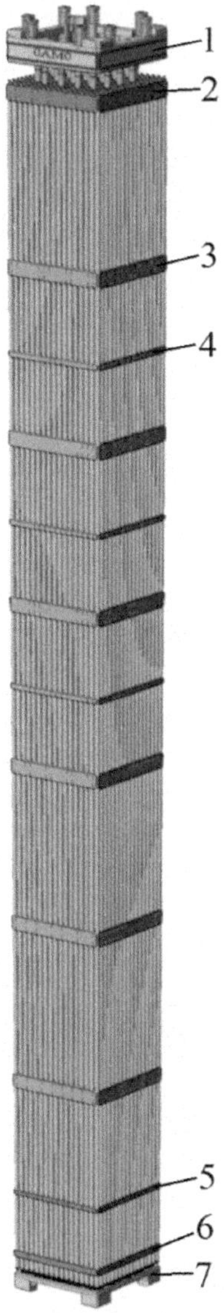

1 – top nozzle; 2 – upper grid; 3 – spacer grid; 4 – IFM grid;
5 – anti-fretting grid; 6 – anti-vibration grid; 7 – bottom nozzle

Fig. 1. TVS-K fuel assembly.

3 Hydraulic and Lifetime Tests

Hydraulic and lifetime tests of the TVS-K mockup fabricated according to design technology were performed on the Afrikantov OKBM RGS test facility.

Schematic diagram of the RGS test facility is shown in Fig. 2.

Main parameters of the RGS test facility:

– pressure	up to 15.7 MPa;
– inlet temperature	20–300 °C;
– flow rate	200–550 m^3/h;
– Reynolds number	up to 500000

Pressure drops of the total TVS-K and individual sections (top and bottom nozzles, active part and spacer grids) were determined during tests.

The lifetime test (2000 h) of the TVS-K mockup was performed. Visual inspection after test showed that there are no defects of FA elements and fretting of fuel rod cladding.

4 Critical Heat Flux Investigations

Critical heat flux investigations using 4 × 4 and 5 × 5 test sections with SG and IFM grids were carried out on two thermal test facilities. Schematic diagram of the Afrikantov OKBM thermal test facility is shown in Fig. 3.

Main parameters of thermal test facilities:

– pressure	up to 18 MPa;
– inlet temperature	up to 340 °C;
– mass velocity	up to 5500 $kg/(m^2 \cdot s)$

Experimental CHF data on the test sections with IFM grids and without IFM grids was obtained. CHF was investigated on the test sections with typical cells and guide tube cells as well as with different types of non-uniform axial power shapes (cosine type and with maximum of heat generation at upper part of test sections). Number of test sections is 14.

Investigation of coolant temperature in the cells of the test section with increased stepwise radial power distribution was also conducted for determination of the thermal diffusion coefficient.

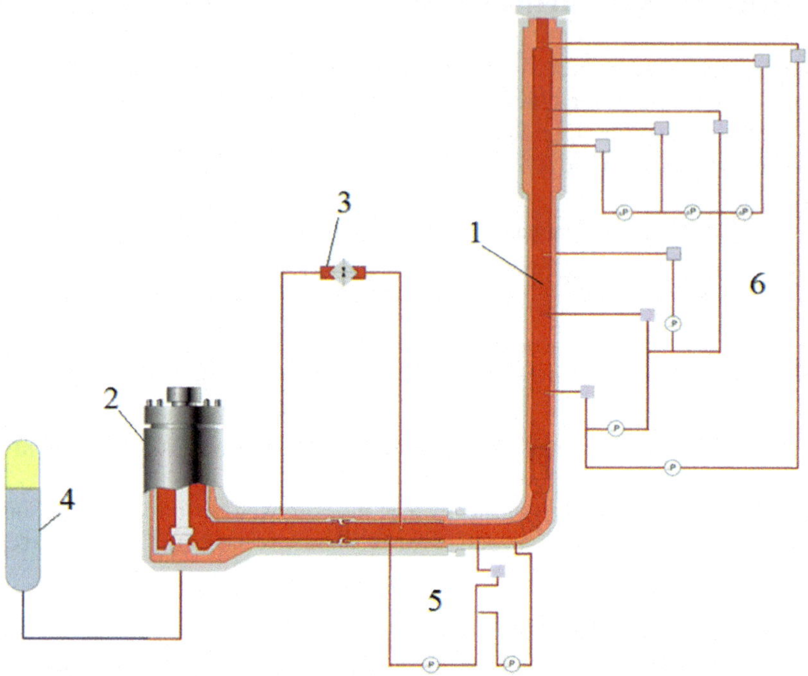

1 – TVS-K mockup; 2 – circulation pump; 3 – electric heater; 4 – pressurizer; 5 – flow rate measuring system; 6 – pressure drop measuring system

Fig. 2. Schematic diagram of the RGS test facility.

5 Thermal-Hydraulic Design Methodology

5.1 Thermal-Hydraulic Design Criteria

Provision of Primary Coolant Flow Rate

Hydraulic characteristics of the reactor core with TVS-K shall ensure coolant flow rate through the reactor in normal operation within the design limits: $Q_{min} < Q < Q_{max}$.

Prevention of Departure from Nucleate Boiling

Exclusion of heat transfer crisis on the hottest fuel rod with probability not less than 95% at confidence level of 95% in normal operation and anticipated operational occurrences: MDNBR > SALDNBR.

Prevention of TVS-K lifting

Exclusion of TVS-K lifting with probability not less than 95% in limiting cold and hot conditions.

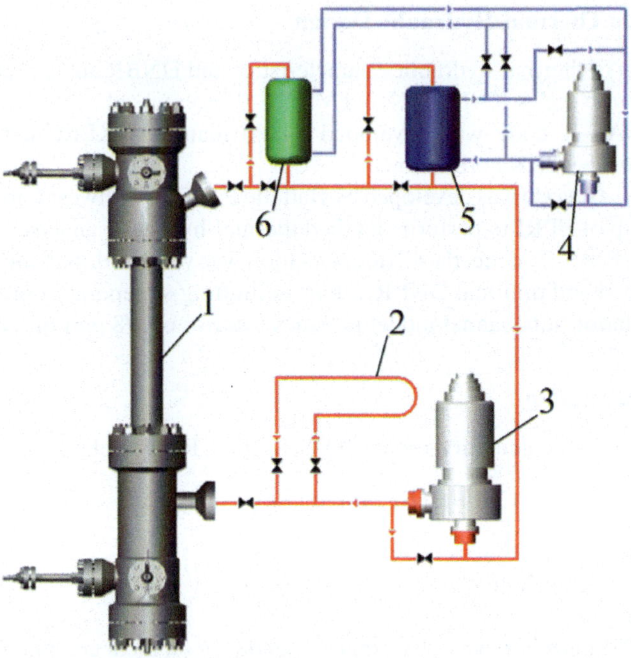

1 – TVS-K test section; 2 – electric heater; 3 – primary circulation pump; 4 – secondary circulation pump; 5 – heat-exchanger; 6 – condenser

Fig. 3. Schematic diagram of the thermal test facility.

Hydraulic compatibility

Pressure loss coefficients of the total TVS-K and individual sections shall be close to pressure loss coefficients of reference FA and shall not lead to:

- unacceptable coolant transverse velocities according to the conditions of vibration loading on fuel rods;
- unacceptable decrease of coolant flow through any FA in mixed core according to the conditions of fuel rods cooling.

5.2 TVS-K Hydraulic Characteristics

Hydraulic characteristics and pressure loss coefficients of the TVS-K fuel assembly and individual sections were determined based on hydraulic test results.

Bypass flow through guide tubes and instrument tube was calculated.

Analysis of flow distribution between fuel assemblies having different hydraulic characteristics in mixed core was performed.

Estimation of transverse velocities of coolant flow around non-identical spacer grids in mixed core is carried out.

5.3 Codes for Thermal-Hydraulic Design

For calculation of thermal-hydraulic characteristics and DNBR analysis, the subchannel thermal-hydraulic codes KANAL-K and VIPRE-01 were used.

The KANAL-K code was developed by Afrikantov OKBM and was used for development of the CRK correlation.

The VIPRE-01 code was developed by Battelle Pacific Northwest Laboratories under the sponsorship of EPRI to perform detailed thermal-hydraulic analyses of PWR cores. By using the VIPRE-01 code, the CRK correlation was validated, statistical design limits were determined and minimal DNBR values in limiting statepoints were calculated.

The calculation subchannel model includes a sector of 1/8 part of the PWR core.

5.4 CHF Correlation

For critical heat flux calculations in the TVS-K, the CRK correlation was developed:

$$\mathrm{CHF} = Q(P, \rho w, x) \cdot F_{SC} \cdot F_{MG} \cdot F_F, \quad (1)$$

P—pressure;
ρw—local mass velocity;
x—local quality.

For the CRK correlation a polynomial dependence on local coolant parameters with empirical coefficients was used.

The subchannel factor F_{SC} takes into account the effect of hydraulic inequality of subchannels.

The mixing grid factor F_{MG} takes into account the dependence on distance from spacer (IFM) grids.

The form-factor F_F taken into account the non-uniform axial heat flux distribution was based on the Tong form-factor for the W-3 correlation [2].

The CRK correlation combines CHF data for the test sections with and without IFM grids.

The applicable range of the CRK correlation:

– pressure	9.0–17.2 MPa;
– local mass velocity	1000–5300 kg/(m^2·s);
– local quality	–0.2–+ 0.5;
– fuel rod outer diameter	9.5 mm;
– guide tube outer diameter	12.6 mm;
– fuel rods pitch	12.6 mm;
– axial grid spacing	261–535 mm

Statistical characteristics of the CRK correlation:

- mean ratio of measured to predicted CHF - M/P = 1.0;
- standard deviation of measured to predicted CHF - s = 7.3%;
- correlation limit - CL = 1.14.

5.5 Statistical Design Limits and Minimal Departure from Nucleate Boiling Ratios

The Statistical Design Limit is calculated with 95% probability at the 95% confidence level using statistical procedure that is similar to STDP [3]. When performing calculations, deviations of the following reactor core and local parameters are taken into account: core power, core pressure, reactor coolant flow, core inlet temperature, core bypass flow, fuel rod power, local subchannel coolant heating, code uncertainty, transient uncertainty and CHF correlation uncertainty.

The Statistical Design Limit:

$$SDLDNBR = \frac{1}{\mu_C \cdot \left(1 - 1.645 \cdot \sqrt{\sum_i \left(S_i \cdot \frac{\sigma_i}{\mu_i}\right)^2 + \left(\frac{\sigma_C}{\mu_C}\right)^2}\right)}, \quad (2)$$

σ_i—standard deviation of the i-th parameter;
μ_i—mean value of the i-th parameter;
S_i—sensitivity coefficient for the i-th parameter;
μ_C—lower conservative value for mean value of measured to predicted CHF;
σ_C—high conservative value for standard deviation of measured to predicted CHF.

The Safety Analysis Limit takes into account additional margins on rod bow U_{RB} and fuel assembly bow U_{AB} and also additional design margin U_{DM}:

$$SALDNBR = SDLDNBR \cdot (1 + U_{RB} + U_{AB}) \cdot (1 + U_{DM}). \quad (3)$$

The SALDNBR for reference PWR is 1.33 (including $U_{DM} = 0.05$).

Calculations of MDNBR for set of reference limiting statepoints corresponding to normal conditions and anticipated operational occurrences (including statepoints with decrease of coolant flow, with change of power and pressure, with change of absorber rods position) were carried out according to the specified boundary conditions.

Thermal characteristics of the TVS-K fuel assembly provide thermal power of 3 loops PWR 3135 MW with MDNBR for nominal conditions more than 2.2 in homogenous core.

5.6 TVS-K Holddown Forces

Holddown forces acting on the TVS-K fuel assembly were calculated for limiting cold and hot conditions.

The nominal value of hold down force is defined as the difference between the pressing force and resultant lift force.

Statistical method was used for taking into account of deviations of the following parameters: reactor coolant flow; core bypass flow; TVS-K pressure loss coefficient; core inlet flow maldistribution; buoyancy lift force; weight of the dry TVS-K; force of the spring block.

Minimal value of holddown force is calculated with 95% probability.

TVS-K holddown forces allow excluding the excessive pressing forces on FA which negatively affect on FA deformation, accumulation of FA bow and increase of interaction forces between assemblies during core operation.

6 Conclusions

The TVS-K 17 × 17 fuel assembly with 12 feet fuel stack has been developed for operation in PWR cores.

To substantiate the TVS-K thermal-hydraulic characteristics, the experimental investigations of the TVS-K mockup and test sections were carried out on hydraulic and thermal test facilities, computational justification was performed and design documentation was developed.

Calculated values of minimal DNBR confirm a fulfilment of the design criterion for prevention of heat transfer crisis and ensure reliable operation of PWR cores with TVS-K for normal operation and anticipated operational occurrences.

Characteristics of the spring block of the TVS-K top nozzle ensure a fulfilment of the design criterion for prevention of TVS-K lifting in homogeneous and mixed PWR cores for limiting conditions.

Operation of the TVS-K fuel assemblies in the core does not change the design reactor flow rate and the distribution of flow in the core and does not have a noticeable effect on the operating conditions of the surrounding fuel assemblies.

Hydraulic compatibility with reference fuel assemblies is ensured by close values of their hydraulic characteristics.

The TVS-K fuel assembly meets modern requirements for PWR fuel for technical, economic and operational characteristics.

References

1. K. Lafchiev, K.D. Johnson, D. Jadernas, P. Askeljung, V.V. Novikov, V.A. Markelov, V.I. Kuznetsov, A.Y. Shevyakov, E.N. Mikheev, A.A. Kabanov, A.V. Ugryumov, A.A. Shishkin, A.F. Radostin, A.I. Romanov, I.E. Simanovskaya, E.V. Sholin, A.A. Enin, M.A. Shustov, A.V. Strukov, S.V. Lozitsky, M. Hemlin, D. Schrire, R. Waginder. TVS-K operation experience under PWR conditions at Ringhals-3 NPP. In: Top Fuel 2021 Conference on Proceedings, Santander, Spain, October 24–28 (2021).
2. L.S. Tong. Boiling crisis and critical heat flux. Westinghouse Electric Corporation. Published by U.S. Atomic Energy Commission, USA (1972).
3. J. Robeyns, F. Parmentier, G. Peeters. Application of a statistical thermal design procedure to evaluate the PWR DNBR Safety Analysis Limits. In: 9th International Conference on Nuclear Engineering (ICONE 9) on Proceedings, Nice, France, April 8–12 (2001).

Study on the Application of Intelligent Storage of Fuel Assembly Based on RFID Technology

Guo Shuang(✉)

CNNC Jianzhong Nuclear Fuel Co., Ltd. (CJNF), Beijing, China
384810948@qq.com

Abstract. With the rapid rise of a new round of industrial change and technological revolution, intelligent storage can integrate warehouse resources, ensure the accuracy of goods management and circulation, and play an important role in the process of intelligent nuclear fuel manufacturing. CJNF mainly applies radio frequency identification and Internet of things technology to realize intelligent storage management of important raw materials, semi-finished products and finished product warehouses in the production line, meet the intelligent storage management of the production line and the process control application requirements of some key production processes, and lay the foundation for the integrated management and control of the whole process of the production of nuclear fuel components.

Keywords: Fuel assembly · Intelligent storage · RFID

1 Introduction

With the continuous and orderly development of nuclear power, higher requirements are put forward for the production capacity of nuclear fuel plants, so as to meet the growing demand for nuclear fuel components. The production process of nuclear fuel assembly is long and complicated, and there are a lot of semi-finished products in the process. Nuclear fuel assembly production plants rely on manual input and storage operations, physical items do not correspond to shelf numbers, and there is no storage management system to manage raw materials, semi-finished products and finished products in production.

With the rapid rise of a new round of industrial reform and technological revolution, the development of modern industrial informatization has entered a new historical stage of intelligent manufacturing construction. Under the background of the national deployment and implementation of the strategic layout of manufacturing power, enterprises accelerate the continuous integration of information technology and industrial technology, relying on the "integration of the two", "Industry 4.0", "intelligent storage", "big data", "Internet of things" and "digitalization" and other technologies, combined with RFID wireless transmission technology, to achieve the unity of material flow and information flow on the production site. To improve the material management level of nuclear fuel assembly production site.

J. Liu and Y. Jiao (Eds.): WRFPM 2023, SPPHY 299, pp. 243–250, 2024.
https://doi.org/10.1007/978-981-99-7157-2_25

Ultra-high FrequencyRadio Frequency Identification (UHF RFID) technology is one of the core technologies of the Internet of Things. It is a non-contact automatic identification RFID technology working in the 860–960MHz frequency band. It integrates semiconductor, electronic information, cryptography and other fields of technology.

As shown in Fig. 1, the typical block diagram of common UHF RFID systems is mainly composed of electronic tags, readers, communication networks and upper computer servers. Among them, the reader is mainly composed of power management module, MCU, radio frequency module, antenna, USB and serial communication module, while the electronic tag is mainly composed of antenna and radio frequency module. The basic principle of the system is as follows: wireless communication is carried out through the electromagnetic backscattering coupling between the reader and the electronic tag antenna, so as to achieve the purpose of automatically collecting data, identifying, reading and writing label information.

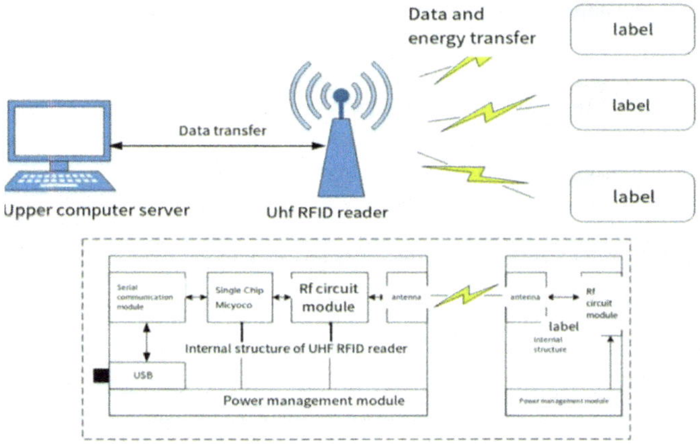

Fig. 1. Uhf RFID system block diagram

2 Requirement Analysis

2.1 A Subsection Sample

The storeroom of the nuclear fuel production workshop in CNNC mainly includes stereo storeroom and manual storeroom. The manual warehouse for storing parts, wastes, materials, semi-finished products and other items mainly relies on the traditional means of manual trolley storage and manual data recording. There are wrong materials, wrong placement, manual recording errors, inventory data deviation and other phenomena. Nuclear fuel production process is long, all kinds of intermediate products do not correspond to the physical material and material, storage management efficiency is relatively low, cannot check the quantity and status of materials. The logistics and transfer process of nuclear fuel production is shown in Fig. 2.

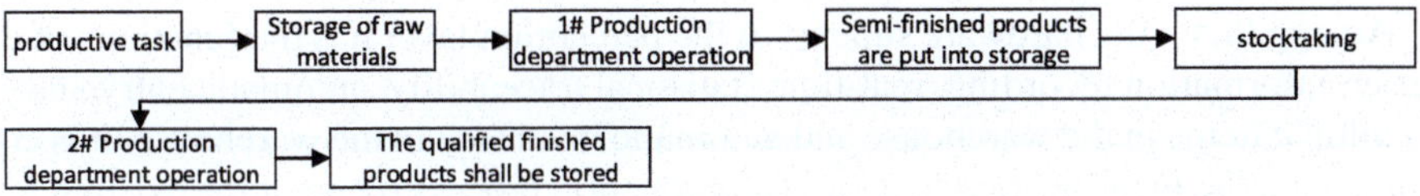

Fig. 2. Flow chart of materials on site in the production workshop

Passive UHF RFID technology has the advantages of non-contact, high accuracy, rapid batch automatic identification, no battery power supply and so on. It can realize wireless data transmission in the nuclear fuel production workshop and optimize the work flow of materials in and out of storage. In the process of designing intelligent production and storage management system, the focus is on the introduction of UHF RFID technology as the main means of information acquisition of the sensing layer. By arranging RFID key application points in each production process point and warehouse location point in the workshop, the alarm logic mechanism is added to ensure the interrelation and operation standard among the subsystems, so as to realize the real-time and fast tracking of materials. Improve the level of material storage management. Based on RFID technology, the data can be transmitted to the antenna at a fixed station through the label equipped with radio frequency chip. After the antenna reads the antenna information, it can be processed to realize the transmission of material information, and finally realize the intelligent control of the storage system.

3 System Design

According to the actual application scenario of fuel assembly production, the design and development of the intelligent warehouse management system based on the Internet of Things technology is mainly composed of pellet and assembly department subsystem and system integration platform. The system integrated platform includes the intelligent warehouse management platform and the interface platform with MES. The system design is shown in Fig. 3.

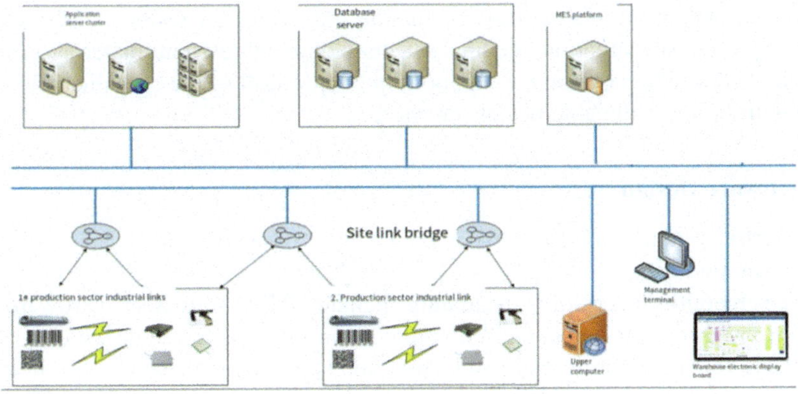

Fig. 3. Overall system design drawing

Among them, the hardware support of the perception layer has the functions of data carrier, information recording, real-time statistical traceability, automatic entry and exit of multiple items in the warehouse and automatic inventory in the warehouse. It is compatible with the identification label of the production process scene of the existing MES system based on the barcode in each link, realizing the integration functions of information collection, item identification, positioning and tracking and data communication. The transmission layer realizes the transmission and exchange of the item identification information collected by the system perception layer and the instruction information of the application layer through the network. The application layer realizes the direct transmission of the acquired data to the data service center, or the first associated packaging upload to the node data server, and then summary to the data service center. Through the high-speed data processing of the database, it supports the application management services of the subsystem of the 1# and 2# production departments, such as item registration, item entry and exit warehouse, alarm warning, information query and so on.

3.1 Hardware Design

From the perspective of the Internet of Things system, the hardware platform architecture designed in this paper basically includes the acquisition layer, the transmission layer and the application layer. Among them, the acquisition layer is the core of the hardware platform of the system, integrating the information perception and acquisition terminal such as readers and the information circulation carrier terminal such as tags to provide hardware support and data sources for the system. The transmission layer takes routers, switches and other devices as the intermediate medium, which mainly realizes the data collected by the collection layer is accessed and transmitted to the upper layer through network. The application layer includes application server cluster, database cluster, intelligent warehouse application service platform, to achieve the application goal of intelligent information management.

Containers and fuel assemblies in the production site are mostly metal components, and metal has an impact on RFID chip signal transmission. In order to judge the impact of metal on RFID transmission, several metal drums were placed on the site to simulate the storage situation of the production site, and the data reading of the label by the antenna was studied. According to the test results, the following test results can be obtained. According to the site situation, the final selection of metal resistant material screw fixed RFID tag UTAG-80, paste on the top, bottom or side wall of the container, can also be bound to the sling with cable ties or screws.

3.2 Software Design

The software system is divided into perception layer, base layer, data layer, service layer and application layer.

The perception layer software mainly includes RFID initializer, hardware-driven interface program and embedded communication control program, which is mainly responsible for the identification, collection and supervision of warehouse items. The basic layer software is mainly network communication software composed of the Internet and AD hoc network, which provides data channel for the communication from the perception layer to the upper layer. The data layer software is mainly database and data access interface program, used to store and process the information from the hardware bottom. Based on the design idea of micro-service architecture, the service layer software encapsulates the core functions into independent services such as scheduling, instruction and data, which can be called by the upper layer to access the data layer and realize the data processing function. The top layer software is mainly the system function business program, providing warehousing information tracking, item identification and positioning, statistical analysis, alarm warning and other applications.

The software architecture of the management system is determined to be B/S architecture, based on HTTP application layer protocol and TCP/IP transport layer protocol, based on J2EE framework construction, through WebService network communication mode for communication, database using MySQL relational database.

The overall network platform design is shown in Fig. 4. Implement the overall network according to the system design drawing.

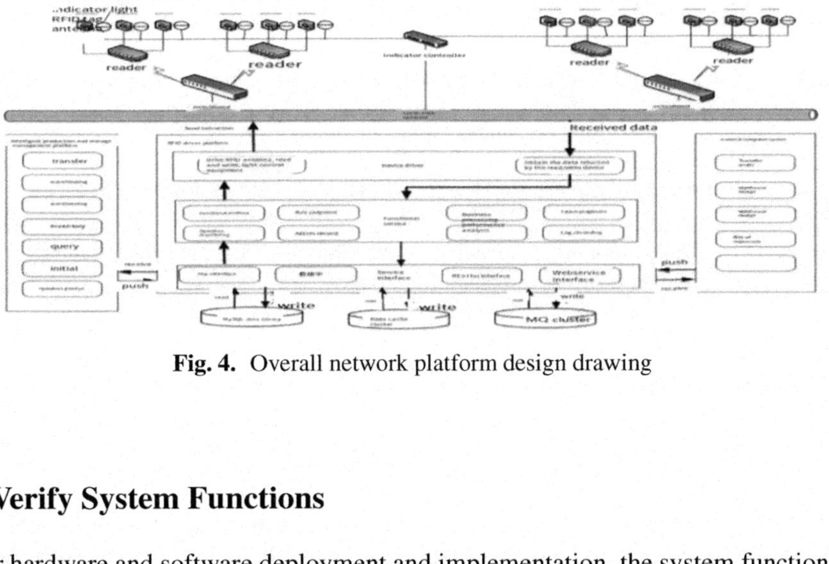

Fig. 4. Overall network platform design drawing

4 Verify System Functions

After hardware and software deployment and implementation, the system functions are verified.

4.1 Hardware Testing

4.1.1 Tag Antenna Test

Test the field signal strength of hardware labels and antennas. The item location program uses Fourier curve fitting simulation to get the sensitivity curve of recognition times. The curve shows that, when the recognition distance is given, the greater the absolute value of slope within a certain range, the greater the change of recognition times, that is, the more sensitive it is. It is of great significance to find out the known transmitting power with the maximum sensitivity to judge the range value in the positive position.

100 groups of samples were tested, and the accuracy rate reached 99%, meeting the requirements of the system.

The container positioning method uses the buried antenna number positioning, antenna number binding library location, label binding container, by adjusting the transmission power in 25 dBm, the 100 container labels for 3 times of accurate identification number test, the reader accurate identification number of labels accumulated to 298, 300, the tag identification accuracy rate meets the 99% requirements. At the same time, the container positioning method based on the label container binding and antenna number positioning ensures the one-to-one correspondence between storage locations, containers and materials. Combined with the simulation test results of the item positioning program proposed in this paper, 100 containers are repeated 3 times warehousing operation, the cumulative number of successful warehousing reaches 298, and the positioning accuracy of the single storage location of product containers can reach 99%.

4.1.2 Database Testing

MySQL database is selected as the system database, which is only used to store structured data, without unstructured data such as pictures and binary files. According to the data in Table 1, if the storage cycle of the system business resources exceeds 10 years, the storage requirement of the database, including other records, needs to reach 21 GB.

Table 1. Database storage requirements calculation description

Contents	Daily amount of data	Annual amount of data (GB)	Maximum usage	Storage cycle (years)	Storage requirement (GB)
Business data	5000	0.870228	60%	10	15
Other records	2000	0.348091	60%	10	6
Total	/	/	/	/	21

By setting 500 virtual users to access the system at the same time, the online test results show that the system can run normally, indicating that the system in this paper can meet the performance index. In addition, the system introduces the measures of cache server and database optimization. Combined with the stress test, the system response time in the actual normal test scenario is about 3 s, which satisfies the performance index that the system response time is less than 5 s in 95% application scenarios.

4.2 Functional Test

4.2.1 Basic Data Configuration Functions

Basic data configuration functions include warehouse management, warehouse location management, read and write device interface verification, station management, container management, reader management, antenna management, material management, dictionary management. Through the system test found that during the period of the interface involved are no problems, front-end, background, database can carry out normal data interaction (Fig. 5).

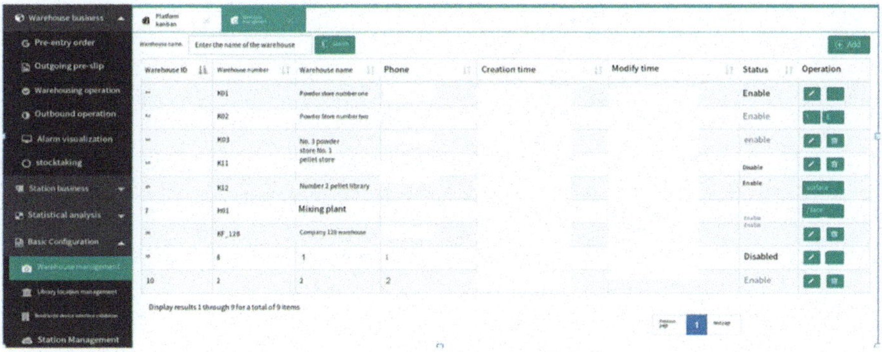

Fig. 5. System test page

4.2.2 In and Out of the Database Function Test

After the completion of basic data configuration, the test results of simulated MES system to create pre-documents and automatic warehousing operation passed.

4.2.3 Querying Statistical Function Tests

This function has realized the statistical test results for the inventory, warehousing, warehousing and inventory taking of a single container after the storage, and the query statistical function test is normal.

5 Conclusion

1. The application of RFID radio frequency technology in multi-metal environment is realized.
2. The system design is stable and reliable, which can realize the control of the material flow process in the production of nuclear fuel assembly, and has positive significance for improving the material management level of nuclear fuel production.

References

1. Guo Yajun, Wang Liang, Wang Caimei. Fundamentals of Internet of Things [M]. Beijing: Tsinghua University Press, 2013.
2. Anas M. Atieh, Hazem Kaylani, Yousef Al-abdallat, et al. Performance improvement of inventory management system processes by an automated warehouse management system[J]. Procedia CIRP on Manufacturing Systems, 2016:568–572.

3. Li M. Design of intelligent drug manufacturing management system based on RFID technology [D]. Chengdu: University of Electronic Science and Technology of China, 2014.
4. Yuan Xingxing. Research on Intelligent Storage Management System of Hydraulic Components Digital Workshop [D]. Hefei: Hefei University of Technology, 2019.

LWR Fission Gas Behavior Modeling Using OpenFOAM Based Fuel Performance Solver OFFBEAT

J. Xie, N. He, Q. Wang, and T. Zhang(✉)

Harbin Engineering University, Harbin, China
{junhao_xie,henan,wangqingyu,tian.zhang}@hrbeu.edu.cn

Abstract. Fission gas release is one of the most important limitations in reactor fuel designs. Low soluble Xenon and Krypton gas generated in fission events may either retained in the fuel pellet, leading to fuel swelling and causes Pellet-Cladding Mechanistic Interaction (PCMI), or releases to fuel gas gap and results in a higher fuel temperature even crack of the cladding due to their relative low conductivities. The behavior of fission gas is mechanistically complicated and hard to model as it is coupled to fuel neutronic and thermal responses. Therefore, the capability of fission gas behavior modeling is one of the top-ranked criteria while evaluating the reliability of a fuel performance code. The OFFBEAT code is a multi-dimensional fuel performance code based on the OpenFOAM@ C++ library. It contains models that are required by LWR fuel analysis and can build from 1.5-D to 3-D. By coupling with the 0-D fission gas solver SCIANTIX, the code can offer mechanistic modeling of fission gas behavior. However, the performance of the coupled code has not yet been fully validated. In this work, preliminary validation of the OFFBEAT code in fission gas modeling is conducted. From open literature, Halden experiment IFA-716 rod 1&6, IFA-677 rod 1&5 are modeled and analyzed. These experiments indicate the capability of the OFFBEAT solver in modeling large grain size. The capability of the code in modeling high burnup fuel is validated against HBEP BK365 rod. The results show that the OFFBEAT code has quite good accuracy in modeling regular grain fission gas release as well as high burnup fuel. However, while dealing with doped large grain fuel, the code overestimates the release at low burnup but underestimates it for relatively high burnup. Nevertheless, the cluster dynamic model applied in SCIANTIX shows a better result than the Speight model normally applied in fuel performances codes.

Keywords: Light water reactor · Fuel modeling · Code validation · Fission gas behavior · OFFBEAT solver

1 Introduction

Light water reactors (LWR) commonly use sintered UO_2 pellets as nuclear fuel. During the reactor operation, The fission of UO_2 produces a certain amount of fission gas, and due to the low solubility of these gas atoms in the fuel matrix, they form bubbles within and between the grains through diffusion, nucleation, and re-solution behaviors. These

J. Liu and Y. Jiao (Eds.): WRFPM 2023, SPPHY 299, pp. 251–260, 2024.
https://doi.org/10.1007/978-981-99-7157-2_26

bubbles not only affect the diffusion rate of gas atoms to the grain boundaries, but also lead to swelling of the fuel pellet, accelerating the pellet-cladding gap closure and pellet-cladding mechanic interaction (PCMI). The intergranular gas bubbles will gradually grow larger through grain boundary diffusion, grain growth, bubble nucleation, bubble growth polymerization, and finally form the intergranular fission gas pathway leading to fission gas release (FGR).

Due to the low thermal conductivity of fission gas, the conductance of the fuel cladding gap is reduced, resulting in a further increase in the core temperature and intensifying the fission gas release, and the swelling of the fuel pellet caused by fission gas will further increase the internal pressure on the cladding, which will seriously affect the physical properties of the cladding and even lead to cladding rupture, seriously endangering the normal operation of the reactor. Thus, the capability and accuracy of FGR modeling is a crucial criterion for fuel performance code.

Motivated by the fuel failure accident in a Swiss BWR and the analysis of its causes, based on OpenFOAM@ C++ library, a multidimensional open-source fuel performance code, the OpenFOAM Fuel Behavior Analysis Tool, or OFFBEAT, has been developed by LRS at EPFL and PSI [1]. The OFFBEAT code can simulate the fission gas behavior by coupling with the 0-D fission gas solver SCIANTIX [2]. The details of this work will be explained in Sect. 2 of this paper.

In order to verify the performance of the OFFBEAT code for modeling and solving the fission gas behavior, as well as the ability to model normal and large grains, the Halden experiment IFA-716 rod 1 and rod 6 [3], IFA-677 rod 1 and rod 5 [4], and HBEP BK365 rod [5] were selected for modeling and the details of the modeling and results will be explained in Sect. 3.

2 FRG Modeling in OFFBEAT Code

The control equations of the OFFBEAT code are discretized on an unstructured mesh using modern finite volume techniques and the solver is fully parallelized by a geometric domain decomposition.

OFFBEAT uses the finite volume method (FVM) for solving, which is a numerical scheme that is very popular in the CFD domain and mainly determined by the development platform (OpenFOAM). Second order solution of the FVM method accelerates the calculation while keep the accuracy within a reasonably good range [1]. The geometric domain of OFFBEAT is divided into several computational meshes composed of continuous polyhedral cells or control bodies [1].

Since the OpenFOAM library on which OFFBEAT development is inherently 3D, the geometry solved by OFFBEAT is likewise 3D [1]. However, OFFBEAT can perform 2D or even 1D simulations by using appropriate methods, mainly by imposing so-called null boundary conditions on opposite sides of the directions to be ignored to force 2D or 1D simulations. The 1.5D model based on OFFBEAT is used for this validation. By adopting a small-angle wedge shape and setting the sides, top and bottom of the wedge as empty boundaries. Although many models have been applied in the OFFBEAT solver, there are limitations so far on modelling MOX fuel, or metallic fuel etc. However, for light water reactors, the solver generally gives satisfying results.

OFFBEAT uses the modular programming [1] (thermal sub-solver, small-strain mechanical sub-solver and solver modules for simulating the fuel envelope gap and the gas evolution inside the rod free volume). During the solving process, OFFBEAT goes through three main loops and operates on different modules separately during the process. When all the iterations are completed, the loop convergence criterion is checked to see if it is met, and the solution result is output when the convergence criterion is satisfied.

OFFBEAT can solve the thermo-mechanical behavior of fuel rods by semi-empirical models and consider the effects of phenomena specific to nuclear, such as swelling or fission gas release, which is the focus of this paper. Fission gas behavior can be solved by coupled SCIANTIX solver. SCIANTIX is an open source zero-dimensional stand-alone computer code developed as a dedicated intermediate scale code linking low-length scale calculations with engineering scale simulations of fuel performance codes [2].

The model used in SCIANTIX considers the basic processes of diffusion of individual gas atoms, namely bubble nucleation, redissolution, and trapping of gas atoms in the bubble. The model is derived from the cluster dynamics formulation, which ultimately models the behavior of fission gas atoms and bubbles within the grain by the following three differential equations [2]:

$$\begin{aligned} \frac{\partial N}{\partial t} &= +\upsilon - \alpha_{\overline{n}} N \\ \frac{\partial m}{\partial t} &= +2\upsilon - \beta_{\overline{n}} N - \alpha_{\overline{n}} m \\ \frac{\partial c_1}{\partial t} &= yF + D\nabla^2 c_1 - 2\upsilon - \beta_{\overline{n}} N + \alpha_{\overline{n}} m \end{aligned} \tag{1}$$

where c_1 is the concentration of single gas atoms, v is the anisotropic bubble nucleation rate, N is the bubble density, m is the concentration of gas in the bubble, α is the anisotropic gas redissolution rate, β is the gas atom capture rate, and F is the fission rate.

For the process of intergranular bubble evolution, SCIANTIX uses the model proposed by Pastore et al. [6], which takes into account the effect of grain boundary microcracking under transient conditions and calculates the fission gas concentration q on the grain boundary by the following equation:

$$\frac{\partial q}{\partial t} = -\left[\frac{3}{r_{gb}} \frac{\alpha}{\alpha + \beta} \frac{\partial}{\partial r}(c_1 + m)\right]_{r=r_b} - R \tag{2}$$

The first term on the right-hand side of this equation is the fission gas flux diffusing from within the grain to the grain boundary, while the second term R represents the fission gas release from the grain boundary.

SCIANTIX is currently designed to be used either as a fission gas behavior module coupled within a particular fuel performance code or as a stand-alone code that solves for fission gas behavior on its own. A number of fuel performance codes that have incorporated SCIANTIX as a fission gas behavior module, including the TRANSURANUS fuel performance code [7], and the OFFBEAT fuel performance code in this paper.

3 Modeling Approach

3.1 Halden IFA-716

The Halden IFA-716 project is an experiment at the Halden Boiling Water Reactor (HBWR) in Norway to study the fission gas release from chromium-doped large grain uranium dioxide fuel [3]. The experimental setup consists of six fuel rods, of which rod 1 and rod 6 are chromium-doped large grain fuels, and both rods have a hole 110 mm deep in the center, which is designed to allow the insertion of thermocouples to measure the fuel centerline temperature. Based on the geometric parameters of the fuel rods in open literature, mesh of rod 1 and rod 6 is built based on OFFBEAT, and in Fig. 1, (a) shows the mesh structure and (b) shows the operational power histories of the two fuel rod models.

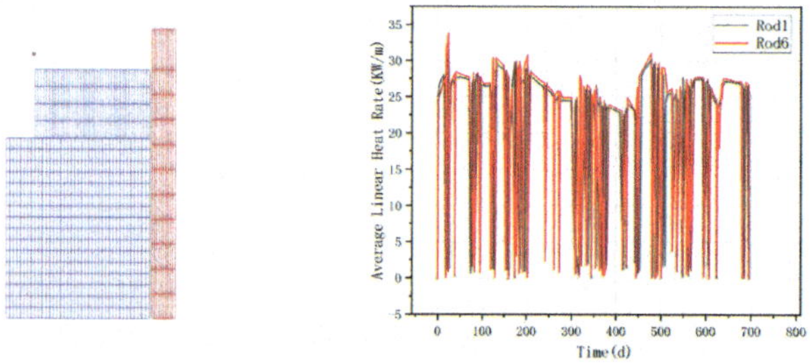

(a) (b)

Fig. 1. Mesh of rod 1&6 (a) and the average linear heat rate history (b).

3.2 Halden IFA-677

The IFA-677 experiment is another project carried out in the Halden reactor. Since the experiments in the cryogenic region were completed by the IFA-716 experiment, the IFA-677 project focused on the thermal behavior, dimensional changes (densification and swelling), FGR, and PCMI of the fuel rods at high initial power [4], the experiment completed six cycles of irradiation under HBWR conditions in September 2007, ultimately achieving an average fuel burnup of about 26.3 MWd/kgU. Two of six rods were doped UO2-containing fuel (rod 1&5). Unlike rod1&6 in IFA-716, rod1&5 in IFA-677 have a borehole of about 110 cm depth in both the upper and lower parts. Based on the geometric information, the mesh built is shown in Fig. 2 (a), and (b) shows the operational power histories of the two fuel rod models.

3.3 HBEP BK365

BK365 rods of the High Burnup Effects Programme (HBEP) Task 3 was to study FGR and fission product distribution at high burnup. The BK365 rods were operated for a long time at lower power in the BR-3 pressurized water reactor, with a final burnup of

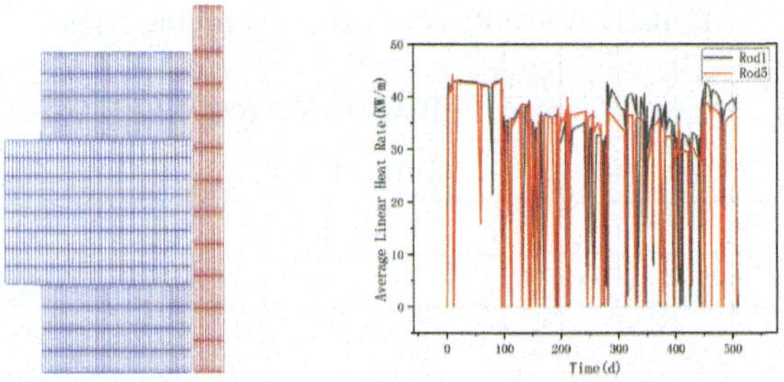

(a) (b)

Fig. 2. Mesh of rod 1&5(a) and the average linear heat rate history(b).

69.4 MWd/kgU. The fuel consists of ring-shaped pellets stacked 1 m long. The mesh built from the geometric information of the BK365 fuel rod is shown in Fig. 3 (a), and (b) shows the operating power history of the fuel rod. A high-burnup structure model is included in the SCIANTIX, but in this case we did not use it because it leads to unreasonable results (over 30% FGR).

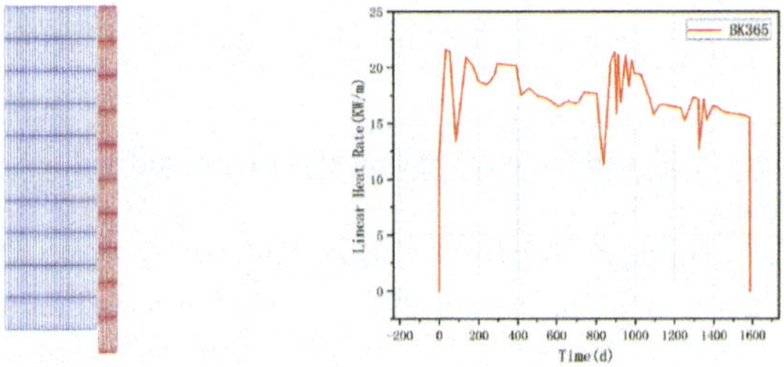

(a) (b)

Fig. 3. Mesh of BK365(a) and the average linear heat rate history(b).

4 Results and Discussions

For the IFA-716 and IFA-677 experiments, the selected benchmark parameters are fuel centerline temperature, FGR, and fuel rod internal pressure. For the HBEP experiments, the selected benchmark parameter is the FGR by the end of irradiation. The diffusion coefficient for the different cases is shown in Table 1.

4.1 IFA-716 rod1&6

Figure 4 (a) and (b) show the variation of rod1 and rod6 fuel center temperature with irradiation time, respectively, where red represents the measured value, green represents the BISON predicted value, and black represents the OFFBEAT predicted value. Comparing the three, it can be seen that compared to BISON, OFFBEAT agrees well with

Table.1. The diffusion coefficient for the cases.

	IFA-716 and 677
Case A	Turnbull [8]
Case B	Matzke [9]

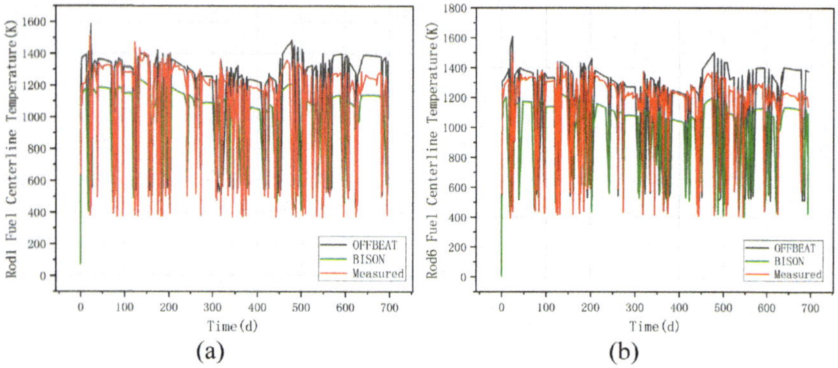

Fig. 4. IFA-716 fuel centerline temperature of rod 1(a) &6(b)

the experimental measured values, except that the measured values are larger at the end of the operating cycle.

The FGR versus burnup for rod 1&6 is shown in Fig. 5. As it's shown in the following Table 1, where Case A uses the Turnbull's diffusion coefficient model [8], which considers the relationship between diffusion coefficient and temperature in different temperature intervals and make the diffusion coefficient more sensitive to temperature. Case B uses the Matzke's model [9], which uses only one equation to show the relationship between diffusion coefficient and temperature.It can be seen that the results of Case A are higher than those of Case B for both rod 1&6, and its results are closer to the measured values.

However, the situation is somewhat different when it comes to the internal pressure of the fuel rods. The variation of the rod internal pressure with time is shown in Fig. 6, and it can be seen that both Case A and Case B have low internal pressures in the early stage relative to Measured and BISON, which is due to the underestimation of the fuel expansion and overestimation of the gasgap between the fuel and the cladding. However, due to the higher FGR, the OFFBEAT's internal pressure at the end of operation period gradually increases to BISON or even higher levels.

4.2 IFA-677 rod1 and rod5

The variation of the fuel center temperature with time for rod 1&5 is shown in Fig. 7, where it can be seen that the predicted values of OFFBEAT are significantly higher than the experimentally measured values for both rod 1&5 at the beginning of the operating

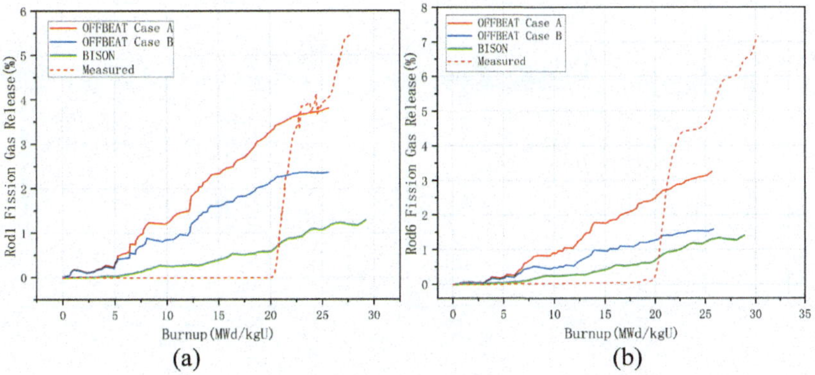

Fig. 5. IFA-716 FGR of rod 1(a) & 6(b)

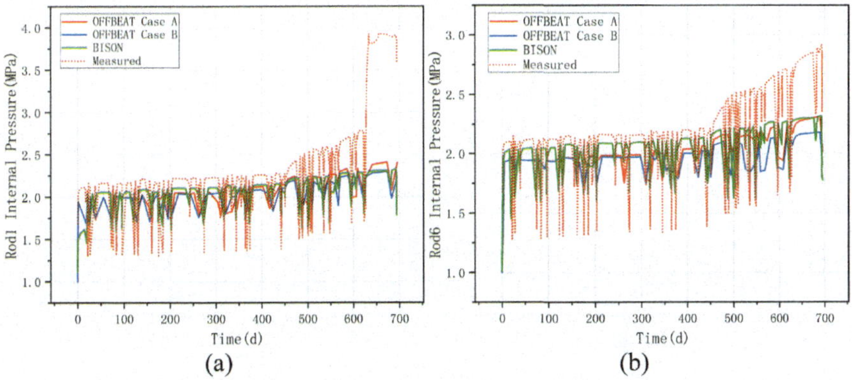

Fig. 6. IFA-716 rod internal pressure of rod 1(a) & 6(b)

cycle, but they match better in the later operating cycles. The reason for this situation is the underestimation of the fuel heat flux, which leads to a high central temperature.

For the FGR, both Case A and Case B exhibit lower results, as shown in Fig. 8. The result fit better in the early part of the operating cycle, but significantly lower than the measured value after the burnup reaches 17 MWd/kgU, which is the same as the BISON before its diffusion coefficient correction.

Compared with FGR, the rod internal pressure for rod 1&5, fit better with the experimental measurements, remaining essentially the same as the measured values in the early stages, but significantly lower at the end of the operating cycle due to the lower fission gas release. The variation of the internal pressure with time for rod1&5 is shown in Fig. 9.

4.3 HBEP BK365

The FGR of BK365 is shown in Fig. 10. The fission gas release of BK365 is shown in Fig. 10. For the intergranular defect diffusion coefficient, OFFBEAT was selected from

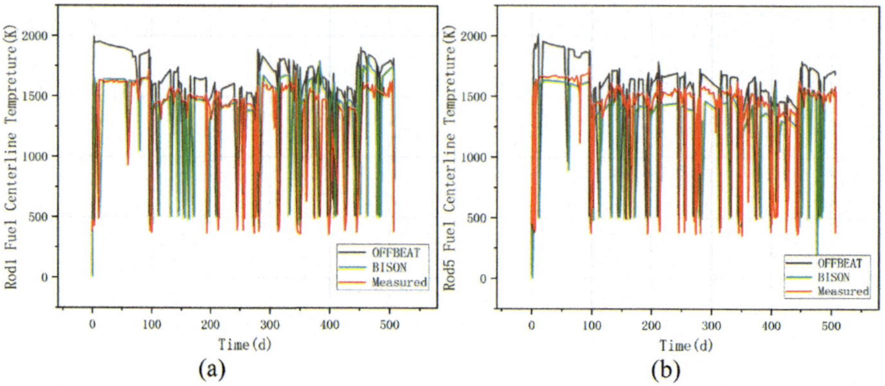

Fig. 7. IFA-677 fuel centerline temperature of rod 1(a) &5(b)

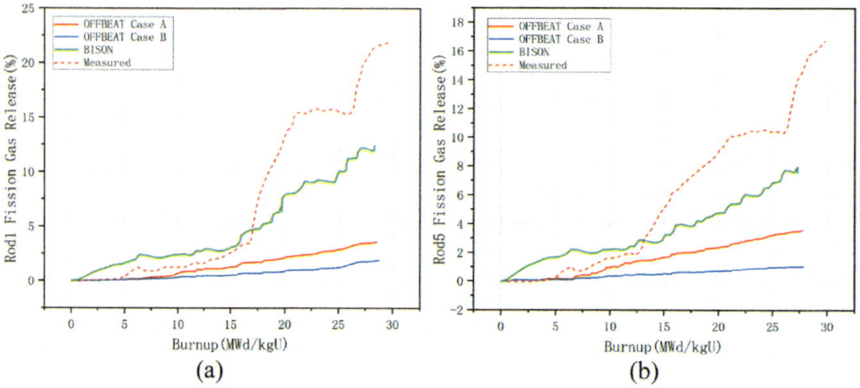

Fig. 8. IFA-677 FGR of rod 1(a) &5(b)

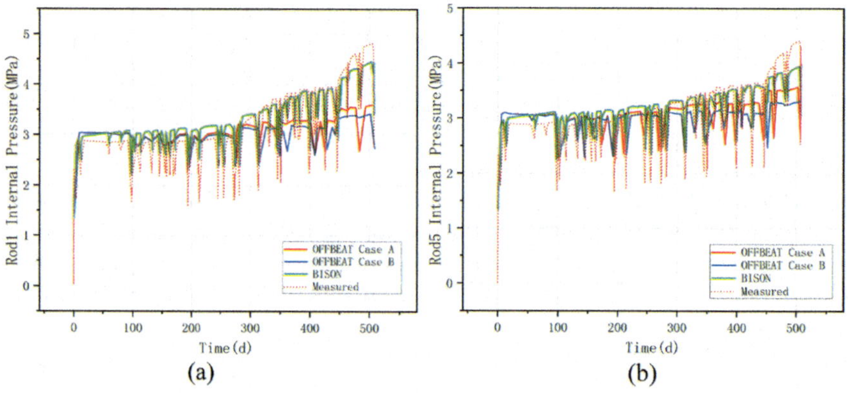

Fig. 9. IFA-677 internal pressure of rod 1(a) &5(b)

the physical model of G. Pastore et al. [6]. It can be seen that OFFBEAT has a higher FGR compared to BISON, and compared to Copernic and Transuranus, OFFBEAT is closest to the measured. As we can see, OFFBEAT has good reliability in predicting FGR at high burnup.

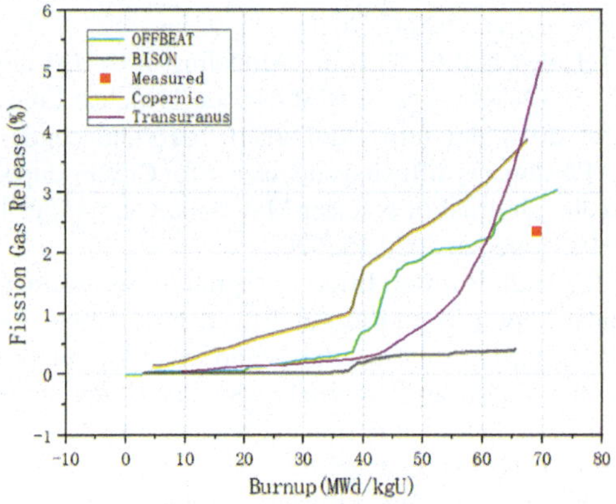

Fig. 10. BK365 FGR

4.4 Discussions

In all IFA cases, offbeats intends to overpredict the fuel center temperature. This may be due to the use of 1.5D modelling and the misestimation of the axial power distribution, which may be inconsistent with the experimental value.

For fission gas release simulation, all cases overestimate FGR at low burnup, which is because we did not use the model for the doped large grain fuels, and this is also the reason for underestimation of FGR in IFA-677 and IFA-716. However, compared to BISON (with a modified fission gas diffusion coefficients for doped UO2), the results obtained by OFFBEAT have a similar accuracy. This indicates that not only the diffusion coefficient matters, but also other parameters have influences on doped large grain UO2 fission gas behaviors.

5 Conclusions

The capability of the OFFBEAT solver fission gas behavior modeling has been preliminarily validated by the development team. In this work, further validation work is done with high burnup case and large grain UO2 experiments.

Overall, the code shows generically good agreement with experimental data in both fuel temperature and FGR. For the case of high burnup fuel with regular grain size, result from the OFFBEAT solver is closer to the measurement value than other codes. For fuel rods with doped UO2, OFFBEAT gives reasonable results but also shows that to have more accurate results, new models with modified fission gas behavior coefficient is necessary.

References

1. A. Scolaro, I. Clifford, C. Fiorina, and A. Pautz. The offbeat multi-dimensional fuel behavior solver. Nuclear Engineering and Design, 358:110416, 2020.
2. D. Pizzocri, T. Barani, and L. Luzzi, "SCIANTIX: A new open source multi-scale code for fission gas behaviour modelling designed for nuclear fuel performance codes," Journal of Nuclear Materials, vol. 532, p. 152042, 2020.
3. Y. Che, G. Pastore, J. Hales, and K. Shirvan, "Modeling of Cr2O3-doped UO2 as a near-term accident tolerant fuel for LWRs using the BISON code," Nuclear Engineering and Design, vol. 337, pp. 271–278, Oct. 2018, doi: https://doi.org/10.1016/j.nucengdes.2018.07.015.
4. M. W. Cooper et al., "Fission gas diffusion and release for Cr2O3-doped UO2: From the atomic to the engineering scale," Journal of Nuclear Materials, vol. 545, p. 152590, Mar. 2021, doi: https://doi.org/10.1016/j.jnucmat.2020.152590.
5. HBEP. Summary of the High Burn-up Effects Programme as abstracted from the Programme Final Report. Technical Report, High Burn-up Effects Programme, December 2002.
6. Pastore G, Swiler L P, Hales J D, et al. Uncertainty and sensitivity analysis of fission gas behavior in engineering-scale fuel modeling[J]. Journal of Nuclear Materials, 2015, 456: 398-408.
7. Gonzalez J, Ševecek M. Modelling of fission gas release in UO2 doped fuel using transuranus code. Acta Polytechnica CTU Proceedings, 37: 24-30, 2022.
8. Turnbull J A, White R J, Wise C. The diffusion coefficient for fission gas atoms in uranium dioxide[R]. 1989.
9. Matzke H. Gas release mechanisms in UO2—a critical review. Radiation Effects, 1980, 53(3-4): 219-242.

Development of Fuel Performance Analysis Code for Liquid Metal Coole Fast Reactor Based on MOOSE Platform

Shihao Shao(✉), Zhouyu Liu, Xiaobei Xu, Yufan Zong, Liangzhi Cao, and Hongchun Wu

Xi'an Jiaotong University, Shaanxi, China
1791445402@qq.com, {zhouyuliu,caolz,hongchun}@mail.xjtu.edu.cn, 1309940619@qq.com

Abstract. The liquid metal cooled fast reactor is of great strategic significance for the innovative development of nuclear energy in China, and the development of fuel performance analysis programs suitable for it is of great significance for the design and safety analysis of fast reactors. This article is based on the multi physics coupling platform MOOSE and finite element method, and independently developed the fuel performance analysis program LoongCALF for liquid metal cooled fast reactors. The program adopts a 1.5-dimensional modeling method to perform multi-physical field coupling calculations on the fuel rods, which uses a single-layer thermal-mechanical module and a global gas behavior module. The program is suitable for fast reactor fuel elements with MOX and HT-9 cladding. For the calculation example of a complete fuel rod whose pellet material is MOX and cladding material is HT-9, LoongCALF program and China Institute of Atomic Energy program are respectively used to model and calculate the calculation example, and the calculation results are compared. The results indicate that the LoongCALF program can accurately simulate the spatial distribution and temporal variation of key physical parameters inside fuel elements under steady-state operating conditions of liquid metal cooled fast reactors.

Keywords: Liquid metal cooled fast reactor · MOOSE · Finite element method · Fuel performance analysis

1 Introduction

The advantages of high uranium resource utilization rate and hard neutron spectrum make fast reactor technology an important step in China's "three step" strategy for nuclear power. The liquid metal cooled fast reactor, as one of the highly potential reactor types in the fourth generation nuclear energy system, has significant strategic significance for the innovative development of nuclear energy. Its research and development design is a major national demand in both military and civilian aspects [1].

Fuel element is one of the most basic components of nuclear reactor, and its performance analysis is one of the most challenging tasks in reactor research and development.

J. Liu and Y. Jiao (Eds.): WRFPM 2023, SPPHY 299, pp. 261–270, 2024.
https://doi.org/10.1007/978-981-99-7157-2_27

Fuel element performance analysis is used to predict the evolution of fuel behavior and key parameters, which involves closely coupled multiple physical phenomena, including: heat conduction, solid mechanics, gap heat transfer, expansion, densification, creep, fission gas release, etc. Fuel element performance analysis is one of the important topics to guide fuel element design, predict the service life of fuel elements, and ensure the safe operation of reactors. A lot of research has been carried out on the fuel performance analysis program for pressurized water reactor in China. Compared with pressurized water reactor, the flux level of fast reactor is high, the operating temperature is high, and the core material structure is more seriously affected by radiation than pressurized water reactor [2]. Developing a fuel performance analysis program suitable for fast reactor environment is of great significance for the design and safety analysis of fast reactor.

Since the 20th century, many countries have developed a series of performance analysis procedures for liquid metal cooled fast reactor fuel elements, such as FEAST-OXIDE [3], developed by the Massachusetts Institute of Technology (MIT), is suitable for oxide fuels and metal fuels, and is considered to be the most comprehensive fast reactor fuel performance analysis procedures for considering physical phenomena, GERMINAL [4], developed by the French Commission on Atomic Energy and Alternative Energy (CEA), and FEMAXI-FBR [5], developed by the Japan Atomic Energy Commission, etc. In recent years, China has also developed some fast reactor fuel element performance analysis software, such as LIFEANLS [6] developed by China Academy of Atomic Energy, the first fast reactor fuel performance analysis program in China, and subsequent FIBER program; and KMC-FUEL developed by University of Science and Technology of China.

To achieve performance analysis of liquid metal cooled fast reactor fuel elements. This article is based on the multi physics coupling platform MOOSE, and uses numerical methods to simulate the evolution of the long-term service performance of fuel elements under steady-state conditions, and develops the fuel element performance analysis program LoongCALF. This can provide reference for the design, service life prediction, and safety evaluation of LMFR fuel elements. Through modeling and calculation of the design example, the LoongCALF program is compared with the China Institute of Atomic Energy program Fiber Oxide, and the LoongCALF program is fully verified.

2 Method

2.1 Program Introduction

The fast reactor fuel performance analysis software developed by the program mainly focuses on liquid metal cooled fast reactor fuel components. The commonly used core material for liquid metal cooled fast reactor fuel elements is MOX fuel, and the commonly used cladding materials are HT-9, etc. The program solver is written in C++ language, and the program adopts modular design with separate material modules. It can not only calculate fast reactor core components with clad materials of and HT-9, but also has the architecture conditions to extend to advanced clad materials such as 316 (Ti) SS and T92, as well as U-Zr and U-Pu-Zr metal fuels.

2.2 Mathematical Models

Governing Equation. The temperature of the fuel element is solved through the heat conduction equation:

$$\rho C_p \frac{\partial T}{\partial t} + \nabla(-k\nabla T) - S = 0 \tag{1}$$

where: ρ—density (kg/m^3); C_p—specific heat (J/(kg°C)); T—temperature (K); k—heat conductivity; S—heat source.

The stress-strain state of the fuel element is solved by the solid mechanical balance equation:

$$\nabla \cdot \boldsymbol{\sigma} + \rho \mathbf{f} = 0 \tag{2}$$

where: $\boldsymbol{\sigma}$—stress; ρ—density (kg/m^3); $\mathbf{f}$—Volume force per unit mass.

Fluid Heat Exchange Model. The program adopts the single-channel assumption that the fuel element exchanges heat with the fluid channel surrounding it, and that there is no material and energy communication between the channel and the adjacent channel. The coolant temperature control equation is written as follows [2]:

$$\frac{\partial(\rho c\,\mathrm{T})}{\partial \mathrm{t}} + \frac{\partial}{\partial \mathrm{z}}(\rho c\,\mathrm{T}w) = q_{v,ch} \tag{3}$$

where: ρ—Cooling agent density (kg/m^3); c—Coolant specific heat capacity (J/(kg·k)); T—coolant temperature (K); w—Axial flow rate of the coolant (m/s); $q_{v,ch}$—Volume heat source in the coolant flow channel (W/m^3), Including a small amount of fission energy deposited from the γ radiation zone in the coolant area and the heat transferred by the cladding to the coolant (Introduction in the form of a bulk heat source).

Model of the Fission-Gas Behavior. The common method to simulate the release of fission gas in fuel is two stages: in the first stage, the fission gas is generated in the grain and spreads to the grain boundary; In the second stage, the bubbles of fission gas grow and merge at the grain boundary, and finally connect with the external gas environment, so that the fission gas stored in the grain boundary is released to the external gas environment [8].

The gas diffusion process in the first stage is described by the following equation:

$$\frac{\partial C_{\mathrm{g}}}{\partial t} = D_{\mathrm{eff}} \frac{1}{r^2} \frac{\partial}{\partial r}\left(r^2 \frac{\partial C_{\mathrm{g}}}{\partial r}\right) + \beta \tag{4}$$

where: C_{g}—the crystal fission gas concentration; t—time; D_{eff}—the effective gas diffusion coefficient; r—Coordinates of the one-dimensional spherical coordinate system; β—Gas production rate.

In the second stage, when the grain boundary gas is connected with the outside world (i.e., the grain boundary gas reaches saturation), its surface concentration on the grain boundary is calculated by the following equation:

$$N_{\mathrm{s}} = \frac{4 r_{\mathrm{b}} f(\theta_{\mathrm{dihe}}) V_{\mathrm{c}}}{3 k_{\mathrm{B}} T \sin^2\theta_{\mathrm{dihe}}}\left(\frac{2\gamma}{r_{\mathrm{b}}} + P_{\mathrm{ext}}\right) \tag{5}$$

$$f(\theta_{\text{dihe}}) = 1 - 1.5\cos\theta_{\text{dihe}} + 0.5\cos^3\theta_{\text{dihe}} \tag{6}$$

where, N_s—The saturated surface concentration of the fission gas at the grain boundary, m^{-2}; r_b—Fission gas bubble radius at the grain boundary, m; θ_{dihe}—Dihedral angle of the grain boundary bubble; V_c—Proportion of the grain boundary covered by bubbles at saturation; k_B—Boltzmann constant, 1.3806×10^{-23}J/K; γ—Air bubble surface tension, J/m^2; P_{ext}—External hydrostatic pressure, Pa.

When the fission gas at the grain boundary exceeds the saturation value, the excess gas will be released into the fuel element gas cavity.

Material Behavior and the Physical Property Model.

Heat Conductivity. MOX fuel thermal conductivity [7]:

$$k = F_1 F_2 F_3 F_4 k_o \tag{7}$$

where: k—The MOX fuel thermal conductivity; O/M—Oxygen-to-metal atom ratio; T—temperature (K); β—burn-up (at %); P—Fuel porosity; α—The pore coefficient (1.5 or 2.5). According to the literature, take $\alpha = 2.5$.

The thermal conductivity of HT-9 capadding is calculated by the following Eq. (2):

$$k = \begin{cases} 17.622 + 2.428 \times 10^{-2}T - 1.696 \times 10^{-5}T^2 & T < 1030K \\ 12.027 + 1.218 \times 10^{-2}T & T \geq 1030K \end{cases} \tag{8}$$

where: k—Thermal conductivity of the HT-9 cladding material (Wm^{-1}K^{-1}); T—temperature (K).

Thermal Expansion. The line thermal expansion strain model for MOX fuel is [7]:

$$\varepsilon_i^{th} = \frac{\Delta L}{L_0} = K_1 T - K_2 + K_3 \exp\left(-\frac{E_D}{kT}\right) \tag{9}$$

where: T—temperature (K); E_D—Defect generation energy (J); k— boltzmann constant $\left(1.38 \times 10^{-23}\right)$ (J/K); $K_{1,2,3}$—Model coefficient.

Thermal expansion of the HT-9 clad [2]:

$$\alpha\Delta T = -0.2191 + 5.678 \times 10^{-4}T + 8.111 \times 10^{-7}T^2 - 2.576 \times 10^{-10}T^3 \tag{10}$$

where: $\alpha\Delta T$—Linear thermal expansion strain (%); T—temperature (K).

Relocation. The MOX fuel diameter migration is calculated by the following equation:

$$DR = 0.111GD_{ci} - 45 \tag{11}$$

where: DR—Diameter migration (μm); G—Diameter gap size (μm); D_{ci}—Clad inner diameter (mm).

In order to ensure that the outer diameter of the pellet is consistent with the calculated value of the relocation model, the central cavity of the pellet also produces the relocation strain. The relocation strain is generated only in the R-θ plane, and the line strain is:

$$\varepsilon_l = \frac{DR}{2R_{pout}} \tag{12}$$

where: ε_l—Linear strain in the R-θ plane; R_{pout}—Outer radius of the pellet (μm).

At the same time, the program considers a pore migration model and a heavy structure model to simulate the formation of fuel center pores.

Swelling. Most of the fission products produced by the MOX fuel fission reaction are solid, and a small part are gaseous. These products remain in the fuel, causing swelling of the fuel.

The swelling caused by the solid fission product is called the solid swelling and can be calculated by the following equation [9]:

$$\Delta\varepsilon_{sw-s} = 5.577 \times 10^{-5}\rho\Delta Bu \tag{13}$$

where: $\Delta\varepsilon_{sw-s}$—Solid-state swelling body strain increment; ρ— density; thickness; density; consistency (kg/m^3); ΔBu—Fuel consumption increment (fissions/atoms-U).

The swelling caused by the gaseous fission product is called the gaseous swelling and can be calculated by the following equation [9]:

$$\begin{aligned}\Delta\varepsilon_{sw-g} &= 1.96 \times 10^{-31}\rho\Delta Bu(2800 - T)^{11.73} \\ &\times \exp(-0.0162(2800 - T) - 0.0178\rho Bu)\end{aligned} \tag{14}$$

where: $\Delta\varepsilon_{sw-g}$—Gaseous swollen body strain increment; ρ—density (kg/m^3); ΔBu—Fuel consumption increment (fissions/atoms-U); T—temperature (K).

Densification. During the service of the fuel core, with the accumulation of fission products and the increase of temperature, the pores in the fuel continue to shrink, leading to the overall contraction of the core. This phenomenon is called compaction.

The MOX fuel density fraction [9] is calculated by the following equation:

$$\rho = \rho_o + (0.96 - \rho_o)\left(1 - \exp\left(-\frac{Bu}{0.6}\right)\right) \tag{15}$$

where: ρ—Fuel density fraction; ρ_o—Initial fuel density fraction; Bu: burn-up (at %).

Strain due to densification:

$$\varepsilon_V = \frac{\rho_o}{\rho} - 1 \tag{16}$$

Creep. At high temperature, the fuel core and the clad can slowly undergo irreversible deformation, even if the stress is less than the yield limit of the material. This phenomenon is called creep.

The MOX fuel thermal creep is calculated by the following equation [9]:

$$\dot{\varepsilon} = \frac{3.23 \times 10^9}{d^2}\sigma\exp\left(-\frac{92500}{RT}\right) + 3.24 \times 10^6\sigma^{44}\exp\left(-\frac{136800}{RT}\right) \tag{17}$$

where: $\dot{\varepsilon}$—Fuel creep rate $\left(h^{-1}\right)$; d—grain size (μm); R—universal gas constant, 1.987; σ—stress (MPa).

Irradiation creep:

$$\dot{\varepsilon}_f = 1.78 \times 10^{-26}\sigma\phi \tag{18}$$

where: $\dot{\varepsilon}_f$—Fuel irradiation creep rate $\left(h^{-1}\right)$; σ—stress (MPa); ϕ—Fission rate $\left(\text{Fission/m}^3/\text{s}\right)$.

Thermal creep of HT-9 clad [2]:

$$\dot{\varepsilon}_{TS} = C_5 \exp\left(-\frac{Q_4}{RT}\right)\overline{\sigma}^2 + C_6 \exp\left(-\frac{Q_5}{RT}\right)\overline{\sigma}^s \tag{19}$$

where: $\dot{\varepsilon}_{TS}$—Effective thermal creep strain (%/s); T—temperature (K); $\overline{\sigma}$—equivalent effective stress (MPa); model parameter $C_5 = 1.17\times10^9$; $C_6 = 8.33\times10^9$; $Q_4 = 83142$; $Q_5 = 108276$; R = 1.987.

Irradiated creep of the HT-9 clad:

$$\dot{\varepsilon}_{irr}^{cr} = \left[B_3 + B_4 \exp\left(-\frac{73000}{RT}\right)\right]\phi\sigma_{eff}^{1.3} \tag{20}$$

where: $\dot{\varepsilon}_{irr}^{cr}$—Irradiated creep strain rate (%/s); ϕ—Neutron injection rate $\left(\times10^{22} n/\left(\text{cm}^2 \cdot \text{s}\right)\right)$; model parameter $B_3 = 1.83\times10^{-4}$; $B_4 = 2.59\times10^{14}$; R = 1.987.

3 Result

In order to test the calculation accuracy and applicability of the performance analysis program LoongCALF developed in this paper, this section will use the MOX example for numerical simulation and calculation using the LoongCALF program and Fiber-Oxide program, which are fuel rods with MOX and cladding material HT-9 [10]. Results include the fuel center temperature, gap width, plenum pressure and fission gas release fraction.

3.1 Example Parameters

The examples are fuel rods of MOX and HT-9. The specific structure and operating parameters are shown in Tables 1, 2 and 3.

Table 1. Structural parameters of fuel elements

Parameter	Value
Upper plenum length, mm	48
Lower plenum length, mm	450
Fuel stack height, mm	450
Pellet hole diameter, mm	1.6
Pellet diameter, mm	5.2
Inner diameter of cladding, mm	5.4
Outer diameter of cladding, mm	6
Fuel density, g/cm3	10.5

Table 2. Main parameters of power distribution and cladding temperature

Number	Axial height, m (relative height of the lower end face of the active zone)	Power normalized distribution	Outer surface temperature of the cladding, °C
1	0.0225	0.781	388.968
2	0.0675	0.863	422.099
3	0.1125	0.933	454.274
4	0.1575	0.98	487.69
5	0.2025	1	524.4
6	0.2475	0.998	560.201
7	0.2925	0.963	594.928
8	0.3375	0.899	624.573
9	0.3825	0.813	650.839
10	0.4275	0.702	674.741

Table 3. Operating parameters

Parameter	Value
Core irradiation time, days	240
Maximum line power at the beginning of service life, kw/m	40.5
Maximum line power at the end of life, kw/m	37.8

3.2 Interpretation of Result

The parameters such as center line temperature, gap width, cladding outer surface displacement, air cavity pressure and fission gas release share calculated by LoongCALF program are compared with the calculation results of Fiber-Oxide program. In this paper, only the parameters of axial segment 5 of maximum power change with the time point, the comparison results are shown in the figure below, and the conclusions of other power axial segments are the same.

As can be seen from Fig. 1, the fuel center temperature experienced the process of rise-down-up-down-down-rise, which is caused by the fuel compaction, swelling, deterioration of thermal conductivity, cladding creep and fission gas release. Judging from the relative error line, the maximum relative deviation between the fuel center temperature obtained by LoongCALF program and the Fiber-Oxide program is about 10%, which is within the acceptable range.

As can be seen from Fig. 2, the LoongCALF program and the Fiber-Oxide program predict the contact time of the core coating; however, due to the physical quantities affecting the coating gap, such as thermal expansion, swelling, compaction and creep

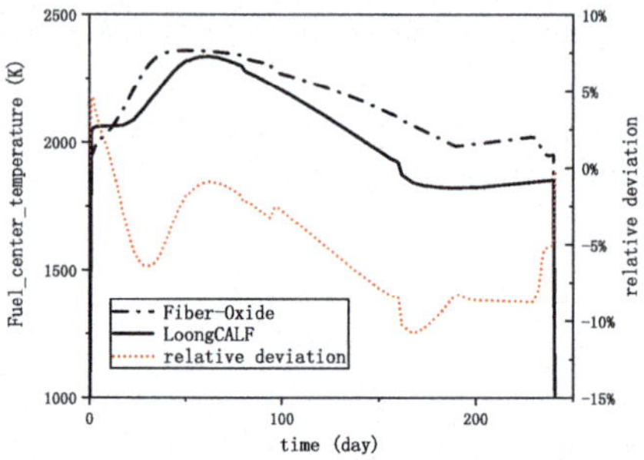

Fig. 1. Change in fuel center temperature in axial section 5 over time

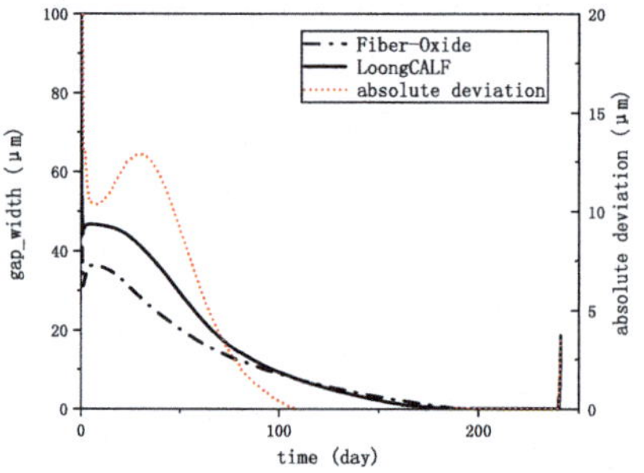

Fig. 2. Change in gap width of axial section 5 over time

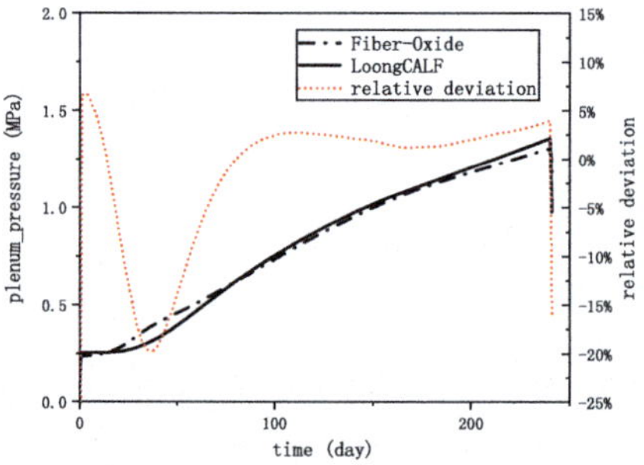

Fig. 3. Change in plenum pressure over time

will produce strain on the coating of the core, it is difficult to achieve the complete agreement between different procedures, and the overall compliance is good.

As can be seen from Fig. 3, the results of the plenum pressure are better, but the results is slightly different due to the different time of fission gas release of the two programs.

As can be seen from Fig. 4, the combustion consumption of fission gas release calculated by the Fiber-Oxide program is significantly earlier than the LoongCALF program, considering the relatively large uncertainty and the differences between different models, the fission gas release calculation is the same, which is acceptable.

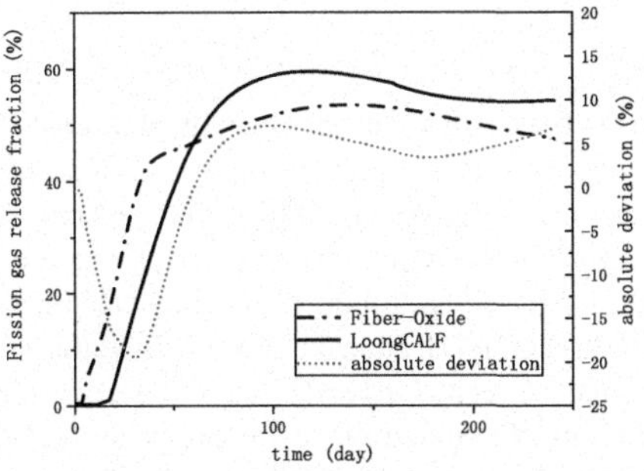

Fig. 4. Changes in fission gas release fraction over time

4 Conclusion

The LoongCALF program for analyzing the performance of fast reactor fuel elements developed in this article is fully functional and has fuel performance analysis functions for fast reactor MOX fuel, including thermal conductivity, gap heat transfer, heavy structure, fission product migration, fission gas release, irradiation swelling, irradiation creep, and other calculation functions. A numerical example was designed to validate the program, and the results showed that: (1) for the validation example, LoongCALF and Fiber Oxide were in good agreement with key parameters such as fuel centerline temperature, gap width, plenum pressure, and fission gas release fraction. The maximum relative deviation of fuel centerline temperature was 11%, the maximum absolute deviation of gap width was 13 (μm), the maximum relative deviation of plenum pressure was 20%, and the maximum absolute deviation of fission gas release fraction was 19%; (2) The maximum deviation displayed mostly occurs on the first day, which is the first time it reaches full power operation. This phenomenon is caused by differences in the processing of repositioning and fracture models between the two programs, and the deviation of various physical quantities is basically maintained within 10% during subsequent operation; (3) At the same time, due to the use of a large number of empirical models for fuel performance analysis, there are inevitably differences between different models. The above differences shown in the calculation results of the LongCALF program and the Fiber Oxide program are within an acceptable range.

The numeric results prove that the LoongCALF program can accurately simulate the spatial distribution and change of the temperature, stress, strain and fission gas release of the fuel element under the steady state condition of the liquid metal cooling fast reactor. In the future, the fast reactor fuel performance analysis program will be fully verified to further verify the accuracy and rationality of the procedure, and the advantages of the finite element method retain the possibility of the program to higher dimension calculation.

References

1. Hongchun W, Hongyi Y, Liangzhi C, et al. Current Status and Prospects of Key Analysis Software for Metal Cooled Fast Reactor[J]. Modern Applied Physics, 2021, 12(01): 4–15.
2. Feipeng Q. Development and Application of Performance Analysis Program for Liquid Metal Cooled Fast Reactor Fuel Elements[D]. University of Science and Technology of China, 2018.
3. Karahan A, Buongiorno J. Modeling of thermo-mechanical and irradiation behavior of mixed oxide fuel for sodium fast reactors[J]. Journal of Nuclear Materials, 2010, 396(2-3): 272–282.
4. Lainet M, Michel B, Dumas J C, et al. GERMINAL, a fuel performance code of the PLEIADES platform to simulate the in-pile behaviour of mixed oxide fuel pins for sodium-cooled fast reactors[J]. Journal of Nuclear Materials, 2019, 516: 30–53.
5. Tsuboi Y, Endo H, Ishizu T, et al. Analysis of fuel pin behavior under slow-ramp type transient overpower condition by using the fuel performance evaluation code 'FEMAXI-FBR'[J]. Journal of Nuclear Materials, 2012, 49(4): 408–424.
6. Chen H. Fast Reactor Fuel Rod Life Analysis Program[M]. Beijing: China Institute of Atomic Energy, 1999
7. A. Karahan. Modelling of thermo-mechanical and irradiation behavior of metallic and oxide fuels for sodium fast reactors[D]. Massachusetts Institute of Technology, 2007.
8. Forsberg K, Massih A R. Diffusion theory of fission gas migration in irradiated nuclear fuel UO2[J]. Journal of Nuclear Materials, 1985, 135(2-3): 140–148.
9. J. D. Hales, R. L. Williamson, S. R. Novascone, et al. BISON Theory Manual: the Equations behind Nuclear Fuel Analysis (BISON Release 1.3)[R]. 2016, INL/EXT-13–29930 Rev. 3.
10. Fuhai G. The development of a computer code for predicting fast reactor oxide fuel element thermal and mechanical behaviour (FIBER-Oxide)[C]. FR17: International Conference on Fast Reactors and Related Fuel Cycles, 2017.

PWR Fuel Cycle Increased Enrichment, Combination of Burnable Absorbers

V. Kuzin(✉)

TVEL, JSC, Moscow, Russian Federation
tvel@info.ru

Abstract. Assessment of the opportunity for development of various durability fuel cycles for PWR-type reactors using the increased enrichment fuel TVS-K. The combination of gadolinium oxide in U-Gd fuel rods and erbium oxide evenly distributed along regular fuel rods is used as a burnable absorber. The paper contains the calculation results for a great number of fuel cycles in the PWR-type reactor that differ in terms of the number of fresh fuel assemblies and average fuel enrichments, the analysis of such fuel efficiency indicators as durability of the fuel inventory operation at full power and burnup of the unloaded fuel. The estimations were obtained using the approximation models, which parameters had been identified based on three-dimensional calculations for a great number of fuel cycles in the PWR-type reactor. Using the combination of burnable absorbers enables to produce the fuel cycles of various durability. Varying the number of U-Gd fuel rods and content of erbium in regular fuel rods facilitates the control of such fuel cycle parameters as power distribution, reactivity coefficients, and critical concentration of boron acid. The completed studies resulted in characterization of properties for 18- and 24-months fuel cycles.

Keywords: TVS-K · PWR · 24-month fuel cycle · Combination of burnable absorbers

1 Introduction

This scientific article contains the results of a computational study of the neutronic characteristics of the fuel cycles of a 3-loop PWR reactor with a thermal power of 2895 MW. The fuel used is TVS-K. The opportunity of using of enhanced enrichment fuel (up to 6.45% by ^{235}U) is considered. The characteristics of fuel cycles when using a combination of burnable absorbers (Gd_2O_3 and Er_2O_3) are determined.

2 Input Data for Calculations

2.1 Characteristics of the Fuel Assembly

The LFA TVS-K is designed for pilot operation in Unit 3 of Ringhals-3 NPP. TVS-K is an array of 17 × 17 fuel rods with a lattice pitch of 12.6 mm, a fuel rod diameter of 9.5 mm and a fuel column height of 12 feet. Table 1 shows the main characteristics of TVS-K.

J. Liu and Y. Jiao (Eds.): WRFPM 2023, SPPHY 299, pp. 271–279, 2024.
https://doi.org/10.1007/978-981-99-7157-2_28

Table 1. Main parameters of TVS-K

Parameter	Value
Fuel rod layout	17 × 17 square
Fuel assembly pitch, mm	215.04
Core height in cold zero power, cm	366
Fuel rods in assembly	264
Lattice pitch, mm	12.6
Outer diameter of cladding, mm	9.5
Fuel pellet diameter, mm	8.19
Number of guide/centre thimbles	24/1
Outer diameter of GT and CT, mm	12.6

2.2 Characteristics of the Reactor Plant

As the reactor plant under study, the standard 3-loop PWR reactor with power 2895 MW was selected. The characteristics of the reactor plant are given in Table 2.

Table 2. Main parameters of reactor plant

Parameter	Value
Reactor type	PWR 157 FA in square grid
Thermal power (100%)	2895 MW
Coolant pressure	15.8 MPa
Coolant temperature at the core inlet (HFP)	282.0 °C
Average coolant temperature at the core	302.2 °C
Core coolant flow rate	71370 m^3/h

3 Characteristics of TVS-K with Various Types of Burnable Absorbers

3.1 Preparing a Constant Library

Based on the data presented above, a set of fuel assemblies was formed that differ by:

- fuel enrichment (4.95; 5.45; 5.95; 6.45 by ^{235}U);
- erbium oxide content (0; 0.4; 0.8; 1.2% Er_2O_3);
- number of U-Gd fuel rods (0; 4; 8; 12; 16; 20; 24, 28, 32 pcs.)

Radial and axial shaping were not used. U-Gd Fuel rods with ^{235}U enrichment of 3.6% and Gd_2O_3 content 8 wt.% were used in the analysis. Equilibrium fuel cycles of various durations were subsequently made up using the obtained set of fuel assemblies.

3.2 Neutronic Characteristics of TVS-K

Burnup dependencies were obtained for the entire set of FAs made up, such as:

- multiplication factor in an infinite medium;
- power distribution;
- burnup distribution;
- distribution of heavy isotopes and fission products.

Figure 1 shows the multiplication factor-burnup dependences for fuel assemblies with 5.95% ^{235}U enrichment and various combinations of burnable absorbers. The combination of burnable absorbers allows to avoid the increase peak pin power (the relative power of the fuel pin when performing two-dimensional calculations), which inevitably occurs when the number of U-Gd fuel rods in a fuel assembly increases (Fig. 2). At the same time, reducing the number of U-Gd fuel rods leads to a slight increase in the average fuel enrichment, which in turn makes it possible to compensate for some of the losses on unburned erbium oxide.

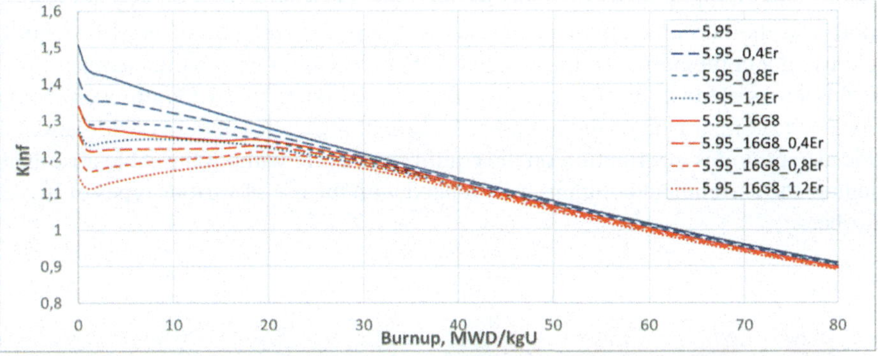

Fig. 1. Dependence of the multiplication factor on burnup for various types of fuel assemblies

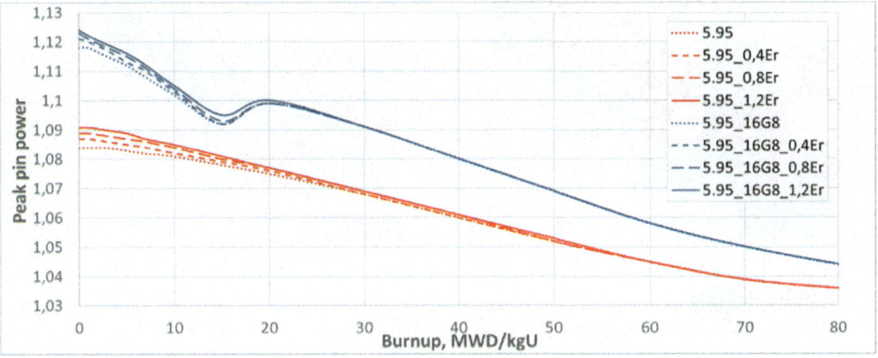

Fig. 2. Dependence of maximum power on burnup for various types of fuel assemblies

4 Characteristics of Equilibrium Fuel Cycles

4.1 18-Month Fuel Cycle

Equilibrium 18-month cycles were generated using the obtained library. Detailed neutronic calculations of 3 variants of equilibrium cycles with 60/61 fresh fuel assemblies were performed to assess the influence of burnable absorbers composition on 18-month fuel cycle characteristics:

- burnable absorber Gd_2O_3 (number of U-Gd fuel rods in fuel assembly is from 8 to 24);
- burnable absorber Er_2O_3 (Erbium oxide content in fuel elements from 0.4 to 1.2%);
- a combination of burnable absorbers $Gd_2O_3 + Er_2O_3$ (number of U-Gd fuel rods in fuel assemblies from 0 to 16, erbium oxide content in fuel elements of 0.4%).

The main criteria for the making up of the cycles were the power distribution in the fuel rod, the maximum linear load less than 1.5 and 1.9 and the negative reactivity coefficient. The arrangement and replace pattern of fuel assembly are the same for all variants. The duration of fuel cycles is ~495 eff. Days. Figure 3 shows the refueling pattern and the burnup of the average fuel assemblies at the end of the 18-month fuel cycle using a combination of burnable absorbers. Figure 4 shows the change in power distribution as a function of burnup for the various types of 18-month cycle. Table 3 includes the comparison of the main parameters of the cycles under consideration. The Table also provides information on the equilibrium cycle without burnable absorbers to calculate the enrichment compensation related with the use of various types of burnable absorbers.

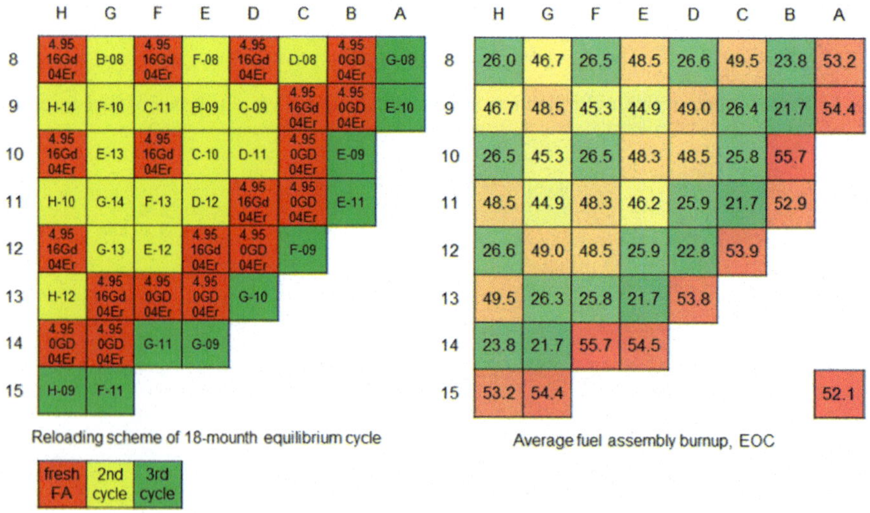

Fig. 3. 18-month fuel cycle pattern and burnup distribution (EOC)

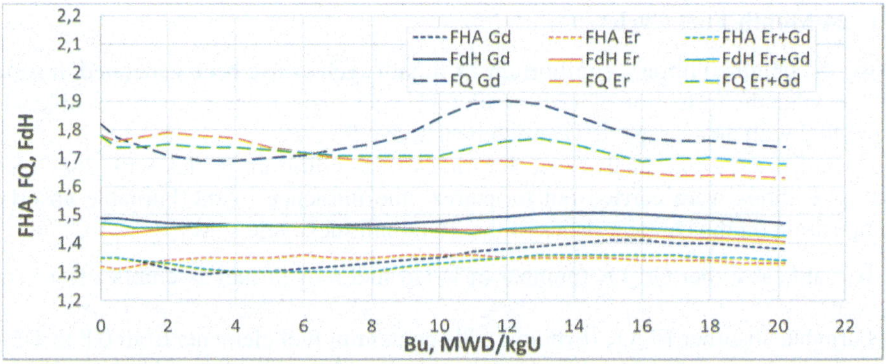

Fig.4. Dependence of power distribution rates on burnup in 18-month fuel cycles

Table 3. Parameters for 18-month fuel cycles

Type of cycle	Duration, eff. Days	U-Gd fuel rods number	ω Er_2O_3, %	^{235}U,%	MTC, 10^{-5} 1/°C	Cb HFP, g/kg	Cb 20 °C keff 0,95, г/кг	Fq (K0)	FdH (Kr)	AO MIN	AO MAX
Without absorber	494.5			4.617							
Gd_2O_3	494.6	1024	0	4.802	0	8.62	16.53	1.9	1.506	−7.6	3.6
Gd_2O_3 + Er_2O_3	495	576	0.4	4.901	−2	8.66	16.87	1.78	1.468	−4.5	0
Er_2O_3	494.3	0	0.72	5.017	−4	8.99	17.39	1.79	1.463	−2.7	−2

Based on the results of these calculations, the following conclusions can be made for fuel cycles with various types of burnable absorbers:

(1) Cycles based on gadolinium oxide require the lowest average ^{235}U fuel enrichment, but their formation may cause problems with the core arrangement to align the power distribution fields, as well as with the Doppler reactivity coefficient, (its value is the maximum for the cycles under consideration).

(2) Erbium oxide-based cycles require the highest average ^{235}U fuel enrichment, their formation is easy to reach low power distribution, the Doppler reactivity coefficient is negative (its value is the minimum for the cycles under consideration). At the same time, problems may arise with such parameter as the standby concentration of boric acid, which is related to the various behavior of the absorption properties of erbium and gadolinium depending on the change of coolant temperature.

(3) The use of a combination of burnable absorbers allows eliminating all of the problems mentioned above while only slightly increasing the average enrichment compared to the gadolinium oxide-based cycle. In addition, in a particular fuel cycle, the use of a combined burnable absorbers allows the ^{235}U enrichment to remain within 4.95%.

4.2 24-Month Fuel Cycle

Using the obtained library, equilibrium 24-month cycles have been generated. It should be noted that the formation of a 24-month fuel cycle in this reactor is only possible when using fuel with an average ^{235}U enrichment above 5%.

Detailed neutronic analysis of 3 variants of equilibrium cycles with 76/77 fresh fuel assemblies were carried out to analyze the influence of the burnable absorbers composition on the characteristics of the 24-month fuel cycle:

- burnable absorber Gd_2O_3 (number of U-Gd fuel rods in fuel assemblies from 16 to 32);
- burnable absorber Er_2O_3 (Erbium oxide content in fuel elements from 0.8 to 1.2%);
- a combination of $Gd_2O_3 + Er_2O_3$ burnable absorbers (number of U-Gd fuel rods in fuel assemblies from 8 to 24, Erbium oxide content in fuel elements of 0.4%).

The cycling criteria are the same as for the 18-month fuel cycle. The arrangement and replace pattern of fuel assembly are the same for all variants. The duration of fuel loads is ~668 eff. Days.

Figure 5 shows the refueling and burnup pattern of the average fuel assemblies at the end of a 24-month fuel cycle using a combination of burnable absorbers. Figure 6 shows the change in power distribution as a function of burnup for the various types of 18-month cycle. Table 4 includes the comparison of the main parameters of the cycles under consideration. The Table also provides information on the equilibrium cycle without burnable absorbers to calculate the enrichment compensation related with the use of various burnable absorbers.

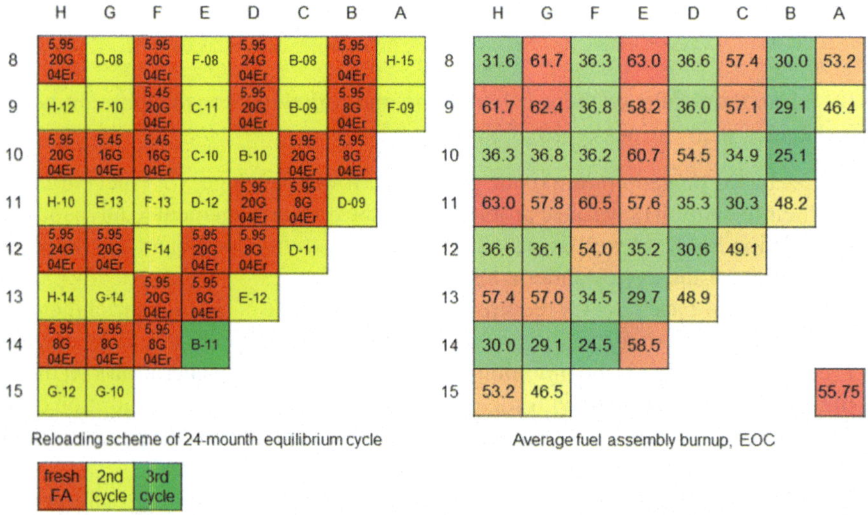

Fig.5. 24-month fuel cycle pattern and burnup distribution (EOC)

The conclusions about the 24-month fuel cycle look similar as for the 18-month cycle. The features of the cycles when using various types of burnable absorbers only

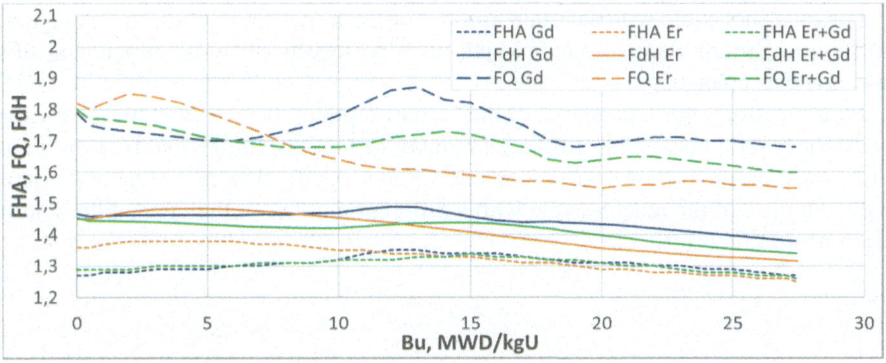

Fig.6. Dependence of power distribution rates on burnup in 24-month fuel cycles

Table 4. Parameters for 24-month fuel cycles

Type of cycle	Duration, eff. Days	U-Gd fuel rods number	ω Er_2O_3, %	^{235}U,%	MTC, 10^{-5} 1/°C	Cb HFP, г/кг	Cb 20 °C keff 0,95, г/кг	Fq (Кo)	FdH (Kr)	AO MIN	AO MAX
Without absorber	669			5,450							
Gd_2O_3	668,3	1792	0	5,621	−0,3	9,90	18,67	1,87	1,488	−9,0	2,6
Gd_2O_3 + Er_2O_3	668,9	1152	0,4	5,742	−4,6	9,65	18,97	1,8	1,454	−4	2
Er_2O_3	668,3	0	1,09	5,950	−5,2	10,37	19,90	1,85	1,486	−6,4	−1

enhanced. In the presence of a large amount of erbium in the core, there is a significant distortion of power distribution to the lower part of the core at the beginning of the cycle, which leads to an increase in the values of relative power compared to the cycle using a combined burnable absorbers.

4.3 Formation of Fuel Cycles of Various Durations

Using the library of constants considered above, a set of equilibrium fuel cycles of various durations was formed with the number of fuel assemblies from 44 to 80/81 and ^{235}U fuel enrichment in fuel rods from 4.95 to 6.45%.

The following restrictions were observed when forming fuel cycles:

(1) Negative Doppler reactivity coefficient;
(2) The boron concentration at the beginning of the cycle at full power is ~1650 ppm;
(3) The maximum power of the fuel rod does not exceed 1.52;
(4) Reactor trip worth without one of the most effective RCCA at power is at least 7%;
(5) The content of erbium oxide in the fresh fuel is on average ~0.6%.

For each fuel cycle with a certain number of fresh fuel assemblies, the arrangement and replace pattern were selected, which were preserved when the enrichment of the fresh fuel was changed.

Figure 7 shows the dependence of the burnup depth of the unloaded fuel on the duration of the reactor cycle with a different number of fuel assemblies and fuel enrichment in fuel rods. Figure 8 shows the dependence of the specific flow rate of natural uranium on the duration of the reactor cycle with a different number of fuel assemblies and fuel enrichment in fuel rods.

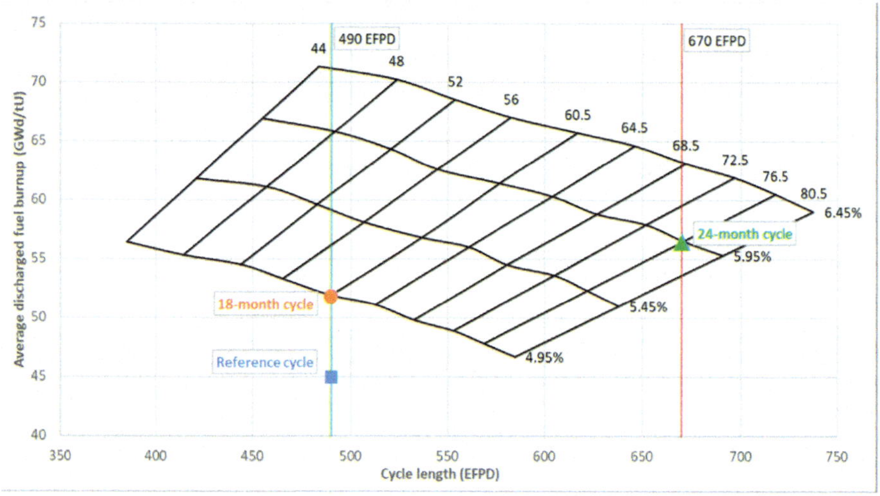

Fig.7. Dependence of the burnup depth of the discharged fuel on the duration of the cycle

5 Results

The use of a combination of gadolinium oxide in U-Gd fuel rods and erbium oxide homogenously distributed over fuel rods as a burnable absorber allows the formation of fuel cycles of various durations while remaining within design limitations. By varying the number of U-Gd fuel rods and the content of erbium in fuel rods, it is easy to control such parameters of the fuel cycle as the power distribution, reactivity coefficients, and the critical and steady shutdown boron concentration.

Neutronic calculations show the possibility of implementing fuel cycles with a duration of 385 to 738 effective days and an average burnup of discharged fuel from 71.3 to 46.7 MWd/kgU. Such ranges are provided with batches from 44 to 80/81 fresh TVS-K with an average enrichment of ^{235}U from 4.95 to 6.45%.

The possibility of forming an 18-month fuel cycle using 60 fuel assemblies with a combined absorber and fuel enrichment of ^{235}U fuel rods up to 4.95% is demonstrated. The use of the considered fuel cycle makes it possible to reduce the specific consumption of natural uranium by 3% compared to the reference fuel cycle. It is shown that the

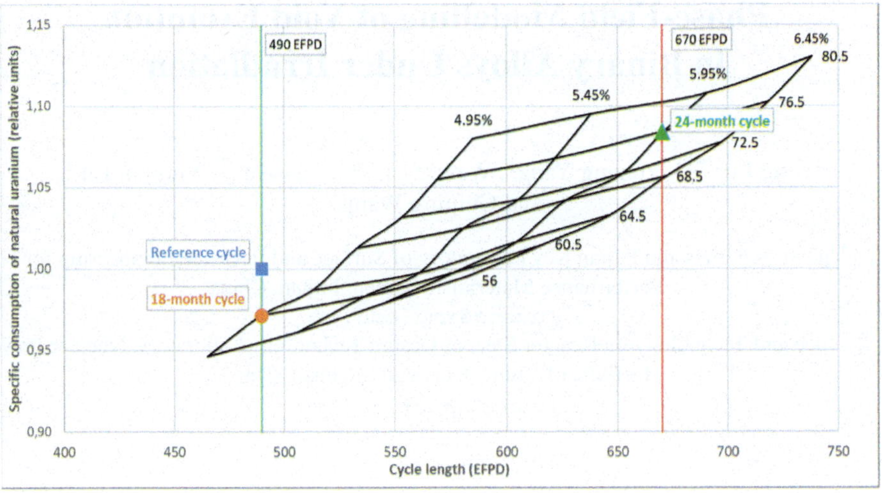

Fig.8. Dependence of the specific consumption of natural uranium on the duration of the cycle

formation of a 24-month fuel cycle is possible with the use of 76 fresh fuel assemblies and ^{235}U fuel enrichment of fuel rods of 5.95%.

Based on the results of the work, it can be concluded that the use of TVS-K of increased enrichment with the use of Gd_2O_3 in U-Gd fuel rods and homogeneously distributed Er_2O_3 in other fuel rods as burnup absorbers is prospective. The use of such fuel, on the one hand, makes it possible to provide reserves for neutronic characteristics important for safety, on the other hand, it makes it possible to improve the technical and economic indicators of the fuel cycle of the PWR reactor.

References

1. -TVSK-DR-RRB-07 TVS-K Design. JSC "TVEL".
2. N-TO-001–2022-TVS-K "Fuel cycles of PWR. Neutronic characteristics" Technical report. JSC "TVEL".
3. Equilibrium Core Design 18-Month, Low-Leakage, 3-Loop PWR Model. Studsvik.

Phase-Field Modelling of Void Evolution in Binary Alloys Under Irradiation

Yong Lu[1(✉)], Yuhang Yang[1], Wenjie Li[2(✉)], Dan Sun[2], Xingjun Liu[1], and Cuiping Wang[1]

[1] College of Materials and Fujian Key Laboratory of Surface and Interface Engineering for High Performance Materials, Xiamen, Fujian, China
luyong@xmu.edu.cn

[2] Science and Technology on Reactor System Design Technology Laboratory, Nuclear Power Institute of China, Chengdu, Sichuan, China
liwenjie@npic.ac.cn

Abstract. The formation and growth of voids have significant influence on the dimensional changes of nuclear materials subjected to irradiation. In this paper, we developed a phase-field model to investigate the evolution of voids in binary alloys under irradiation, which includes the process of irradiation-induced defect formation and annihilation. The relationship of alloy composition and defect formation energy was determined. The nucleation and growth of voids for various alloy compositions were simulated to investigate the effect of solute composition on the void growth kinetics. Our simulation results indicate that, the composition of matrix phase and its influence on the formation energy of vacancy are essential for the growth kinetics and void distribution during irradiation. As an example, the void nucleation and growth in U-Zr alloys are also studied and consistence between the present results and experimental observations were obtained.

Keywords: Irradiation · Voids · Phase-field · Binary alloy · U-Zr

1 Introduction

Metallic nuclear fuels are considered as candidates for advanced fast reactor fuels, due to their high thermal conductivity and burnup [1, 2]. The microstructure of metallic nuclear fuels could have complex evolution under intense irradiation, and consequently lead to major changes in the shape and thermo-mechanical properties [3]. The continuous bombardment of high energetic particles produces numerous vacancies and self-interstitial atoms (SIAs) [4]. These cumulatively generated vacancies and interstitials gradually become supersaturated and facilitate the nucleation and growth of voids, resulting in irradiation-induced swelling of nuclear fuels [5]. To assess the long-term safety of reactors, understanding the growth kinetics of voids and their impact on swelling is essential for the development of advanced swelling-resistant nuclear fuels.

Much effort has been dedicated to investigating the formation and growth of voids in irradiated materials [6, 7]. As an efficient method, computer simulation based on

J. Liu and Y. Jiao (Eds.): WRFPM 2023, SPPHY 299, pp. 280–292, 2024.
https://doi.org/10.1007/978-981-99-7157-2_29

reaction rate theory has been widely used to monitor the void growth under irradiation. The Phase-Field Model (PFM) is recognized as an effective approach to characterize the phase transformation and microstructure evolution in various materials and processes [8, 9]. With respect to irradiation effects, the PFM can well capture the kinetic and morphological evolution of irradiation-induced voids [10, 11], gas bubbles [12–14] and dislocation loops [15]. Millett et al. [16] developed a PFM on the basis of Cahn-Hilliard and Allen-Cahn equations to predict void formation and evolution in irradiated metals. The generation and recombination of point defects and annihilation of defects at sinks are incorporated into their model. Wang et al. [17] studied effects of neutron irradiation and elastic tensile loading on microstructure evolution in polycrystalline tungsten by using PFM, in which the interaction of point defects with grain boundaries (GBs), thermal resistance of GBs, and voids were considered. Rokkam et al. [18] present a PFM for void formation in metals with vacancy concentrations exceeding the thermal equilibrium values, which allows for a unified treatment of void nucleation and growth under the condition of random generation of vacancies. Up to now, typical PFMs of void formation and growth are limited to pure metals, such as Fe, V and W [7, 17, 19]. However, metallic nuclear fuels usually contain several kinds of elements, and the formation energies and activation energies of vacancies and interstitials can be different according to the composition of matrix phase.

For metallic nuclear fuels, Uranium is usually alloyed to stabilize the body-centered cubic γ phase and alleviate swelling behavior [20, 21]. When the chosen element is alloyed into Uranium as an effective metallic fuel, there is often a tradeoff between mechanical properties and swelling resistance [22]. Different atoms own different binding energies between atoms and some typical point defects, and thus cause different defects production rate under irradiation [23]. Therefore, it is necessary to determine the influence of different alloy compositions on the evolution of voids.

In this work, we develop a phase-field model to investigate the void formation and evolution in binary alloys. Thermodynamic models for describing the vacancy formation energy and interaction parameter on the void formation and evolution are proposed. Then the kinetics of void growth under different alloy compositions in binary alloy are systematically studied.

2 Phase-Field Method

In the derivation of the phase-field equations, we first consider a model system consisting of two phases: the matrix phase with vacancies and interstitials, and the void phase. A non-conserved field parameter η is defined to identify the matrix ($\eta = 0$) and void ($\eta = 1$) phases, which vary smoothly from 0 to 1 within the interface between void and matrix phases. The concentrations of vacancy $c_v(r, t)$ and interstitial $c_i(r, t)$ are conserved fields and assumed to be supersaturated at the initial stage. Then the free energy functional F can be written as follows:

$$F = \int_V [(1 - h(\eta))f^m(c_v, c_i) + h(\eta)f^v(c_v, c_i)$$

$$+w_h g(\eta) + \frac{\kappa_v}{2}|\nabla c_v|^2 + \frac{\kappa_i}{2}|\nabla c_i|^2 + \frac{\kappa_\eta}{2}|\nabla \eta|^2]dV \tag{1}$$

where f^{m} and f^{v} represent the free energies of the matrix phase and void phase, respectively. $h(\eta) = \eta^3(6\eta^2 - 15\eta + 10)$ is a function to interpolates the free energy curves of the two phases. $g(\eta) = \eta^2(1-\eta)^2$ is the double-well potential, w_h is a scalar that governs the barrier height of the double-well potential. κ_v, κ_i and κ_η are the gradient energy coefficients. An isotropic surface energy is imposed in the simulation to generate the spherical void. The free energy of the matrix phase as a function of vacancy and interstitial concentrations are given by [18],

$$f^m(c_v, c_i) = E_v c_v + E_i c_i + k_B T[c_v lnc_v + c_i lnc_i$$

$$+(1 - c_v - c_i)ln(1 - c_v - c_i)] \tag{2}$$

where E_v is the vacancy formation energy, E_i is the interstitial formation energy, k_B is the Boltzmann constant, and T is the absolute temperature.

Considering an A-B binary alloy, the vacancy formation energy and interstitial formation energy as a function of alloy composition are expressed as:

$$E_v = E_v^A(1-c) + E_v^B c + E_v^{AB} c(1-c) \tag{3}$$

$$E_i = E_i^A(1-c) + E_i^B c + E_i^{AB} c(1-c) \tag{4}$$

Here c is the mole fraction of B atoms. E_v^A and E_v^B are the vacancy formation energy of atom A and atom B, respectively. E_i^A and E_i^B are the interstitial formation energy of atom A and atom B, respectively. E_v^{AB} and E_i^{AB} are the vacancy and interstitial atom formation energy interaction parameters, respectively. The free energy of the void phase is related to the vacancy, and the interstitial concentration is given by

$$f^v(c_v, c_i) = k_B T\left[(c_v - 1)^2 + {c_i}^2\right] \tag{5}$$

The spatial and temporal evolutions of concentration fields of vacancy and interstitial atom, as well as the phase field parameter can be expressed by the Cahn-Hilliard equation [24] and the Allen-Cahn equation [25] as

$$\frac{\partial c_v}{\partial t} = \nabla\left[M_v \nabla \frac{\delta F}{\delta c_v}\right] + \xi_v + P_v - R_{iv} - S_v \tag{6}$$

$$\frac{\partial c_i}{\partial t} = \nabla\left[M_i \nabla \frac{\delta F}{\delta c_i}\right] + \xi_i + P_i - R_{iv} - S_i \tag{7}$$

$$\frac{\partial \eta}{\partial t} = -L\frac{\delta F}{\delta \eta} + \xi_\eta \tag{8}$$

where $M_v = D_v c_v / k_B T$ and $M_i = D_i c_i / k_B T$ are the mobilities of vacancy and interstitial, respectively, D_v and D_i are the diffusivities of vacancy and interstitial atom, respectively, and L is the interface mobility. ξ is the thermal induced fluctuations, P_v and P_i are the generation of vacancies or interstitials due to irradiation, subscripts v and i stand for vacancy and interstitial respectively. R_{iv} in Eqs. (6) and (7) is the vacancy-interstitial recombination rate, which can be expressed as $R_{iv} = R_r c_v c_i$ and $R_r = (1\text{-}\ h(\eta))R^{bulk} + h(\eta)R^{surf}$ [16, 26]. S_v and S_i are the sink rate of the vacancies and interstitial atoms. The excess vacancies and interstitials will be absorbed by grain boundary and dislocation. The sink terms can be described as $S_v = {k_v}^2 D_v c_v$ and $S_i = {k_i}^2 D_i c_i$. k_v and k_i are the sink strengths of grain boundary and dislocation for vacancies and interstitials [27].

The nucleation is a statistical process during phase transformation. The explicit nucleation algorithm is a proper treatment for the nucleation of voids within the present metastable system [28]. The nucleation probability in a sufficiently small-time interval Δt can be obtained by

$$P_T = 1 - e^{-J\Delta V\Delta t}, \tag{9}$$

where J is the nucleation rate per phase-field cell, ΔV is the volume of phase-field cell. At every time step, a random number P_r $(0 < P_r < 1)$ is generated for each phase-field cell. The void nucleation event can only take place in the location where P_T is great than P_r. For the two-dimensional phase-field simulations, the traditional rate equations for the steady-state nucleation in matrix can be written as,

$$J = J_0 e^{-\pi\sigma^2/k_B T\Delta f} \tag{10}$$

where J_0 is the nucleation coefficient, σ is the interface energy per unit area, Δf is the driving force for nucleation. In the present work, the free energy difference between the void and the matrix phases (i.e., $\Delta f = f^m - f^v$) is used to calculate the driving force for nucleation, which is a function of vacancy concentration, alloy composition and temperature.

In the present work, the governing equations (Eqs. (6)– (8)) in the phase-field model are solved numerically in two dimensions using the explicit finite difference method. Periodic boundary condition is applied to the simulation domain. We use the following normalizations for the numerical calculation: $r^* = r/l$, $t^* = tD/l^2$, $\Delta^* = \Delta \cdot l$, $M_v^* = M_v k_B T/D$, $M_i^* = M_i k_B T/D$, $L^* = Lk_B T/D$, $F^* = F/k_B T$, $\kappa^* = \kappa_v/k_B T l^2$, ${\kappa_i}^* = \kappa_i/k_B T l^2$, ${\kappa_\eta}^* = \kappa_\eta/k_B T l^2$, ${\xi_v}^* = \xi_v l^2/D$, ${P_v}^* = P_v l^2/D$, ${R_{iv}}^* = R_{iv} l^2/D$, ${S_v}^* = S_v l^2/D$, ${J_0}^* = J_0 l^4/D$. The dimensionless grid size is $\Delta x^* = \Delta y^* = 1.0$, and the simulation cell is $512\Delta x \times 512\Delta y$ with a grid length $\Delta x = \Delta y = 10^{-9}$m. The time step is $\Delta t^* = 0.0005$. Thus, Eqs. (6)–(8) can be transformed into dimensionless forms

$$\frac{\partial c_v}{\partial t^*} = \nabla^*\left[M_v^*\nabla^*\frac{\delta F}{\delta c_v}\right] + {\xi_v}^* + {P_v}^* - {R_{iv}}^* - {S_v}^* \tag{11}$$

$$\frac{\partial c_i}{\partial t^*} = \nabla^*\left[M_i^*\nabla^*\frac{\delta F}{\delta c_i}\right] + {\xi_i}^* + {P_i}^* - {R_{iv}}^* - {S_i}^* \tag{12}$$

$$\frac{\partial \eta}{\partial t^*} = -L^*\frac{\delta F^*}{\delta \eta} + \xi_\eta^* \tag{13}$$

3 Results and Discussion

3.1 Vacancy Formation Energy and Vacancy Evolution

The change of vacancy formation energy with concentration in binary alloy is demonstrated in Fig. 1. The vacancy formation energies of atom A and B are taken as $E_v^A = E_v^B = 1.0$ eV. It can be seen that the vacancy formation energy interaction parameter has great influence on the vacancy formation energy of binary alloy. When the vacancy formation interaction parameter E_v^{AB} is equal to zero, the vacancy formation energy has a liner relationship with alloy concentration. When E_v^{AB} greater than zero, the vacancy formation energy reaches its maximum value at A-50at.% B alloy. When E_v^{AB} is less than zero, the vacancy formation energy will reach its minimum value at A-50at.% B alloy. Figure 2 shows the vacancy distribution evolution with time versus alloy concentration. The initial vacancy concentration was set as 0.02 at time $t^* = 0$. When the vacancy formation interaction parameter $E_v^{AB} = -1.5$, as the simulation time $t*$ increases, the vacancy concentration decreases in both pure A and pure B sides, and has a maximum value at $c_B = 50$ at.%. However, the vacancy diffused inversely from gradient chemical concentration to pure A and pure B sides, when $E_v^{AB} = 1.5$. The main reason for the evolution of vacancy is the effect of alloy concentration on the vacancy formation energy, thus reducing the total free energy of the vacancy.

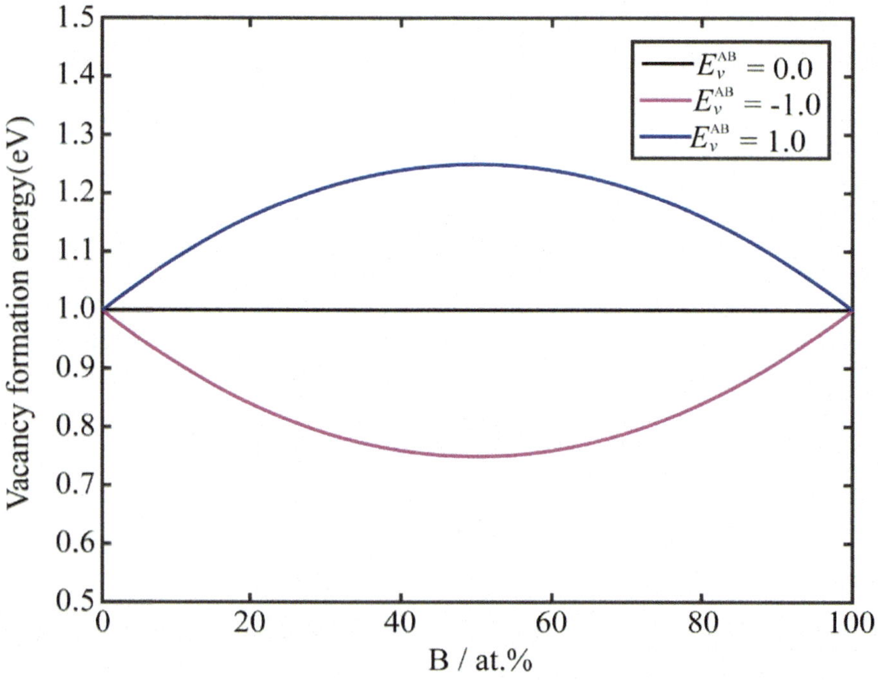

Fig. 1. The vacancy formation energy as a function of alloy composition in binary system.

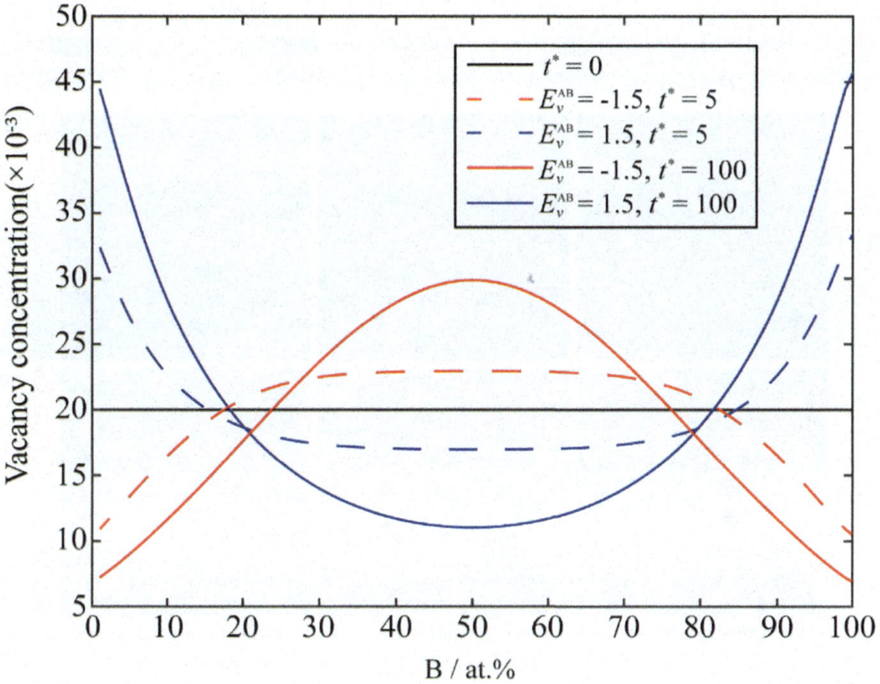

Fig. 2. Vacancy evolution with time versus concentration in binary alloy.

3.2 Void Formation and Growth Under Different Alloy Composition in Binary Alloy

In this section, we introduce the effect of alloy concentration on the void formation and growth in binary alloy. It is known that the local vacancy concentration of matrix has great influence on the nucleation and growth of voids. The aforementioned calculation results have shown that vacancy formation interaction parameter E_v^{AB} and alloy concentration can cause changes of vacancy concentration in the matrix. To examine the effect of alloy concentration on void evolution, the void formation and growth in the matrix with horizontal gradient alloy concentration are simulated. The results of temporal microstructure evolution for the three cases ($E_v^{AB} = 0, E_v^{AB} > 0\text{and}E_v^{AB} < 0$) are shown in Fig. 3. Due to the irradiation effect, the concentrations of vacancies and interstitials are generated stochastically in both time and space. The continuous production of vacancies and interstitials eventually increases the supersaturation level and promotes the nucleation of voids. In these plots, red color indicates the void phase, while blue color represents the matrix phase. As can be readily seen, the voids tend to nucleate randomly in the matrix when the vacancy formation interaction parameter $E_v^{AB} = 0$. When E_v^{AB} is positive, voids tend to nucleate and grow near the pure A or B range due to the accumulation of vacancies. When E_v^{AB} is less than zero, voids prefer to nucleate and grow in the gradient chemical composition range.

Figures 4 and 5 display the morphologies of different initial compositions in A-B binary alloy under the irradiation intensity of 3×10^{-4} dpa/s with $E_v^{AB} = 2.0$ and

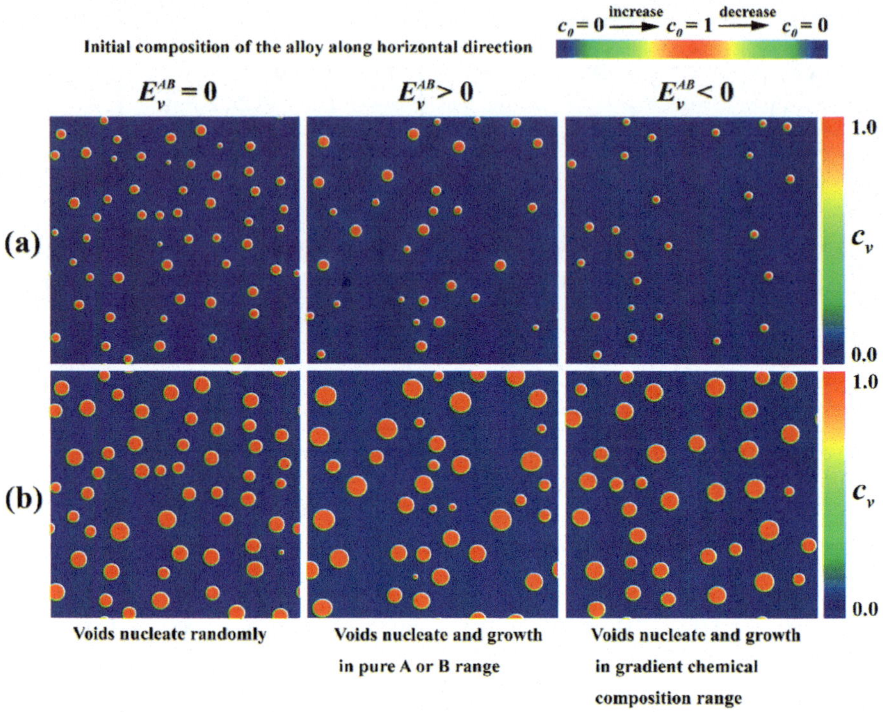

Fig. 3. The results of temporal microstructure evolution for the three cases ($E_v^{AB} = 0, E_v^{AB} > 0andE_v^{AB} < 0$), (a) $t^* = 80$, b) $t^* = 200$.

E_v^{AB} = -2.0, respectively. It can be seen that, when E_v^{AB}= 2.0, the increasing atom B can lead to decrease of void size and increase of void density. While the density of voids decreases and the size of voids increases with the increase of B concentration when E_v^{AB} = -2.0. Figure 6 shows the dependence of void porosity and density on alloy composition. As shown in Fig. 6(a), the voids porosities show an increase from 0.108 to 0.121 for E_v^{AB} = -2.0 and decrease from 0.108 to 0.096 for E_v^{AB}= 2.0 as the concentration of B increase from 0 to 50%, respectively. Figure 6(b) illustrates the relationship between alloy composition and void density. The void density increases from 1.5×10^{-4} nm^{-2} to 2.8×10^{-4} nm^{-2} for E_v^{AB} = 2.0 and decreases from 1.5×10^{-4} nm^{-2} to 0.8×10^{-4} nm^{-2} for E_v^{AB} = -2.0, as the concentration of B increases from 0 to 50%, respectively.

The above simulation results demonstrate that alloy composition and vacancy formation interaction parameter E_v^{AB} have significant impact on void nucleation and growth. The increase of vacancy formation energy leads to the increase of the free energy difference between the void phase and the matrix phase, and consequently accelerates the nucleation rate of voids. Meanwhile, the increase of vacancy formation energy causes harder diffusion of atoms away from their lattice sites and reduces the generation rate of vacancies and interstitials under irradiation. Therefore, under the same alloy concentration, the void porosity decreases with increase of E_v^{AB}, while void density increases with

$0, E_v^{AB} > 0$ and $E_v^{AB} < 0$), (a) $t^* = 80$, b) $t^* = 200$.

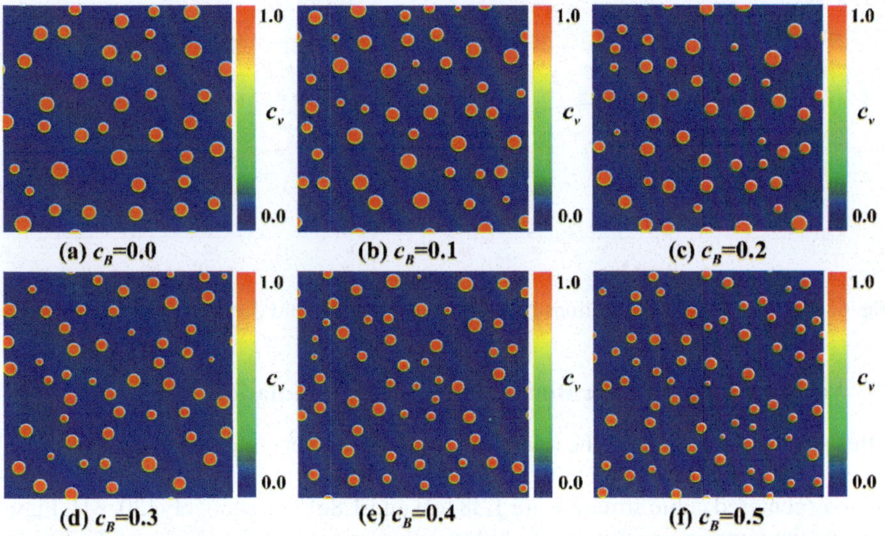

Fig. 4. Effect of alloy compositions on void morphology in A-B binary alloy when $E_v^{AB} = 2.0$.

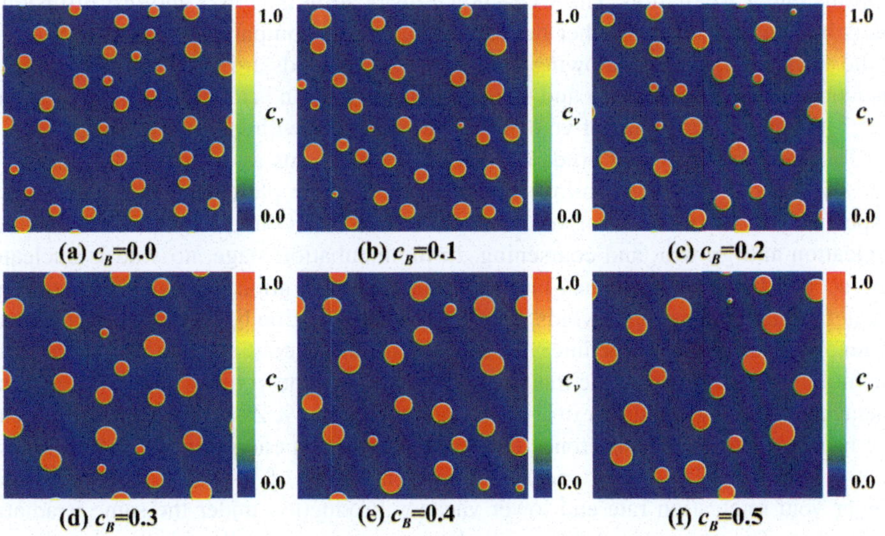

Fig. 5. Effect of alloy compositions on void morphology in A-B binary alloy when $E_v^{AB} = -2.0$.

increase of E_v^{AB}. The present results are consistent well with the experimental observations, e.g., the addition of a small amount of chromium in iron matrix can strongly reduce void swelling and largely increase void density [29].

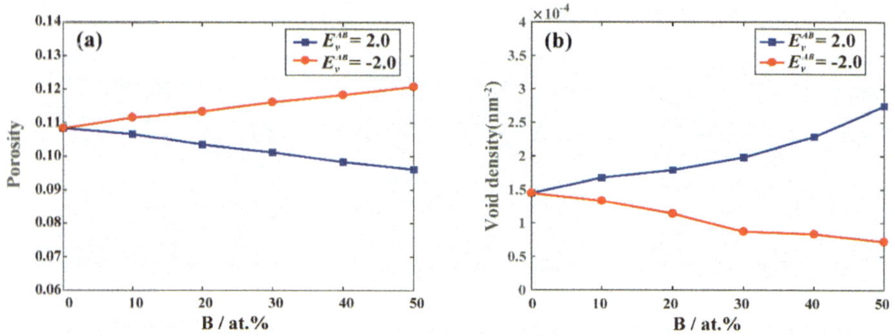

Fig. 6. Effect of alloy compositions on (a) porosity and (b) void density in A-B binary system.

3.3 Void Evolution in U-Zr Binary Alloy Under Irradiation

In this section, we introduce the related material parameters into the model and study the void evolution in U-Zr binary alloy. The vacancy formation energy of U and Zr atoms of the body centered cubic structure are 1.384eV and 1.8eV, respectively [30, 31]. Figure 7 displays the temporal evolution of void morphologies in U-5at.% Zr and U-70at.% Zr alloys under irradiation rate of 5×10^{-4} dpa/s. With the increase of time, the on-going generation of vacancies and interstitials in the matrix eventually result in the nucleation of voids under irradiation. The well-known dislocation bias [32] and fast diffusion of interstitials ensure that a higher net flux of vacancies compared to interstitials arrives at the void surfaces, thus allowing void growth. The voids are spherical shaped due to the isotropy of surface energy and their spatial distribution is random. As can be seen in Fig. 7, the number and size of voids increase obviously with increasing time.

The volume fraction of voids and average void radius as a function of time in U-5at.% Zr and U-70at.% Zr alloys under irradiation are plotted in Fig. 8(a) and (b), respectively. The evolution of voids can be divided into three stages, i.e., incubation, nucleation and growth, and coarsening. In the incubation stage, no void is nucleated. The accumulation of vacancies increases the supersaturation level of the system and accelerates the nucleation of voids. Once the void nucleation begins, multiple nuclei are formed within a narrow time interval. The existing voids continue to grow by absorbing the mobile vacancies from the matrix. By comparison, the volume fraction of void and average void radius are higher in U-5at.% Zr than U-70at.% Zr alloys. This phenomenon can be explained by the difference of vacancy formation energy in U-5at.% Zr and U-70at.% Zr alloys. The higher vacancy formation energy of U-70at.% Zr alloy leads to higher void nucleation rate and lower vacancy production under the same irradiation conditions. The present simulation results are consistent well with the experimental observations in irradiated U-Zr alloy. Ahn et al. [33] studied the cavity nucleation and growth in the U-0.1wt.% Zr and U-40wt.% Zr alloys under He^+ ion irradiation. Their results show that U-0.1wt.% Zr is more sensitive to irradiation and forms larger spherical voids, while numerous smaller voids were observed in U-40wt.% Zr alloy.

Figure 9 shows the temporal evolution of voids in U-Zr two-phase microstructure under irradiation. The initial microstructures are bcc U-rich and Zr-rich phases, as shown in Fig. 9(a). With the increase of time, the irradiation-induced vacancies diffuse into the

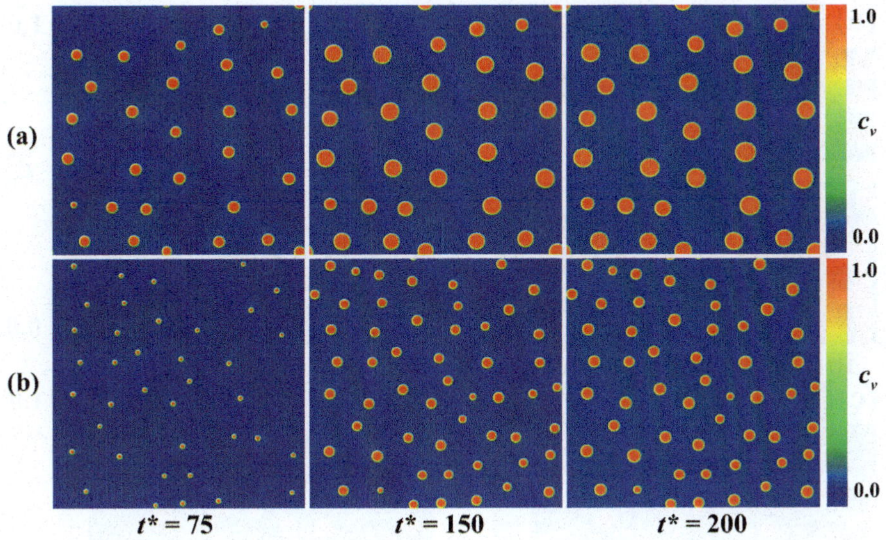

Fig. 7. Temporal evolution of vacancy concentration for (a) U-5at.% Zr and (b) U-70at.% Zr alloys under irradiation.

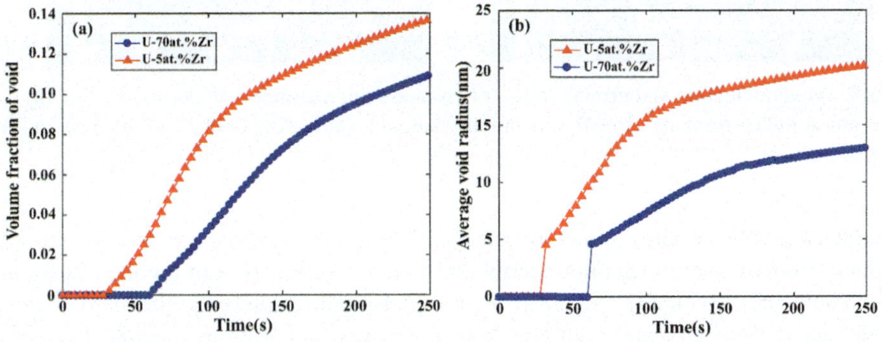

Fig. 8. Temporal evolution of (a) volume fraction of voids and (b) average void radius for U-5at.% Zr and U-70at.% Zr alloys under irradiation.

U-rich phase to reduce the total free energy of the system and lead to the nucleation of voids in U-rich phase under irradiation. The newly formed void nuclei continue to absorb the nearby saturated vacancies and grow inside the U-rich phase in U-Zr two-phase microstructure.

4 Conclusions

The relationship of alloy composition and defect formation energy was proposed to describe the total free energy of the binary system. A Phase-field model based on reaction rate theory is used to simulate the accumulation process of the irradiation-induced

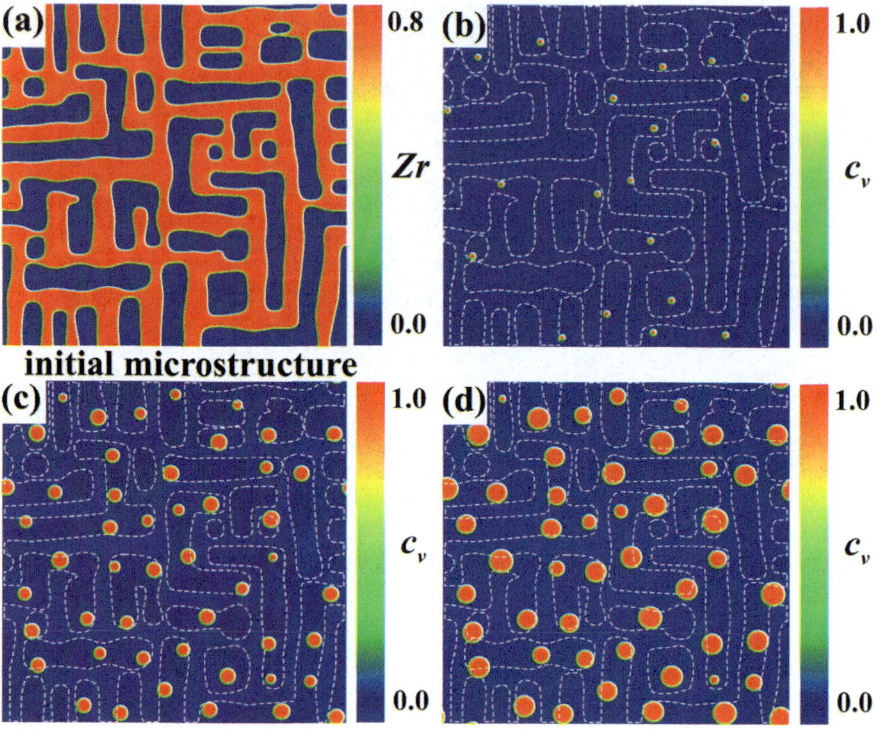

Fig. 9. Void nucleation and growth in U-Zr two-phase microstructure under irradiation, (a) U-Zr two phase microstructure, (b)-(d) simulation results of voids growth at $t^* = 40$, 100 and 200, respectively.

vacancies in binary alloy under irradiation. The alloy composition has an important influence on vacancy formation energy and vacancy production in the matrix. Simulation results indicate that the increase of vacancy formation energy in the matrix can lead to the decrease of size and volume fraction of voids. In addition, specific study of the voids evolution in U-Zr alloys demonstrates the volume fraction and average radius of the voids in U-5at.% Zr are higher than that in U-70at.% Zr alloys, which agrees well with experimental observations. Voids seemingly prefer to grow inside the bcc U-rich phase rather than Zr-rich phase.

Acknowledgments. The authors would like to thank the support for this research by the National Natural Science Foundation of China (12205286), National Key R&D Program of China (2017YFB0702401) and the Fundamental Research Funds for the Central Universities (20720170038).

References

1. V.P. Sinha, P.V. Hegde, G.J. Prasad, G.P. Mishra, S. Pal, Development of high density uranium compounds and alloys as dispersion fuel for research and test reactors, Trans. Indian Inst. Met. 61 (2008) 115–120.
2. S.A. Jonah, K. Ibikunle, Y. Li, A feasibility study of LEU enrichment uranium fuels for MNSR conversion using MCNP, Ann. Nucl. Energy 36 (2009) 1285–1286.
3. Y. Li, S. Hu, S. Xin, M. Stan, A review: applications of the phase field method in predicting microstructure and property evolution of irradiated nuclear materials, npj Comput. Mater. 3 (2017) 1–17.
4. J.A. Brinkman, On the nature of radiation damage in metals, J. Appl. Phys. 25 (1954) 961–970.
5. F.A. Garner, M.B. Toloczko, B.H. Sencer, Comparison of swelling and irradiation creep behavior of fcc-austenitic and bcc-ferritic/martensitic alloys at high neutron exposure, J. Nucl. Mater. 276 (2000) 123–142.
6. L.K. Mansur, Void swelling in metals and alloys under irradiation: an assessment of the theory, Nucl. Technol. 40 (1978) 5–34.
7. Y. Wang, J. Ding, S. Huang, J. Zhao, Y. Wang, Mesoscale modeling of irradiation damage evolution in bcc iron and vanadium: A comparative study, Fusion Eng. Des. 137 (2018) 303–311.
8. Y. Sun, J. Luo, J. Zhu, Phase field study of the microstructure evolution and thermomechanical properties of polycrystalline shape memory alloys: Grain size effect and rate effect, Comput. Mater. Sci. 145 (2018) 252–262.
9. P.H. Leo, W. Johnson, Spinodal decomposition and coarsening of stressed thin films on compliant substrates, Acta Mater. 49 (2001) 1771–1787.
10. Y. Wang, J. Ding, W. Liu, S. Huang, X. Ke, Y. Wang, C. Zhang, J. Zhao, Irradiation-induced void evolution in iron: A phase-field approach with atomistic derived parameters, Chin. Phys. B 26 (2017) 026102.
11. W. Liu, N. Wang, Y. Ji, P. Song, C. Zhang, Z. Yang, L. Chen, Effects of surface energy anisotropy on void evolution during irradiation: A phase-field model, J. Nucl. Mater. 479 (2016) 316–322.
12. Z. Xiao, Y. Wang, S. Hu, Y. Li, S. Shi, A quantitative phase-field model of gas bubble evolution in UO_2, Comput. Mater. Sci. 184 (2020) 109867.
13. S. Hu, Y. Li, X. Sun, F. Gao, R. Devanathan, C.H. Henager, M.A. Khaleel, Application of the phase-field method in predicting gas bubble microstructure evolution in nuclear fuels, Int. J. Mater. Res. 101(4) (2010) 515–522.
14. Y. Li, S. Hu, R. Montgomery, F. Gao, X. Sun, Phase-field simulations of intragranular fission gas bubble evolution in UO_2 under post-irradiation thermal annealing, Nucl. Instrum. Methods Phys. Res. Sect. B 303 (2013) 62–67.
15. H. Rouchette, L. Thuinet, A. Legris, A. Ambard, C. Domain, Numerical evaluation of dislocation loop sink strengths: A phase-field approach, Nucl. Instrum. Methods Phys. Res. Sect. B 352 (2015) 31–35.
16. P.C. Millett, A. El-Azab, S. Rokkam, M. Tonks, D. Wolf, Phase-field simulation of irradiated metals: Part I: Void kinetics, Comput. Mater. Sci. 50 (2011) 949–959.
17. H. Wang, S. Biswas, Y. Han, V. Tomar, A phase field modeling based study of microstructure evolution and its influence on thermal conductivity in polycrystalline tungsten under irradiation, Comput. Mater. Sci. 150 (2018) 169–179.
18. S. Rokkam, A. El-Azab, P. Millett, D. Wolf, Phase field modeling of void nucleation and growth in irradiated metals, Modell. Simul. Mater. Sci. Eng. 17 (2009) 064002.
19. Y. Wang, J. Ding, Y. Chen, J. Zhao, Y. Wang, Three-dimensional phase field simulation of intragranular void formation and thermal conductivity in irradiated α-Fe, J. Mater. Sci. 53 (2018) 11002–11014.https://doi.org/10.1007/s10853-018-2376-3

20. M. Steiner, E. Garlea, S. Agnew, Modeling solute segregation during the solidification of γ-phase U-Mo alloys, J. Nucl. Mater. 474 (2016) 105–112.
21. J. Rest, Kinetics of fission-gas-bubble-nucleated void swelling of the alpha-uranium phase of irradiated U-Zr and U-Pu-Zr fuel, J. Nucl. Mater. 207 (1993) 192–204.
22. D.E. Burkes, R. Prabhakaran, J. Jue, F.J. Rice, Mechanical properties of DU-xMo alloys with x= 7 to 12 weight percent, Metall. Mater. Trans. A 40 (2009) 1069–1079.
23. M. Yoo, L. Mansur, Distributions of point defects in bounded media under irradiation, J. Nucl. Mater. 62 (1976) 282–292.
24. J.W. Cahn, On spinodal decomposition, Acta Metall. 9 (1961) 795–801.
25. J.W. Cahn, S.M. Allen, A microscopic theory for domain wall motion and its experimental verification in Fe-Al alloy domain growth kinetics, J. Phys. 38 (1977) C7–51-C7–54.
26. J.J. Carter, W.H. Howland, R.W. Smith, A rate-theory approach to irradiation damage modeling with random cascades in space and time, Metall. Mater. Trans. A 46 (2015) 93–101.
27. B. Li, S. Hu, C. Li, Q. Li, J. Chen, G. Shu, C. Henager, Y. Weng, B. Xu, W. Liu, Simulations of irradiated-enhanced segregation and phase separation in Fe–Cu–Mn alloys, Modell. Simul. Mater. Sci. Eng. 25 (2017) 065007.
28. T.W. Heo, L. Chen, Phase-field modeling of nucleation in solid-state phase transformations, JOM 66 (2014) 1520–1528.
29. A. Bhattacharya, E. Meslin, J. Henry, A. Barbu, S. Poissonnet, B. Décamps, Effect of chromium on void swelling in ion irradiated high purity Fe–Cr alloys, Acta Metall. 108 (2016) 241–251.
30. B. Beeler, C. Deo, M. Baskes, M. Okuniewski, First principles calculations of the structure and elastic constants of α, β and γ uranium, J. Nucl. Mater. 433 (2013) 143–151.
31. A. Moore, B. Beeler, C. Deo, M. Baskes, M. Okuniewski, Atomistic modeling of high temperature uranium–zirconium alloy structure and thermodynamics, J. Nucl. Mater. 467 (2015) 802–819.
32. C. Woo, B. Singh, Production bias due to clustering of point defects in irradiation-induced cascades, Philos. Mag. A 65 (1992) 889–912.
33. S. Ahn, S. Irukuvarghula, S.M. Mcdeavitt, Microstructure of α-U and δ-UZr_2 phase uranium–zirconium alloys irradiated with 140-keV He+ ion-beam, J. Alloys Compd. 681 (2016) 6–11.

Digital Transformation of Fuel Pellet Production Facilities

Xiaoyu Guo(✉)

CNNC Jianzhong Nuclear Fuel Co., Ltd. (CJNF), Sichuan, China
guoxiaoyu@cnnc.com.cn

Abstract. Many companies in the nuclear sector have undertaken a digital transformation during the past few years owing to the rapid development of new technologies. As the leader in production of nuclear fuel in China, CJNF has been focusing on the technical advances in the field and has a strong commitment to the digital transformation plans. Automation of production facilities and digital optimization of management were studied and applied in the pellet plant. The results showed a good performance and experience on the pellet production, process and quality management, logistics and inventory management with the application of automation and digital technologies.

Keywords: Digital transformation · Fuel pellet · Production management

1 Introduction

At present, a new-round scientific and technological revolution and industrial transformation thrive in the world, which greatly and deeply impacts on global economic development, social progress and human civilization. Digital technologies such as big data, cloud computing, 5G, Internet of Things (IoT) and artificial intelligence (AI) continue to emerge. The digital economy has increasingly become the core driving force for economic growth, playing an increasingly important role in promoting economic and social development. In the new era, promoting the digital and intelligent transformation and upgrading of the nuclear sector is an inevitable requirement for achieving the high-quality development of the nuclear industry.

Nuclear fuel cycle is an important foundation for the development of the nuclear industry. Major nuclear fuel suppliers in the world have been carrying out modernization program and digital transformation in recent years to improve production efficiency, reduce manufacturing costs and maintain their competitive advantages in the international market.

CJNF is the largest manufacturer of nuclear fuel assemblies for PWRs in Asia, and has been working on digital transformation for many years. This paper introduces the principles, objectives, contents, effects and related experiences of the transformation work to realize real-time data acquisition and data analysis in production lines, taking fuel pellets production as an example.

J. Liu and Y. Jiao (Eds.): WRFPM 2023, SPPHY 299, pp. 293–299, 2024.
https://doi.org/10.1007/978-981-99-7157-2_30

2 Transformation Principles

2.1 Design Ideas

As per hardware, the hardware such as equipment and instruments are adapted and transformed to realize internet of things and automation control by arranging data acquisition and control network, production information network, workshop monitoring center, data center and auxiliary terminals.

As per software, production management, material tracking, quality management, equipment management and other modules are built in the manufacturing execution system to realize the integration of production management functions.

As per application, through real-time data collection and analysis of production, quality and equipment operation, visual program is applied to promote the digital transformation of fuel pellet manufacturing.

2.2 Design Work

To build equipment data collection and control network and production information network to realize networking of equipment and production operation terminals.

To build workshop production scheduling and monitoring center to realize data collection and storage sharing, production scheduling and visual monitoring of equipment operation.

To deploy production post operation terminals and systems to realize the informationized and real-time task receiving, operation data processing and material tracking on site.

To deploy warehouse management terminals and systems to realize informationized, paperless and visualized management of material.

To establish digital models of workshop resources, products, processes and quality to realize digital control of resources, products, processes and quality by use of terminals.

To establish data interface standards to realize the transfer of planning, production, process and quality data between the workshop and the management departments.

To develop the industrial control network security strategy to ensure communication security.

3 Implementation

3.1 Implementation Process

Network construction.

Field equipment. The In order to realize digital production, all on-site industrial equipment is interconnected with the production monitoring layer acquisition and execution server, and the main data transferred is the status information of the industrial equipment and the execution commands issued by the server.

Production monitoring. The security protection of the production monitoring layer is mainly for the host protection of the industrial server and the boundary protection of the production monitoring layer and the field equipment layer. The industrial control security monitoring platform is connected to the industrial network, and perceives early warning and dynamic protection for the whole industrial network.

Production Management. The information devices in the production management layer include application servers, data servers, and a large number of computer terminals. The industrial boundary isolation gateway is deployed between the production management MES server and the production monitoring layer collection and execution server, configured with a whitelist security protection policy based on OPC protocol, which only allows OPC communication between the two servers of MES server and collection and execution server and blocks all other network access behavior between the two networks to guarantee the logical isolation between the two networks and avoid mutual influence (Fig. 1).

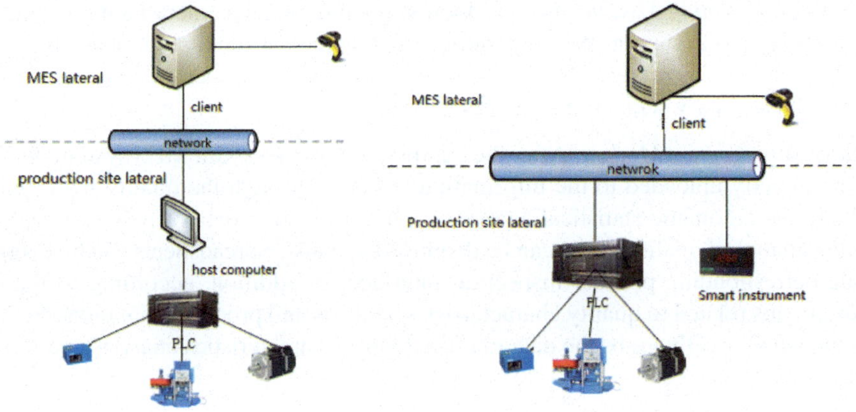

Fig. 1. Integration by use of host computers and Ethernet

Backup and recovery. Full data backup of data information related to MES system and collection and execution system with off-site storage. Regular configuration backup of important hardware devices in the device layer network. The key data of the industrial control system, stored in the industrial control equipment, is backed up periodically or after the configuration and function of the equipment has changed.

Equipment interface modification

Powder granulation system. The monitoring and control of the pelletizing system of rolled chips is realized by using the host computer of the single cone mixing system, and the equipment parameters and operation data of the pelletizing system are collected by integrating the OPC of the host computer Wincc software with the MES system.

Pressing and loading boat system. The rotary presses are equipped with their own network port or with additional network interface modules to enable the MES system to collect equipment parameters and operating data.

Add network interface module to the automatic boat loading device to realize the collection of equipment parameters and operation data.

Material transfer system. Integration with MES system is realized through the Web Service interface of the host computer to complete data acquisition.

High temperature sintering system. Integration with the MES system is achieved through the OPC interface of the sintering furnace host computer to complete data acquisition.

Pellet grinding system. Add the upper system software to the grinding system to complete the online measurement of pellet diameter and data acquisition.

Pellet oxidation system. Add a network interface module to the oxidation unit to enable the collection of equipment parameters and operational data.

Management module

Production Planning Module. The production planning management module is built in the MES system to support the preparation and issuance of production plans, including information such as production plan start date, end date, and total output. The module has certain auxiliary scheduling functions.

Quality Management Module. After the equipment is docked with MES system, the data can be directly uploaded to the information system. Through the quality management module, the automatic statistical analysis of various testing data of pellets is realized, and the abnormal quality status can be displayed visually. Spreadsheets can be exported in the corresponding product inspection interface. In addition, according to the data requirements related to quality characteristics analysis and process evaluation, the MES system automatically grabs the data to form quality characteristics analysis and process analysis reports.

Device management module. The main content of the equipment management module is to take asset equipment and spare parts as the basic management object, cover all aspects of the equipment life cycle (selection, installation, maintenance, repair, analysis and disposal), provide various maintenance modes such as preventive maintenance and fault maintenance, and provide equipment management business with maintenance task planning, review, execution and analysis as the main business line. It provides various maintenance modes such as preventive maintenance and breakdown maintenance, and provides equipment management business with maintenance task planning, review, execution and analysis as the main business line.

Certificate management module. According to the requirements of certificate validity management, the certificate management module is added. Through color differentiation, the system can visually display the validity of various certificates such as personnel qualification certificate, operation certificate, equipment calibration certificate, process certificate, etc. Before the certificate is about to expire, an email will be sent automatically to notify the relevant person in charge, eliminating the need for manual daily inspection.

Production statistics module. Construction of production statistics management module, according to the implementation of product production tasks, to achieve automatic statistics and visualization of pellet product production data, the results support direct export of spreadsheets.

3.2 Performance

Real-time data acquisition. Completed the adaptive transformation of production line equipment to meet the data interface requirements. Built an industrial control data collection network and established a company-wide unified real-time database. Realized real-time collection and storage of equipment operation parameters, process data and production data, which can provide data basis for production monitoring, equipment management, quality management and big data analysis (Figs. 2 and 3).

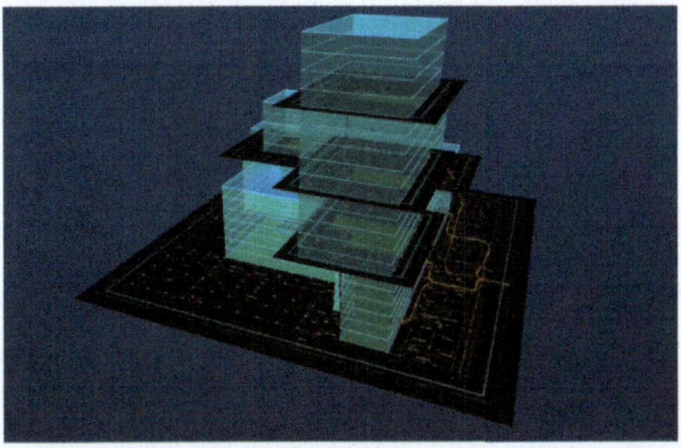

Fig. 2. The model of the production lines

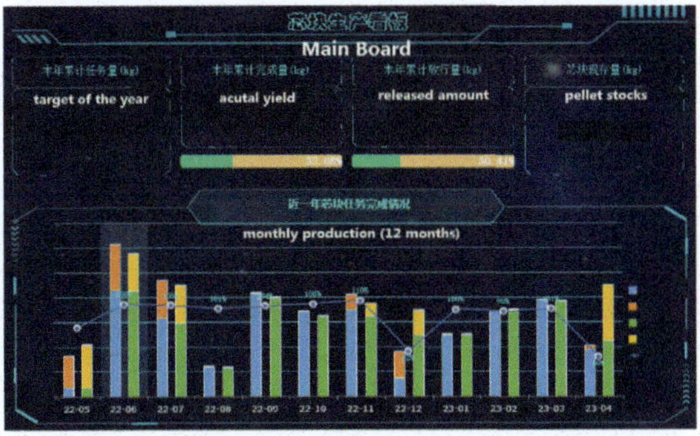

Fig. 3. The main board of pellet production

Production monitoring. Material tracking: With the functions of material preparation plan management, material basic information management and logistics information collection, it realizes the material information tracking and traceability of the production process and supports to provide optimization suggestions for material scheduling.

Process monitoring: with process indicators, equipment operation, production progress monitoring and other functions, support multi-trend monitoring and trend graph parameter customization, can provide intuitive and fast interface support for production scheduling and exception management.

Data large-screen: Graphically display the production operation and equipment status, click on the graph to quickly view the detailed data of production and equipment, and support different professions to customize the interface of Large-screen according to their needs, to achieve visualized and transparent production and manufacturing management. Data large-screen has become an auxiliary tool for decision making and execution.

Quality Management. With the function of real-time collection of production process, online inspection and physical and chemical inspection data, it can meet the business needs of dynamic quality analysis, quality control plan, full quality management, quality SPC analysis and quality traceability.

Equipment Management. Equipment status monitoring: With real-time equipment operation status data collection and monitoring function, it realizes the visualization display of equipment operation status.

Equipment ledger management: A basic database of equipment is established, containing data of equipment acceptance information, equipment change information, and equipment identification history, supporting data retrieval, access, revision, and update.

Equipment operation and maintenance management: It supports automatic data collection of equipment operation records, equipment maintenance records and equipment fault repair records, etc. By managing the spare parts ledger, it can provide support for carrying out preventive maintenance of equipment.

Error-proof management. Error-proof information entry: For manually entered information, set data and format verification rules, and send alarm alerts when information is wrong to ensure the accuracy of information to the greatest extent.

Production operation error prevention: for the production of multi-enrichment pellets, one is to interact with the MES system information through the industrial control network, according to the production tasks issued by the MES system, to achieve control of equipment feeding operation; two is to automatically read the barcode or label of materials and containers, to achieve error prevention control of feeding, refilling and changing operations; three is to exceed the limit value in the production process, the system automatically gives The system can automatically give an alarm to remind, and the system can be forced to stop in case of emergency to avoid personnel misoperation.

4 Conclusion

The digital transformation program of pellet production in CJNF is implemented in phases with a master plan. By doing information network construction and equipment data interface modification on the basis of industrial control network security, it provides the foundation for the improvement of professional management modules in the upper systems with the full realization of automatic data acquisition and multi-function integration. It is proved that the pellet production line has been modernized with the upgraded

MES, where the advanced production module, quality module, equipment module and others were launched and optimized.

References

1. Chen Baoshan. Fuel assemblies for light water reactors [M]. 1st edition. China: Chemical Industry Press, 2007
2. Li Yongzhan. Manufacturing Execution System structure [J]. Automation Expo, 2008;25(3):57–59
3. Wei Y. Zhou, Wu Y., Li F. Bo. Latest nuclear fuel cycle [M]. 1st edition. China: Shanghai Jiaotong University Press, 2016
4. Li Hu. Analysis and design of quality management module in MES [D]. Wuhan University of Technology, 2008

Application of WANO Fuel Reliability Indicator in CGN Operating PWR Units

Pengtao Fu(✉) and Zhijun Li

China Nuclear Power Technology Research Institute Co, Ltd, Shenzhen, China
{fupengtao,lizhijun}@cgnpc.com.cn

Abstract. The fuel cladding is the first barrier to prevent the release of radioactive material into the environment in the reactor. The industry has focused on the improvement of fuel reliability in the past decades. The WANO fuel reliability indicator (FRI) is defined by the iodine volume activities in the primary loops adjusted by purification, the tramp uranium, and the power level of the unit. The WANO FRI has been widely used as the performance indicator of fuel integrity for peer review across the world for several decades. According to the CGN operation experience, the WANO FRI can reflect the fuel integrity at most of the operating cycles in CGN 1000 MWe PWR units. However, the FRI may become invalid at the following phenomenon: the calculated FRI may be less than the threshold when the defect size of failed rods is very tiny, or the calculated FRI may be higher than the threshold in the cycles without fuel failure but in which the primary loops is contaminated by the disseminated actinides in the previous cycles with fuel failure. It is found that the root cause of the phenomenon is that the release of Iodine-131 is significantly affected by the characteristics, e.g., defect size, the location of defects, and the linear power of failed rods which cannot be predicted before shutdown. One improvement of the current FRI has been proposed by CGN that way is to define FRI by the Xe-133 activity in the primary loops because the noble gas can release readily even though for the tiny fuel defect. It has been verified by the measured fission product in 58 cycles without fuel failure and 17 fuel cycles with fuel failure in CGN operating PWR units.

Keywords: Fuel failure · WANO FRI · Iodine-131 · Xenon-133 · Secondary degradation

1 Introduction

Fuel rods are the first barrier to confine the radioactive material in the nuclear power plant. For operational states (normal operation and anticipated operational occurrences), it is required to maintain the integrity of fuel elements; for design basis accidents it is required to ensure that any damage to fuel elements is kept to a minimum [1]. Once the fuel rods fail, some parts of the noble gases and the volatile fission products accumulated between the fuel pellet and fuel cladding will release to primary loops and lead to radiation dose and radioactive waste. In addition, the disseminated actinides in the fuel pellets can also be released to the primary loops due to secondary hydriding. When the fission

J. Liu and Y. Jiao (Eds.): WRFPM 2023, SPPHY 299, pp. 300–308, 2024.
https://doi.org/10.1007/978-981-99-7157-2_31

product activity in the coolant reaches some given extent, the plant must shut down within several hours to prevent further deterioration of the fuel rods. Therefore, it is important and necessary to diagnose the status of fuel failure in the core according to the monitored fission products.

A fuel reliability indicator (FRI) has been proposed by the World Association of Nuclear Operators (WANO) to monitor fuel reliability and the industry's progress in achieving high fuel integrity. This paper introduces the definition of WANO FRI in Pressurized Water Reactors (PWRs), and its application in 1000 MWe operating units in CGN (China General Nuclear Power Corporation). One suggestion to improve the FRI has been proposed based on the operation experience in CGN.

2 Definition of WANO FRI

In the early nuclear plant designed by Westinghouse, the ratio of the measured I-131 activity in the primary loops to the limit in the operation technical specification has been considered in the statistics of fuel failure before the 1980s and then has been substituted by the average measured I-131 activity during the 1980s. With the operation of more PWRs of different types, the WANO fuel reliability indicator has been gradually widely used for peer review since the 1990s in the nuclear industry. It is defined as the steady-state primary coolant I-131 activity, corrected for the tramp uranium contribution and power level, and normalized to a common purification rate. The monthly value of the PWR indicator is calculated as follows [2–4]:

$$FRI = \left[(A_{131})_N - k \times (A_{134})_N\right] \times [(L_n/LHGR) \times (100/P_0)]^{1.5} \quad (1)$$

where:

FRI = the average steady-state primary coolant I-131 activity corrected for the tramp contribution and power level and normalized to a common purification rate and LHGR (Bq/g).

$(A_{131})_N$ = the average steady-state activity of I-131 in the coolant normalized to a common purification rate (Bq/g).

k = the tramp correction coefficient 0.0318 which is based on a tramp material composition of 30% uranium and 70% plutonium.

$(A_{134})_N$ = the average steady-state activity of I-134 in the coolant normalized to a common purification rate (Bq/g).

L_n = the linear heat generation rate used as a basis for normalization 18.0 kW/m (kW/m).

$LHGR$ = the average linear heat generation rate at 100% power for the unit (kW/m).

P_0 = the average reactor power at the time activities are measured (%).

Tramp contribution is caused by fissionable material that has been deposited on reactor core internals from previous defective fuel elements or is present on the surface of fuel elements from the manufacturing process. For this indicator, tramp material composition is assumed to be uranium (30%) and plutonium (70%). The purification rate is the fraction of the reactor coolant system mass passing through the clean-up system each second. Linear heat generation rate is the average core thermal power in kilowatts per meter or foot at 100% power.

Based on information obtained from fuel vendors, a reactor core containing one or more defects is likely to produce indicator values (under steady-state conditions) 19 Bq/g (5.0E-04 μCi/g) for EPRs, PWRs, and PHWRs. Cores that have fuel reliability indicator values equal to or less than these values have a high probability of containing no steady-state fuel defects.

3 The Application in CGN Operating Units

CGN's first nuclear station uses reactors designed and built by Framatome, specifically the M310 plants at Daya Bay Plant. Based on the M310 type reactor, CGN has developed an improved Generation II pressurized water reactor CPR1000. The characteristics of these reactors, e.g., configurations, purification, and operating mode are similar. These reactors take up a large proportion of all the reactors being built in China. The typical reactor design, fuel design, and operating parameters in CGN 1000 MWe PWR units have been listed in Table 1. Most of these units have loaded AFA 3GAA fuel assemblies and operated with 18 months fuel management in recent years.

Table 1. The typical characteristics of CGN 1000 MWe PWR

Parameter	Value
Core power	2895 MW
Electricity output	1089 MWe
Loops	3
Linear power	186 W/cm
Assemblies in core	157
RCP pressure	15.5 MPa
RCP flow rate	3 × 23790
Water mass in RCP	~180 ton
RCP pressure	15.5 MPa
Fuel type	AFA 3GAA
Active height	365.76 cm
Fuel cladding	M5 alloy
Assembly configuration	17 × 17
Rods per assembly	264
^{235}U enrichment	4.45 wt%
Nominal letdown flowrate	13.6 t/h
Max letdown flowrate	27.2 t/h
Inlet coolant temperature (°C)	293.0
Cycle length	~18 months

The annual failure rate of fuel rods in CGN 1000 MWe PWR units from 2001 to 2020 has been presented in Fig. 1. Most of these fuel rod failures happened at the first three cycles and the average value of fuel rod failure rate in these cycles is 3.2E-06. After the third cycle, the average value of fuel rod failure rate in the fuel cycles has decreased to 1.2E-06. The examinations show that debris and fabrication are the main causes of these fuel cycles. To improve fuel reliability, several improvements have been used in CGN operating units, e.g., the use of the anti-debris device at the bottom nozzle of the fuel assembly, the strict management of debris, and the improvement of welding technology.

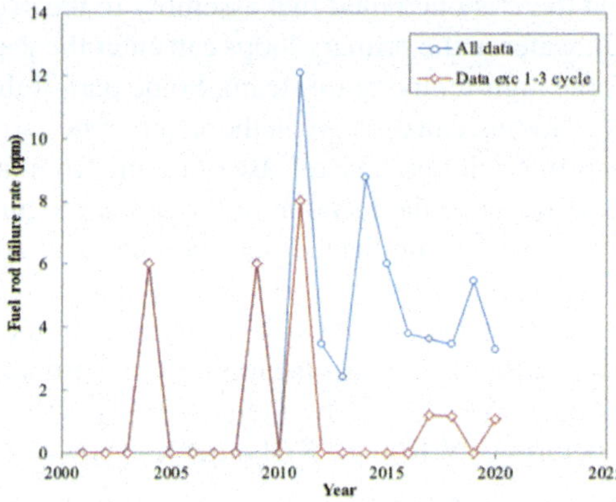

Fig. 1. Trend of fuel rod failure rate in CGN

The statistical results of fuel rod failure rate in the USA, Europe, Japan, Korea, France, and other counties in PWR have been listed in Table 2 [5]. The average fuel rod failure rate from 2006 to 2015 in the world is 3.7E-5. Therefore, the fuel rod failure rate in CGN 1000 MWe units is much better than the average level in the world.

Table 2. Results of fuel rod failure rate in the world

Region	Furl rod failure rate (ppm)	
	1994–2006	2006–2015
World	86.8	36.7
USA	131.6	49.1
France	56.9	31.3
Rest of Europe	108.1	34.7
Republic of Korea	40.5	16.5
Japan (2006–2010)	3.7	10.3

In most cycles of CGN 1000 MWe units, the integrity of fuel rods can be well evaluated by the calculated WANO FRI, i.e., FRI $<$ 19 Bq/g for cycles without fuel failure and FRI > 19 Bq/g for cycles with fuel failure. However, some phenomena have been found that the WANO FRI cannot reflect the real fuel integrity and the two typical exceptions have been introduced as follows.

- ***One typical exception***

At one cycle with fuel failure, it is difficult to detect I-131 in the primary loops although the detection limit is as low as the order of magnitude of several Bq/g. According to the measured Xe-133 and Xe-135, ratios Xe-133/Xe-135 have continuously reached 1.0 which indicates the fuel failure in the core. After the outage, the sipping test and other examinations show that there is one failed fuel assembly in the cycle.

When fuel rods fail, water in the primary loops can enter the gap between pellets and fuel cladding and become high-temperature steam. Some part of the iodine deposited in the inner surface of the cladding can dissolve in the steam or be carried over by the steam and then release to the primary loops. The release of iodine from failed rods depends on some factors, e.g., the defect size, the location of the failed rod and the power of failed rods. The phenomenon is very complicated and the chemical species of iodine may change in the condition. There isn't some formula to quantitatively rule the mitigation of iodine from failed rods. When the defect size is very tight or the relative power of the failed rod is very low, it is difficult to detect the iodine in the primary loops and therefore it is impractical to calculate WANO FRI by iodine activity. The phenomenon has also been found in other operating PWR, e.g., Krško nuclear power plant [6] and Siemens nuclear power plant [7].

- ***The other typical exception***

In some cycles with fuel failure, secondary degradation can lead to large defects with several centimeters. Some parts can release some amount of disseminated actinides can release to the primary loops and deposit on the surface of fuel rods and the primary loops. Figure 2 presents the trend of WANO FRI in continuous three cycles in one CGN unit. Fuel failure is suspected in the N cycle owing to high FRI and one failed fuel rod has been identified by the sipping examination and post-irradiation examinations. At the N + 1 cycle, fission product volume concentration is high and WANO FRI at most of the period is greater than criterion 19 Bq/g, but all the fuel rods are intact which has been proved by the ratio of I-131/ I-133 and Xe-133/Xe-135. The cause is that the secondary degradation of failed fuel rod happened and the composition of the disseminated actinides from the failed fuel rod at the N cycle doesn't remain the assumption with the composition of 30% uranium and 70% plutonium. The post-irradiation examinations (PIE) confirmed that six defects on the failed fuel rod had been inspected and the maximum defect size reached the order of centimeters in diameter. In addition, the results of gamma scanning of the defective rod have confirmed that several grams of disseminated fissile material have been lost in the failed fuel rod [8, 9].

4 The Improvement and Verification

The release of fission product from the failed rods is very complicated and volume activities of iodine and noble gases are affected by many factors, e.g., the relative power of failed rods, the number of failed rods, defect size, the distribution of defects in the

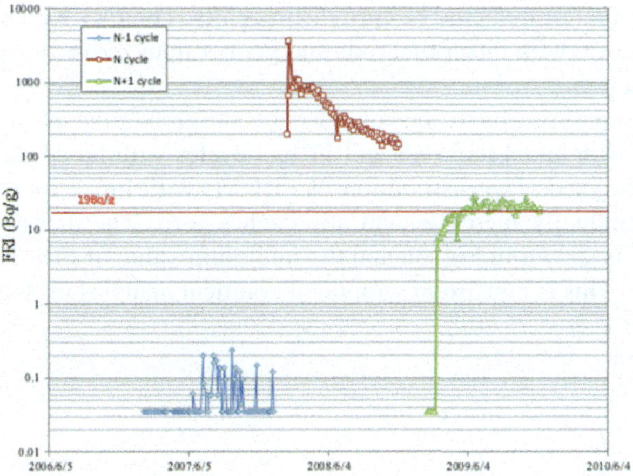

Fig. 2. Comparison of WANO FRI in three cycles in CGN

failed rods and the occurrence of secondary degradation. It is impossible to determine these key factors according to the measured radiochemical data in the primary loops before the examination after shutdown.

The improvement of WANO FRI has been launched and discussed in the industry in recent years. For the above two typical situations, the following suggestion has been proposed and the new FRI can be defined by the noble gas activities.

In comparison with the release of iodine, the noble gases can readily release from failed rods due to their chemically inert behavior even if the defect size is tiny. One improvement for the fuel reliability indicator is to replace the I-131 and I-134 with Xe-133 and Xe-138 respectively. The improved fuel reliability indicator FRI-XE can be defined as follows.

$$\text{FRI - XE} = \left[(A_{133})_N - k_{\text{XE}} \times (A_{138})_N\right] \times [(L_n/LHGR) \times (100/P_0)]^{1.5} \quad (2)$$

where:

FRI - XE = the average steady-state primary coolant Xe-133 activity corrected for the tramp contribution and power level and normalized to a common purification rate and LHGR (Bq/g).

$(A_{133})_N$ = the average steady-state activity of Xe-133 in the coolant normalized to a common purification rate (Bq/g).

k_{XE} = the tramp correction coefficient which can envelop the fuel composition of 0%-5% U-235 enrichment according to the recoil release.

$(A_{138})_N$ = the average steady-state activity of Xe-138 in the coolant normalized to a common purification rate (Bq/g).

The definitions of other symbols are the same as that in the formula (1).

The correction coefficient k_{XE} is affected by the relative composition of U-235 and Pu-239. According to the accumulative yields Xe-138 by U-235 and Pu-239, when the correction coefficient k_{XE} is 1.4, the influence of contaminated uranium can be deducted from the FRI-XE value to the greatest extent. The maximum value of FRI-XE in a fuel cycle with intact fuel rods should be zero theoretically. The FRI-XE should be less than zero in the fuel cycle with intact fuel rods and greater than zero theoretically in the fuel cycle with failed rods.

To verify the analysis above, the volume concentration of the measured fission product in 58 cycles with intact fuel rods and 17 fuel cycles with failed fuel rods have been collected and analyzed. The distribution of FRI-XE with intact fuel rods is shown in Fig. 3a and it gradually increases and approaches zero. The distribution of FRI-XE with fuel failure is shown in Fig. 3b and the minimum value of FRI-XE in these cycles approaches zero. This is consistent with the theoretical analysis in Sect. 3.

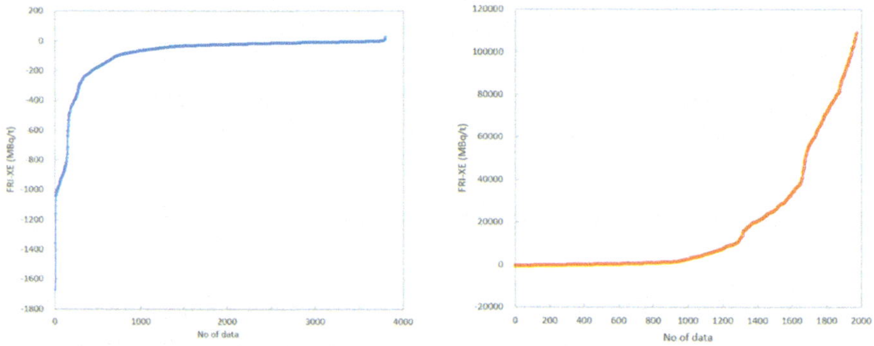

(A) Cycles without fuel failure

(B) Cycles with fuel failure

Fig. 3. Distribution of FRI-XE in the fuel cycles

Figure 4 presents the local distribution of FRI-XE for the fuel cycle without intact fuel rods and failed fuel rods and the overlapping area of a few discrete data exist which accounts for about 2% of the total data based on the statistical analysis. When the threshold of FRI-XE 37 MBq/t has been taken, all of the fue failures can be readily distinguished. It will also help to partially eliminate the misjudgment caused by data sampling and measurement.

According to fuel reliability, the operating PWR units can be the following four levels:

- **Excellent level**: The fuel rods loaded in the reactor are intact and the uranium contamination in the primary circuit is extremely low. It is usually called "the clean reactor" in the industry.
- **Advanced level**: The fuel rods loaded in the reactor are intact, but the reactor has been more contaminated by the severe fuel damage in the previous cycles.
- **Intermediate level**: A small number of fuel rods loaded in the reactor are defective.
- **Worst level**: A relatively large fraction of the fuel rods load in the reactor has failed or a small number of high-power fuel rods has failed.

It is noted there are no strict boundaries for the four levels, especially between the intermediate level and the worst level, because the higher FRI-XE doesn't necessarily result from more failed fuel rods. The relationship between four levels of fuel reliability

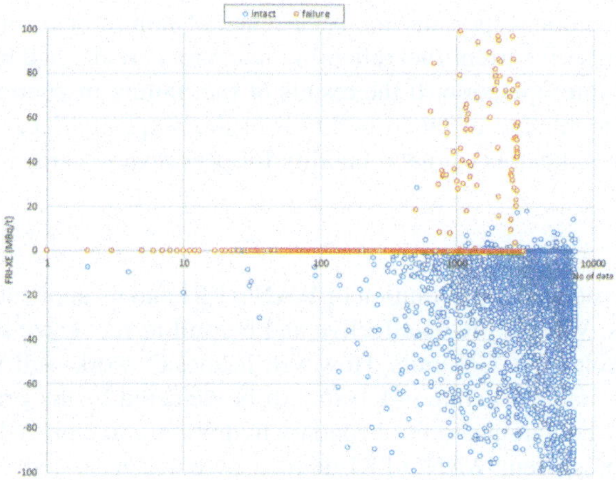

Fig. 4. The local distribution of FRI-XE in the fuel cycle with and without fuel failure

and the status of fuel integrity has been illustrated in Fig. 5 and these boundary values are preliminarily proposed based on the operation experience of PWR in CGN. The prediction of fuel integrity for all the operating PWRs in CGN is consistent with the sipping examinations and post-irradiation examinations.

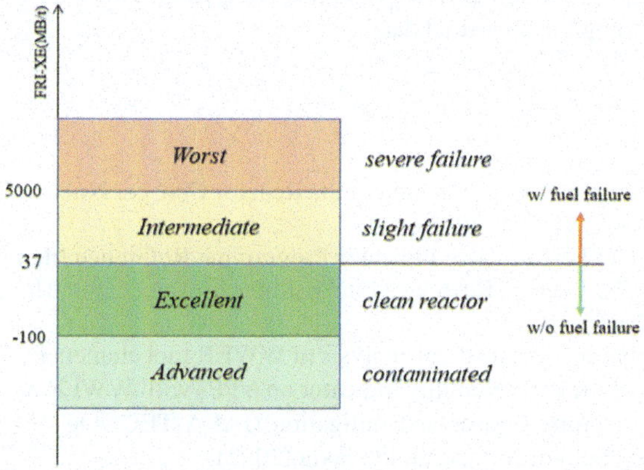

Fig. 5. Schematic diagram of FRI-XE and status of fuel reliability

The degree of fuel failure can be evaluated from many aspects, e.g., defect size and the release of activity of fission products. The number of failed fuel rods in the fuel cycles may be the most important characteristic of fuel failure for the utilities of operating reactors because it will impact the plan for fuel assemblies loading at the following cycle and the economic assessment of nuclear fuel. Therefore, the corresponding failure percentage in one fuel cycle may be much more effective for the evaluation of fuel failure in PWR reasonably. However, it is still difficult to make an accurate prediction of the number of failed fuel rods based on the measured activities of fission products during power operation, especially the occurrence of failure of several dozens of fuel rods. It is noted that all current fuel reliability indicators are defined by coolant activity measured during unit operation. If the results of fuel failure inspection after shutdown are acceptable, the number of failed fuel rods or the proportion of failed fuel rods are also very suitable indicators of fuel reliability for peer review in the industry.

5 Summary

The paper has introduced the definition of WANO FRI and its application in CGN 1000 MWe PWR units. WANO FRI is effective and helpful to recognize the fuel reliability in most of the cycles in CGN units. However, it doesn't work well in other cycles in which the release rate of iodine is too low to be detected or the primary loops have been significantly contaminated by secondary hydrating in the preceding cycles. In the comparison of the current WANO FRI defined by corrected I-131 activity, CGN has proposed one suggestion to improve the current FRI, using the corrected Xe-133 activity in the primary loops because the noble gas can readily release even though the defect size in the failed rods is tiny. It has been verified by the measured fission product in 58 cycles without fuel failure and 17 fuel cycles with fuel failure in CGN operating units.

Acknowledgments. Thanks to Peng Xu and Yaoxin Ye in China Nuclear Power Technology Research Institute Co. Ltd., Xiao Yuan in Fangchenggang Nuclear Power Co., Ltd., Shan Liang in Yangjiang Nuclear Power Co., Ltd., Anlong Zhang in Daya Bay Nuclear Power Operation Management Co., Ltd., and Jingjing Liu in Liaoning Hongyanhe Nuclear Power Co., Ltd. for their discussion and providing the measured data.

References

1. Anegawa, T., Brighton, P., et al., Design of the Reactor Core for Nuclear Power Plants (NS-G-1.12), pp.5–6, IAEA, Vienna (2005)
2. R. Chiarelli, WANO Performance Indicator Programme Reference Manual (2014)
3. Alvarez, L., Daniels, T., et al., Review of Fuel Failures in Water Cooled Reactors, No. NF-T-2.1, IAEA, Vienna (2010)
4. I. Chestakov, Probabilistic-statistical analysis of WWER fuel element leaking causes and comparative analysis of the fuel reliability indicator on NPPs with WWER and PWR reactors, Fuel failure in water reactors: Causes and mitigation (IAEA-TECDOC-1345), Proceedings of a Technical Meeting held in Bratislava, Slovakia (2002)
5. Alonso Pacheco J.M., Armstrong, J., et al., Review of Fuel Failures in Water Cooled Reactors (2006–2015), pp.9–10, IAEA, Vienna (2019)
6. Aleksandra Antolovič, Bojan Kurinčič: NPP Krško Experiences with Spent Fuel Assembly Inspection Methods. In: International Conference of Nuclear Energy for New Europe 2008(PORT2008), pp.1011.1–1011.11, Nuclear Society of Slovenia, Slovenia (2008)
7. Schuster, E., Von Jan R., Fischer, G.: Evaluation of Fuel Performance from Coolant Activity Data. In: Proc. Int. Top. Mtg LWR Fuel Performance, Avignon, 1991, Vol. 1, SFEN, Paris (1991)
8. X. Wang, F. Chu, Z. Liang, PIE and Cause Analysis of the Failed AFA-3G Fuel Rods (in Chinese), Annual report for China Institute of Atomic Energy (2016).
9. Fu, Pengtao. Determination of escape rate coefficients of fission products from the defective fuel rod with large defects in PWR. Nuclear Engineering and Technology (2023)

Creation of Smart Polydimethylsiloxane Sponge for Selective Recovery of Lanthanides from Radioactive Wastes

A. Yiwei Zhang, B. Ki Chul Park, C. Naokazu Idota, and D. Takehiko Tsukahara(✉)

Tokyo Institute of Technology, Tokyo, Japan
{zhang.y.bf,idota.n.aa,tsukahara.t.ab}@m.titech.ac.jp

Abstract. Mutual separation of lanthanides and minor actinides from high-level liquid wastes can effectively improve the efficiency of partitioning and transmutation strategy, reducing the potential risks of high-level liquid wastes to the environment. Conventional separation techniques have been shown efficiency in lanthanides separation but are limited by their long operation time and potential hazards to environment. In our research, a low-cost adsorbent was developed by impregnating lipophilic extractant bis(2-ethylhexyl) phosphate (HDEHP) in the polydimethylsiloxane (PDMS) sponge via hydrophobic interaction, aimed to selectively and efficiently remove lanthanides from high-level liquid wastes. The obtained PDMS-HDEHP sponge can rapidly adsorb lanthanide ions and showed the best lanthanide adsorption performance in the pH3 nitric acid solution. Via nonlinear fitting the experimental results by Langmuir adsorption isotherm model, we found that the maximum lanthanide adsorption capacity of PDMS-HDEHP can reach 8.06–21.51 mg/g depending on the ionic radii and charges, higher than many adsorbents. Furthermore, nearly all of adsorbed lighter lanthanide ions can be recovered into the solution after ethylenediaminetetraacetic acid (EDTA) treatment. It is noteworthy that synthesized PDMS-HDEHP sponge can still maintain lanthanide adsorption ability after desorption, indicating that the obtained sponge owns tremendous potential in nuclear waste management due to its high adsorption capacities and regeneration abilities.

Keywords: PDMS sponge · Lanthanides · Adsorption · High-level liquid wastes

1 Introduction

Nuclear energy, as an important sustainable energy source, plays an indispensable role in achieving the goal of a zero-carbon society. With the development of advanced nuclear power techniques such as small modular nuclear-fission reactors (SMRs), the attention and worries about the potential risks of nuclear wastes and related environmental problems also augment [1, 2]. High-level liquid wastes (HLLWs) easily cause potential risks to operators and environments due to the existence of minor actinides (MAs) and radioactive lanthanides [3, 4]. To overcome these issues, partitioning and transmutation strategy is hopeful to minimum the negative impacts to the environments via transforming radioactive isotopes to short-lived or stable nuclides [5, 6]. In addition, lanthanides

J. Liu and Y. Jiao (Eds.): WRFPM 2023, SPPHY 299, pp. 309–316, 2024.
https://doi.org/10.1007/978-981-99-7157-2_32

play important roles in catalysis, electric devices, medical imaging agents [7] due to their characteristics such as unique optical and magnetic properties, etc. [8]. Therefore, recovering lanthanides from HLLWs can effectively reduce the potential risks of nuclear wastes and is benefit for the sustainable development of nuclear energy, for another, it also has very high economical value.

Solvent extraction is one of the typical methods to separate them, however, it requires large amounts of organic solvents, easily resulting in secondary wastes [9]. Recently, adsorption was widely considered as an optimistic technology for nuclear wastes management because it is economic, environmentally friendly, and easier to operate compared with solvent extraction [10]

Porous polydimethylsiloxane (PDMS) sponge is expected to become a novel adsorbent with its unique Si-O-Si structure, mechanical performance, high flexibility, large surface area, etc., has been widely used in oil and water separation [11, 12]. In our study, we aimed to develop a functional PDMS sponge via a facile manufacturing approach. We selected bis(2-ethylhexyl) phosphate (HDEHP) as functional ligand via immobilizing it in the sponge to selectively remove existing lanthanides in simulated HLLWs. Physicochemical characteristics and mass spectrometry were further investigated to evaluate the properties of the sponge and its possibilities for lanthanide recovery. Additional adsorption experiments were conducted in the simulated HLLWs to examine the selectivity of PDMS-HDEHP sponge towards other competitive metallic ions for the lanthanides. Moreover, immobilized HDEHP in the sponge was proven to remain and continuously work as functional groups after elution. A novel and functional PDMS-HDEHP sponge with high adsorption performance has been developed.

2 Experimental

2.1 Fabrication of PDMS-HDEHP Sponge

Briefly, 5g sylgard 184 base polymer (Dow Corning Co., Michigan, USA) was mixed with 0.5g its curing agent in a polypropylene cup with 8 mL cyclohexane (Tokyo Chemical Industry Co., Tokyo, Japan). After mixing completely, gradually added the ground citric acid monohydrate (Kanto Chemical Co., Tokyo, Japan) powders under constantly stirring. The mixture was then placed in vacuum enough time to remove air, and finally cured the mixture in the oven at 70 °C for 10 h. Repeatedly soaking the obtained PDMS cube in water and ethanol for certain hours to gradually remove the template and remaining cyclohexane, the bare PDMS sponge could be obtained after washing and drying.

In this study, the adsorbent was fabricated following a facile procedure. First, picking up certain amounts of HDEHP (Tokyo Chemical Industry Co., Tokyo, Japan) and then diluted them in n-hexane (Kanto Chemical Co., Tokyo, Japan), constantly stirred the mixture for 30 min in a round-bottom flask. The bare PDMS sponge was then placed into the flask and gently shake the mixture until the sponge fully contacted with organic solvent. HDEHP can immobilize in the sponge via hydrophobic interactions by gradually evaporating the excess hexane. Washing the sponge several times to remove the remaining hexane and HDEHP, then dried the resulting sponge and labelled the obtained adsorbent as PDMS-HDEHP sponge.

2.2 Adsorption Experiments

We chose lanthanum, samarium and ytterbium as representative light, medium and heavy rare earth elements to analyze the adsorption performances of PDMS-HDEHP sponge in aqueous solutions. Stock solutions were prepared by mixing lanthanide nitrates ($Ln(NO_3)_3 \cdot (H_2O)_n$ (n = 3–5)) in nitric acid ranging from pH1 to pH7, individually. The acidity of solution was adjusted by ultrapure water, nitric acid and sodium hydroxide.

To evaluate the performances of PDMS-HDEHP in HLLWs, several of existed elements were selected to simulate the chemical constitutes in HLLWs. Impacts from uranium are negligible due to its low concentrations in HLLWs, while La (III), Ce (III), Eu (III) and Lu (III) represent the rare earth elements that existed in HLLWs. Each time collecting supernatant by syringe attached a 10 μm filter for further experiments. Concentration of each metal were detected by inductively coupled plasma mass spectrometry (ICP-MS) after dilution. Adsorption capacity (qe, mg/g) was applied for analyzing:

$$q_e = \frac{(C_0 - C_e)V}{m} \quad (1)$$

where C_0 (mg/L) and C_e (mg/L) represent the initial and equilibrium concentration, respectively. V (L) is the solution volume, and m (g) means the mass of the adsorbents.

3 Results and Discussion

3.1 Characterization of PDMS-HDEHP Sponge

The fabricated PDMS sponge is a 3-dimensional white cubic (Fig. 1(a)) with hierarchical porous structure, macro pores distributed in the bare PDMS sponge with 60–70% porosity, as Fig. 1(b)–(c) shows. Wettability of PDMS sponge transformed after HDEHP immobilizing, water contact angle changed from 135° to 76° (DME-211, Kyowa, Japan). Through comparing the attenuated total reflection infrared spectroscopy (ATR-IR; FT/IR-4700, JASCO Co.) results before and after HDEHP attaching, different chemical bonds were observed. As Fig. 2(a) shows, the peaks located around 1664 cm^{-1} and 1186 cm^{-1} are ascribed to the P-OH stretching and P = O stretching, respectively. The strong peak ranged from 1009 to 1077 cm^{-1} are both due to the Si-O-Si and P-O-C stretching vibration band. Another peak at 786 cm^{-1} is Si-CH_3 rocking band from PDMS sponge backbone.

Combustion experiments were conducted to investigate the immobilization rate of HDEHP via DTG-60H (SHIMADAZU Co., Kyoto, Japan). By comparing the Thermalgravimetric (TG) curves of PDMS-HDEHP sponge and bare PDMS sponge in Fig. 2(b), a new region of weight loss can be obviously detected, starting from 144 °C and ending at 231 °C, mainly because of the decomposition of immobilized HDEHP in the sponge. Around 14.6% weight loss is assigned to the coverage extractants from thermalgravimetric curves, calculated impregnated amounts of HDEHP is basically consistent with the mass of theoretical results.

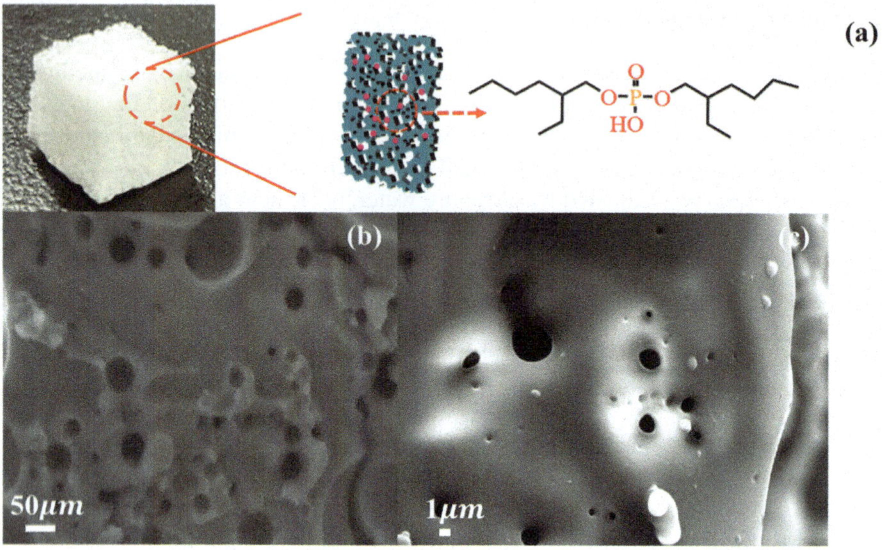

Fig. 1. (a) optical images of bare PDMS sponge, scanning electron microscope results of bare PDMS sponge at (b) 50 μm and (c) 1 μm.

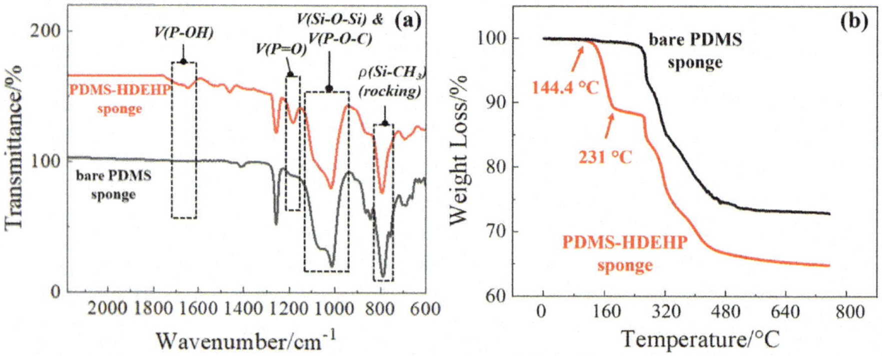

Fig. 2. (a) ATR-IR results of PDMS-HDEHP sponge and bare PDMS sponge, (b) TG results of bare PDMS sponge and PDMS-HDEHP sponge.

3.2 Adsorption Results

Figure 3(a) shows the impacts of nitric acid concentration on lanthanide adsorption performances of PDMS-HDEHP sponge. Lanthanum, samarium and ytterbium are chosen to represent the light, medium and heavy rare earth elements, respectively. Lanthanide adsorption abilities of PDMS-HDEHP sponge gradually increase with the decreasing concentration of nitric acid between pH 1–3, the sponge exhibits the maximum adsorption capacity to lanthanum, samarium and ytterbium in pH3 nitric acid, reaching 7.11 mg/g, 8.68 mg/g and 20.6 mg/g, respectively. Then, the adsorption performances lessen along with the decreasing acidity. This interesting phenomenon is in reference to the properties of HDEHP, an acid with pKa around 1.5, adsorption capacity was negatively affected when the solution acidity increased, and in more neutral solutions, poor protonation of HDEHP resulted in limited adsorption performances. Notably, PDMS-HDEHP sponge showed different adsorption capacity towards lighter and heavier lanthanide ions, inversely proportional to the ionic radii. Lower ionic radii resulted in the stronger interaction between lanthanides and HDEHP, forming the more stable complexation.

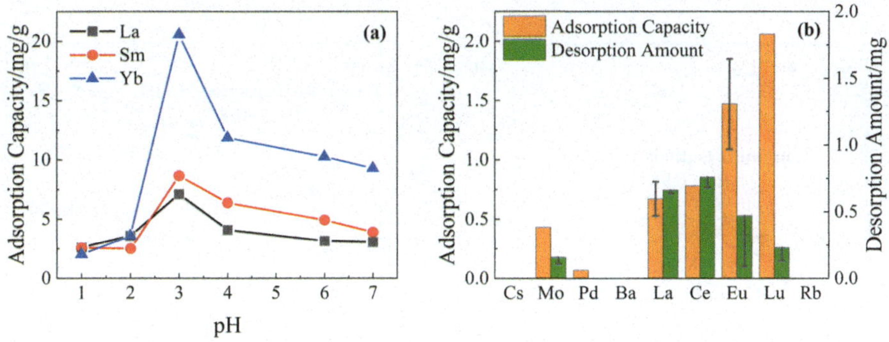

Fig. 3 (a) lanthanides adsorption performances of PDMS-HDEHP sponge in nitric acid solution with different pH, (b) PDMS-HDEHP sponge adsorption & desorption performances in simulated HLLWs.

Fission reaction, reprocessing of spent nuclear fuel (SNF), corrosion of protective materials etc. contributed to the complex chemical environment in high-level liquid wastes. Caesium, molybdenum, palladium, barium, rubidium and lanthanides are selected to prepare the simulated waste solution in pH3 nitric acid to investigate the feasibility of PDMS-HDEHP sponge under the impacts of competitive ions in high-level liquid wastes. As implied in Fig. 3(b), PDMS-HDEHP sponge still exhibited high affinity to lanthanides in the multi-element enviroment, especially to the heavy rare earth elements, adsorption capacity varied from 0.67 mg/g to 2.06 mg/g. Around 50% molybdenum and 7% palladium existed in the simulated HLLWs were also removed by PDMS-HDEHP sponge.

Adsorption isotherm models were introduced to evaluate the maximum lanthanide adsorption capacity of PDMS-HDEHP sponge. La (III) and Yb (III) were chosen as representatives to conduct experiments in room temperature, their concentrations were varied from 0.001 mg/L to 20 mg/L. Typical Langmuir and Freundlich isotherm models were introduced to fit the experimental data. And the equations are shown below.

Langmuir Isotherm model:

$$q_e = \frac{C_e K_L q_{max}}{1+C_e K_L} \quad (2)$$

Freundlich isotherm model:

$$q_e = K_f C_e^{\frac{1}{n}} \quad (3)$$

where q_e (mg/g) and q_{max} (mg/g) mean the equilibrium and maximum lanthanide adsorption capacity of PDMS-HDEHP sponge, respectively. C_e (mg/L) stands for the Ln (III) equilibrium concentration. K_L is Langmuir constant, while K_f and n are Freundlich constants.

Results (Fig. 4(a), (b)) showed that lanthanides adsorption capacity increased with the increasing metal concentration as expected, and gradually reached plateau. Langmuir isotherm model can better explain the adsorption behavior with higher R^2. The calculated maximum adsorption capacity of La (III) and Yb (III) are 8.06 mg/g and 21.51 mg/g, respectively, in agreement with the experimental data.

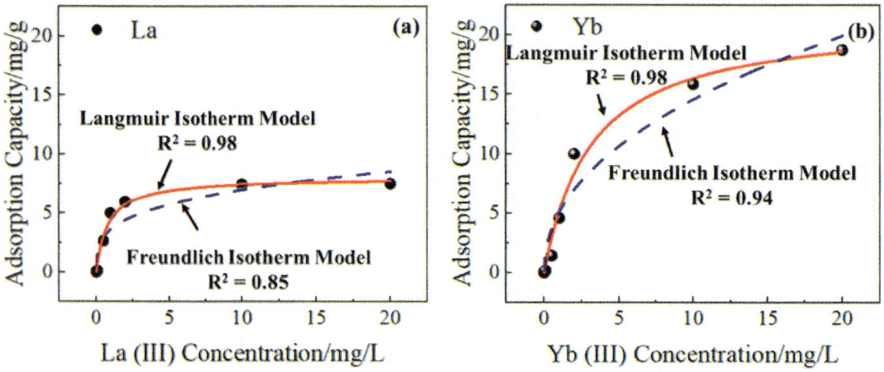

Fig. 4 Langmuir adsorption isotherm model and Freundlich adsorption isotherm model for (a) lanthanum and (b) ytterbium adsorbed in the PDMS-HDEHP sponge.

3.3 Desorption Results

After existed lanthanides in the simulated HLLWs were successfully adsorbed in the sponge, desorption experiments were conducted by immersing the obtained sponge in the ethylenediaminetetraacetic acid (EDTA) solutions and constantly shaking the mixture for certain hours, repeated two times. As Fig. 3(b) shows, the adsorbed lanthanide ions can be successfully separated from the adsorbent after shaking. Nearly 100% of La (III) and Ce (III) can be separated from the adsorbent, 32% of Eu (III) and only 11% of Lu (III) can detach from the adsorbent after EDTA treatment. Adsorbed lanthanide ions can be rapidly detached from the sponge within the first few hours, then the efficiency gradually decreases with the increasing reaction time, which is possibly due to the variations of acidity in the solution caused by the reaction. Poor desorption performance for Lu (III) is possibly due to the stable complexation of heavier lanthanide ions and HDEHP. Moreover, parts of Mo (VI) also can be recovered into the EDTA solutions together with Ln (III). We also explored the regeneration ability of the sponge and found that the sponge still maintained adsorption ability for lanthanides after desorption.

4 Conclusion

In summary, we successfully synthesized a 3-dimensional PDMS sponge named PDMS-HDEHP sponge with porous structure. The resultant PDMS-HDEHP sponge performs maximum lanthanide adsorption capacity in 0.001 M nitric acid solution, ranging from 8.06 mg/g to 21.51 mg/g depending on the lanthanides. The adsorption mainly relied on the complexation, sensitive to the acidity of the solution and inversely proportional to the ionic radii. Moreover, the as-made sponge still exhibited high lanthanide selectivity in simulated HLLWs, adsorbed lighter lanthanide ions can easily be recovered from the adsorbent via EDTA treatment. According to these results, the synthesized PDMS-HDEHP sponge exhibits high potential for efficiently lanthanides removal from waste solutions due to its facile synthesis procedures, high lanthanide selectivity, and low costs.

Acknowledgements. This work was partially supported by JAEA Nuclear Energy S&T and Human Resource Development Project Grant Number JPJA22F 22717857 and JST SPRING, Grant Number JPMJSP2106.

References

1. Krall L M, Macfarlane A M, Ewing R C.: Nuclear waste from small modular reactors. Proceedings of the National Academy of Sciences 119(23), e2111833119 (2022).
2. Goel A, McCloy J S, Pokorny R, et al.: Challenges with vitrification of Hanford High-Level Waste (HLW) to borosilicate glass–An overview. Journal of Non-Crystalline Solids: X 4, 100033 (2019).
3. Zhang H, Li A, Li K, et al.: Ultrafiltration separation of Am (VI)-polyoxometalate from lanthanides. Nature 616(7951), 482-487 (2023).
4. Xu L, Pu N, Li Y, et al.: Selective separation and complexation of trivalent actinide and lanthanide by a tetradentate soft–hard donor ligand: Solvent extraction, spectroscopy, and DFT calculations. Inorganic chemistry 58(7), 4420-4430 (2019)
5. Wang Z, Huang L, Dong X, et al.: Ion sieving in graphene oxide membrane enables efficient actinides/lanthanides separation. Nature Communications 14(1), 261 (2023).
6. Yang X, Su D, Song L, et al.: Extraction and complexation of americium (III) and europium (III) by a N-donor ligand of 3, 5-bis (2-pyridyl) pyrazole. Separation and Purification Technology 232, 115969 (2020).
7. Song A M, Zhang F D, Hu Q H, et al.: Rational design and synthesis of diimide-based metal-organic frameworks for lanthanides recovery from tailing wastewater. Cell Reports Physical Science 3(11), 101120 (2022).
8. Blackburn O A, Edkins R M, Faulkner S, et al.: Electromagnetic susceptibility anisotropy and its importance for paramagnetic NMR and optical spectroscopy in lanthanide coordination chemistry. Dalton transactions 45(16), 6782-6800 (2016).
9. Park K C, Tateno H, Tsukahara T.: Solid phase extraction based on the phase transition of poly (N-isopropylacrylamide): the extraction behaviour of lanthanide (iii) ions in highly acidic solutions. Reaction Chemistry & Engineering 3(1), 48-54 (2018).
10. Tang J, Zhao J, Wang S, et al.: Pre-modification strategy to prepare a novel Zr-based MOF for selective adsorption of Palladium (II) from solution. Chemical Engineering Journal 407, 127223 (2021).

11. Zhao J, Chen H, Ye H, et al.: Poly (dimethylsiloxane)/graphene oxide composite sponge: a robust and reusable adsorbent for efficient oil/water separation. Soft Matter 15(45), 9224-9232 (2019).
12. Zhao X, Li L, Li B, et al.: Durable superhydrophobic/superoleophilic PDMS sponges and their applications in selective oil absorption and in plugging oil leakages[J]. Journal of Materials Chemistry A 2(43), 18281-18287 (2014).

Friction Corrections to Improve Accuracy of Cladding Strength Measurements from the Ring Tension Test

R. S. Hansen(✉), D. W. Kamerman, P. G. Petersen, and F. Cappia

Idaho National Laboratory, Idaho Falls, USA
robert.hansen@inl.gov

Abstract. Evaluation and qualification of accident tolerant fuel (ATF) cladding concepts requires highly robust mechanical testing, particularly for assessing material property variation due to irradiation. The ring tension test (RTT) is a possible test used to provide hoop direction material properties for anisotropic cladding and other tubular materials. Several configurations of the RTT have been implemented, each with different advantages and challenges. The single gauge RTT configuration with a hemicylindrical mandrel is ideal because of excellent stress state uniformity in the gauge region compared to other configurations. However, the primary challenge for this configuration is sensitivity of measured forces to the friction between the ring and mandrel, which lead to inaccurate strength results. This has previously been mitigated through lubrication and assumed to be insignificant, but when testing irradiated cladding in hot cell conditions, these frictional effects should not be ignored. In this work, finite element modelling of the test configuration is used to investigate the relationship between applied mandrel force and the hoop tensile force carried at different locations in the ring. Several potential friction correction methods resulting from this investigation are presented, showing improved accuracy. An extended analysis of the sensitivity of corrections and associated parameters is given during simulated tests with a variety of test conditions and potential material properties. Ultimately, correcting for friction with these methods produces robust strength measurements with significant accuracy improvement, providing a method to obtain hoop direction properties from both fresh and irradiated cladding with high fidelity.

Keywords: Accident tolerant fuel · Cladding · Mechanical testing · Friction · Post-irradiation testing

This manuscript has been authored by Battelle Energy Alliance, LLC under Contract No. DE-AC07-05ID14517 with the U.S. Department of Energy. The United States Government retains and the publisher, by accepting the article for publication, acknowledges that the United States Government retains a nonexclusive, royalty-free, paid-up, irrevocable, world-wide license to publish or reproduce the published form of this manuscript, or allow others to do so, for United States Government purposes.

This work was funded by the Advanced Fuels Campaign of the U.S. Department of Energy's (DOE) Nuclear Fuel Cycle and Supply Chain (NFCSC) program. This

J. Liu and Y. Jiao (Eds.): WRFPM 2023, SPPHY 299, pp. 317–329, 2024.
https://doi.org/10.1007/978-981-99-7157-2_33

research made use of the resources of the High Performance Computing Center at Idaho National Laboratory, which is supported by the Office of Nuclear Energy of the U.S. Department of Energy and the Nuclear Science User Facilities under Contract No. DE-AC07-05ID14517.

1 Introduction

Mechanical testing is an important step in the evaluation and qualification of accident tolerant fuel (ATF) cladding concepts. The circumferential or hoop direction material properties are critical to cladding performance [1], and cladding materials are often anisotropic [2], leading to the need for direct measurement of hoop properties. However, this testing is not straightforward; it is important to maintain the tubular geometry which precludes a traditional ASTM-style dogbone specimen tensile test. Thus, many researchers have turned to the ring tension test (RTT) to assess hoop direction behavior for tubular material testing, with many variations in its implementation [3–8].

This work focuses on the hemicylindrical mandrel, also called "hemi", version of the RTT. This method consists of machining a ring specimen with a reduced-width gauge region which is placed over two hemicylindrical mandrels with diameters closely matching the ring. The mandrel grips are then pulled apart using a standard universal testing machine, inducing a tensile load in the ring specimen. The hemi method offers an important advantage over other RTT variations: by supporting the gauge region with the mandrel, the curvature of the gauge remains constant throughout the range of deformation. This eliminates the non-uniform stress states of combined tension and bending seen in several of the other variations [9]. By preventing the non-uniform bending, premature yielding (which causes inaccurate yield point determination) is also avoided. Additionally, this ensures nearly uniform stress through the cladding thickness with only negligible variations, which is of particular importance for investigations of coated cladding. The hemi test is also less sensitive to small variations in specimen and fixturing geometry [10], making it ideally suited for testing irradiated cladding in a hot cell environment where fine control of test parameters is more challenging. The primary drawback with the hemi method, however, is the friction between the ring specimen and the mandrels.

The RTT generally, and the hemi method in particular, has been shown to be sensitive to the effects of friction [11, 12]. Although support of the specimen's gauge region by the mandrel leads to an improved loading state, it also lengthens the span of the ring in contact with the mandrel. Friction between the two surfaces causes variation in the tensile load carried by the ring [13, 14]; if a greater length of the ring is in contact with the mandrel, there is a more pronounced drop in the internal tensile load. Thus, the hemi method has been linked to overestimation of strengths [15]. Researchers have sought to reduce friction with a variety of lubricants then assume the impact on measured material properties is negligible [16–18]. While a reduced friction coefficient does diminish the impact on measurement accuracy, in many cases the friction cannot be reduced enough to entirely dismiss. Mechanical test samples from irradiated cladding often contain residual fuel or fuel-cladding chemical interaction (FCCI) material, causing a high degree of inner surface roughness which in turn increases friction. Many preferred lubricants aren't compatible with high-temperature testing or are challenging to implement with the remote handling required for hot-cell testing, further complicating efforts to mitigate friction effects. If mitigation strategies cannot be put in place, the friction effect is significant and cannot be ignored.

This work offers a solution to the deleterious effects of friction in the hemi method of the RTT. By analyzing finite element (FE) models, the frictional behavior can be described with analytical methods. This analytical description can be used to precisely model the tensile load as a function of the azimuthal location, allowing for non-empirical corrections to experimental outputs without the need for additional FE modelling. This recently proposed approach is briefly outlined, and the sensitivity and broader applicability of these corrections is presented. The result is a highly robust process for accurately determining the strength properties of cladding and other tubular materials with the RTT method.

2 Finite Element Approach

To enable detailed investigation of the behavior of the ring specimen during a typical RTT experiment, an FE model of the ring and mandrels was generated in Abaqus standard with implicit analysis. The ring was meshed using linear 8-node reduced integration (C3D8R) elements to accurately model contact and prevent shear locking. The mesh was refined using peak mandrel load as the convergence criterion, resulting in a mesh density of roughly 10 elements/mm through the gauge region. Due to the high stiffness and strength of test fixturing relative to the materials being tested, the mandrels were modeled as analytical rigid surfaces; this improves computation time without sacrificing accuracy. Symmetry boundary conditions were also implemented to improve computation time, taking advantage of symmetry about the mid-width plane. The resulting model can be seen in Fig. 1.

The nominal dimensions of the ring are also labelled in Fig. 1. The full width of the ring was 5 mm, with a single 1 mm-wide gauge. The gauge length was set to 3 mm, with a fillet at the end of the gauge of radius 1 mm. Because this mimics the real-world profile from a ring that could be milled in a hot cell, the 3 mm-long gauge is a chord length, yielding a slightly higher arc length of 3.06 mm for the actual gauge length. The ring dimensions were based on a typical 17 $\times$ 17 pressurized water reactor (PWR) fuel pin, with a nominal outer diameter of 9.5 mm and a wall thickness of 0.57 mm. Mandrels were slightly undersized compared to the inner diameter of the ring, with an outer diameter of 8.24 mm, to mimic the clearance needed to slide the ring over mandrels in an actual experiment. The midpoint of the gauge was oriented at a 45-degree angle relative to the loading direction as seen in Fig. 1, which ensures the gauge is fully supported during the entirety of the test. The material property inputs for the model were taken from [19] for moderately cold-worked unirradiated Zircaloy-4.

Nonlinear analysis was used to enable modeling of plasticity and contact, and a hard contact property was used with a "penalty" friction formulation. A friction coefficient of 0.1 was used to simulate a moderately-lubricated surface, similar to previous RTT models [13, 20]. Because damage accumulation was not incorporated into the model, results are only considered accurate up to the increment of peak load (i.e., before necking

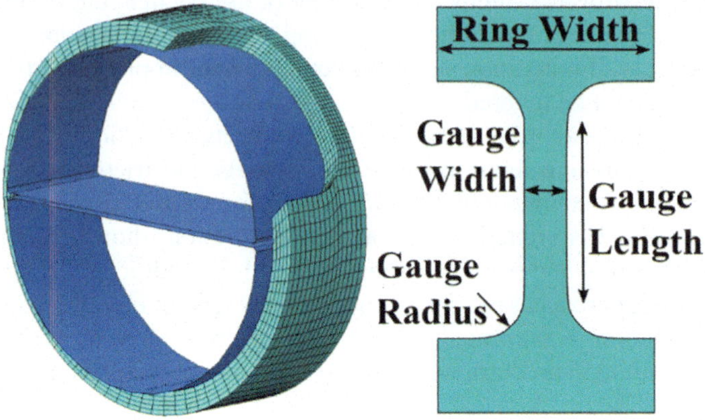

Fig. 1. At left, the FE model of the ring and mandrels, with the gauge oriented at 45 degrees from the vertical loading direction. At right, a side view of the ring with pertinent dimensions labelled.

occurs). As with an actual experiment, the lower mandrel was fixed and the upper mandrel was incrementally displaced, effectively stretching the ring with the deformation concentrated in the gauge region. The load and displacement data were taken from the top grip, as this is the same data collection method in a physical experiment. The result is a model which effectively simulates real-world cladding RTT experiments, with the added benefit of extensive time-resolved monitoring of internal tensile load, pointwise displacements, and distributions of multiaxial stresses and strains. This capability was leveraged in characterizing the angle-dependent internal hoop force to develop friction corrections, discussed in the following section. To better understand the general applicability of the correction factor method, the variable inputs of the correction equations were further investigated by varying the model.

3 Friction Correction Method

As previously discussed, measurements made using raw data from the hemi method will overestimate strength parameters if corrections for friction are not made. An approach for adjusting force values to reflect the actual hoop tension load carried by the ring has been recently proposed, introduced in [10] and discussed with additional comments below. Using the FE model, section cuts can be made through the thickness and width of the ring, both in the gauge and the wider part of the ring. The resultant force at the section cut can be resolved into its components, allowing the model to capture the tensile force aligned in the hoop direction as a function of azimuthal location. This actual internal tension load is plotted in Fig. 2 as the solid blue line, along with other load profiles for different data interpretation schemes which are described below.

The most basic method for converting the force measured by the load cell during the experiment to stress in the gauge is to divide the measured force by two (for the two sides of the ring carrying the load), then divide by the gauge cross-section area. In this method, friction is assumed to be negligible, and the load is not corrected, as shown by the solid horizontal red line in the figure. It is clear that this interpretation will lead to an overestimation of the tensile force experienced in the gauge, and consequent overprediction of yield strength (YS) and ultimate tensile strength (UTS), as has been previously noted [15]. Indeed, when an uncorrected stress-strain curve is used, resulting yield and ultimate tensile strength measurements are 22 and 19% higher than the known material inputs, respectively. It should be noted that this was for the modeled friction coefficient of 0.1 which represents a moderately lubricated surface condition; a lower coefficient of friction will reduce this error. For comparison, the best-case scenario offers a coefficient of roughly 0.02–0.04 with Teflon tape and a pristine inner surface, while a room-temperature unlubricated case is approximately 0.3 [9, 21]. With the significant challenges of achieving these levels of lubrication for in-cell testing of irradiated materials, it is clear some adjustment should be made for accurate property measurement.

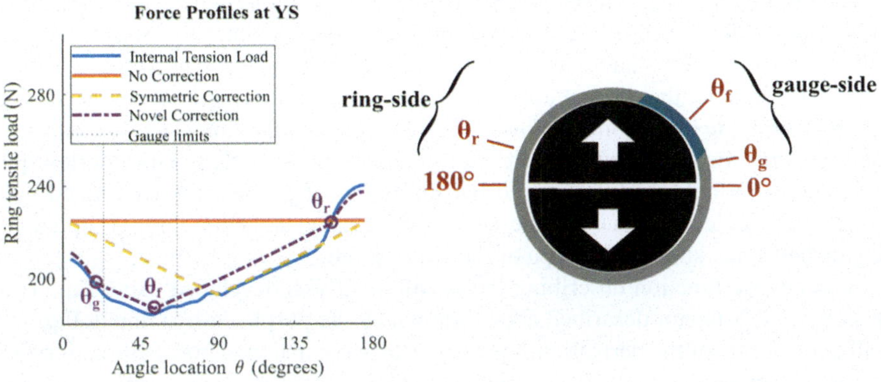

Fig. 2. Plot of force profiles at the YS increment. The internal tension load is taken from the FE model section cuts and compared against the theoretical explanations of the force profile. The angle location of the ends of the gauge are shown with vertical dashed lines, and notable locations for the novel correction are circled. The angle convention is also shown on a schematic of the ring being loaded.

Previous work has noted that force does vary along the ring due to friction [14], as can be clearly seen by the internal tension load shown by the solid blue line in Fig. 2. Past efforts to describe this variation have used the Euler-Eytelwien formula [22], given in Eq. (1), where the tensile load F varies as a function of the angle θ, the coefficient of friction μ, and the tensile load at the gap between mandrels F_g. This tensile load description is paired with the assumption that the load is symmetric about the top of the mandrel (with equal force at $\theta = 0°$ and $\theta = 180°$) [13]. The predicted force using this approach is given by the yellow dashed line in Fig. 2. While this is closer to the actual behavior than not applying any correction, it still doesn't capture the behavior, and leads to overestimating the yield and ultimate tensile strength by 13% and 10%, respectively.

$$F(\theta) = F_g e^{-\mu\theta} \tag{1}$$

However, a deeper understanding of the phenomena causing the load variation in the ring along the top mandrel (θ = 0° to θ = 180°) can improve the predicted forces, leading to highly accurate strength measurements. Using the full-field data from the finite element model, three notable points on the internal tension load of Fig. 2 (designated by circles) were correlated a change in physical behavior. The middle circle in the figure is at the minimum load, near the center of the gauge region. Close inspection showed that at this point the direction of friction changes, indicating that this "fixed" location, θ_f, is stationary relative to the mandrel. This contradicts the previous assumption that the symmetry point for load and deformation is always located at the top of the mandrel (at θ = 90°). Instead, the current results indicate that the gauge location dictates the load profile. If Eq. (1) still describes the local behavior, then a shift in the symmetry point also necessitates a shift in the forces at the gap between mandrels. This leads to the load difference at the two gaps (θ = 0° and θ = 180°) seen in the actual internal tension load of Fig. 2.

The FE model was also inspected at the other two locations indicated by open circles in Fig. 2, where the load profile changes from a "v" shape to a sinusoidal shape: one closer to the gauge, θ_g (≈20°), and one in the bulk of the ring, θ_r (≈160°). At these locations on the RTT sample, the contact condition also changes: the ring is in contact with the mandrel where the load profile follows the "v" shape, and the ring is not in contact with mandrel where the load profile has the sinusoidal shape. Over these unsupported spans, Eq. (1) does not hold, and the carried tension in the ring is instead determined by the cosine of the angle location θ and the ratio α of the radial and hoop force components at the gap, hence the sinusoidal shape of the load profile.

A piecewise function describing the tensile load was developed using Eq. (1) and the sinusoidal behavior described above, shown by the purple dashed line in Fig. 2; for a full explanation of the derivation, see [10]. The accompanying novel correction factor C_{new} for effective load in the gauge (defined by location of the gauge ends, θ_1 and θ_2) is given in Eq. (2).

$$C_{new} = \frac{2 * \left(\cos(\theta_r)\left(\cos(\theta_g) - \alpha \sin(\theta_g)\right)\right)\left(e^{\mu(\theta_2+\theta_g-2\theta_f)} - 2e^{\mu(\theta_g-\theta_f)} + e^{\mu(\theta_g-\theta_1)}\right)}{\mu(\theta_2-\theta_1)\left(\cos(\theta_r) + \left(\alpha \sin(\theta_g) - \cos(\theta_g)\right)e^{\mu(\theta_g+\theta_r-2\theta_f)}\right)} \tag{2}$$

This piecewise function is much closer to the actual behavior, and using the associated correction factor leads to highly accurate yield and ultimate tensile strengths, with errors of 0.7% and -0.4%, respectively. Clearly, this novel method of accounting for friction offers significant improvements in strength measurements from the RTT.

4 Sensitivity of Correction Parameters

With the improvements in strength measurement offered by this novel correction method, it is worth considering the effect of the parameters in Eq. (2) on the correction factor. Although the FE model under investigation in the previous sections was based on a typical Zircaloy-4 cladding RTT with a specific test and sample geometry, ideally it would apply to any RTT experiment with the same general setup. By using a mechanistic, rather than empirical approach, the correction should easily translate to other cladding concepts of varying materials, diameters, wall thicknesses, etc. It is critical to demonstrate that one could apply the correction in Eq. (2) with a rudimentary understanding of the experiment, without requiring any further FE modeling.

The parameters in Eq. (2) can be generally sorted into a few categories. The first category is those which can be easily calculated from the specimen geometry (i.e., the gauge length and location); these include θ_f, which is determined by the center of the gauge, as well as θ_1 and θ_2, which mark the ends of the gauge. The second category is those which cannot directly be calculated, and were identified by inspection of the FE model; these include the contact locations θ_r and θ_g, and the gap force component ratio α. This leaves the friction coefficient μ, which depends on the inner surface condition of the specimen and the mandrel; the sensitivity of strengths to μ is discussed in further detail in other work [10]. Thus, to determine if the correction can be applied generally without any further FE modeling of the specific setup, it is useful to assess the sensitivity of parameters in the second category (θ_r, θ_g, and α); the remainder of this section will discuss these three parameters.

4.1 Sensitivity of Correction to Parameters

First, it is important to consider the sensitivity of the strength correction in Eq. (2) to these identified parameters. To accomplish this, a central finite difference approach was employed, increasing and decreasing one parameter while holding all others to the original values. These sensitivity results are shown in Table 1, where the change in corrected strength is given as a percentage of the strength. Units of change in the angle parameters are given in degrees for ease of use (however, it should be noted that radians are the angle units in Eq. (2)), and the change in α is unitless, as the parameter is a ratio of forces. Examples of change in parameter and change in strength are also given in the table. The sign of the sensitivity is noteworthy; for example, an increase in θ_r results in an increase in corrected strength, while an increase in θ_g will cause a decrease in corrected strength, for a given measured force. Although these sensitivities are local gradient approximations, they indicate the order of magnitude expected for sensitivities of other cladding-type RTT setups with similar parameters. This understanding of parameter sensitivity provides critical context for the following discussions on parameter sensitivity to other factors.

4.2 Sensitivity of Parameters During Deformation

The first aspect of parameter sensitivity is understanding the behavior for the duration of a single test. If the contact angles or radial-to-hoop force ratio are changing drastically as the deformation increases, then a single correction cannot be applied uniformly to the same test. Thus, the area of contact was monitored on both the ring-side and the gauge-side, as in the previous section, but at each model output increment. Similarly, the force ratio alpha was recorded at each increment by performing free-body cuts at the 0° position. These results are shown in Fig. 3, with contact angles in the plot on the left and α

Table 1. Sensitivity of corrected strength measurements to the three parameters θr, θg, and α.

Parameter	Change in corrected strength	Examples	
		Change in parameter	Change in corrected strength
Ring-side contact angle, θ_r	+0.248 (% per degree)	+5° (165° → 170°)	+1.24%
Gauge-side contact angle, θ_g	−0.245 (% per degree)	+5° (15° → 20°)	−1.23%
Radial/hoop force ratio, α	−12.9 (% per alpha)	+0.05 (0.10 → 0.15)	−0.64%

in the plot on the right.[1] Displacements corresponding to the yield strength and ultimate tensile strength are shown by vertical dashed lines; because forces at these locations are used to determine material properties, the correction parameters here become highly important.

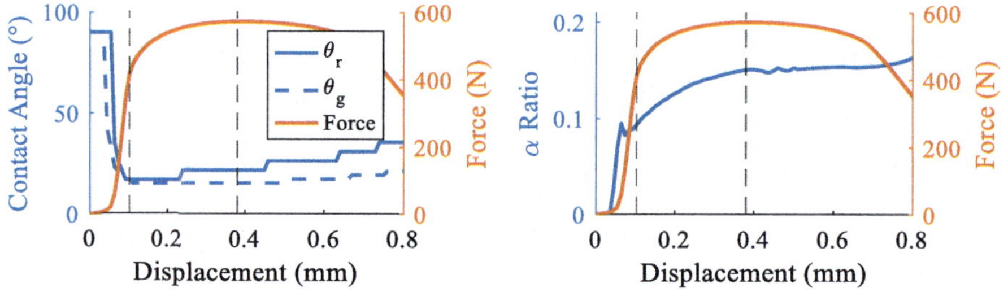

Fig. 3. (Left) Plots of the contact angles θr and θg as a function of grip displacement. (Right) Plot of radial force divided by hoop force at the gap closest to the gauge, α, as a function of displacement. In both, force is plotted as a function of displacement for reference, and displacements associated with YS and UTS are shown with vertical dashed lines.

At initial displacements, the ring is only in contact at 90°, but the contact angles quickly drop to between 15 and 20° as load is applied and the ring conforms to the shape of the slightly smaller mandrel. It is worth noting that from this point through yield and to the peak load, both contact angles are relatively stable. θ_r has a slight increase between yield and peak load, which appears as a discrete step as the contact region moved from one FE element to the next (elements on the ring-side are spaced roughly 4° apart, limiting the angle resolution in this region). Consulting Table 1 reveals that this 4° change in θ_r corresponds to a force correction change on the order of 1%. The right plot shows that although α is likely less stable, it still changes by roughly 0.05 between YS and UTS, with an expected force correction change of <1% per Table 1.

[1] Note: For clarity, in this and following sections a trigonometric transformation of (180°- θ_r) is applied to θ_r in order to directly compare with θ_g (e.g., $\theta_r = 160°$ becomes $\theta_r = 20°$, to compare with $\theta_g = 15°$).

With such minimal changes, the stability of the parameters during a single test is more than sufficient.

4.3 Sensitivity of Parameters to Geometry Differences

After establishing parameter stability during a single test, the investigation moves to sensitivity to changes in test geometry, first by varying gauge orientations. The original geometry featured a gauge centered at 45°; in Fig. 4, the effect of different gauge orientations on the contact angles is shown, at both YS and UTS increments. With this and subsequent plots, the error bars represent the resolution of the contact angle as limited by the mesh size in the vicinity of the contact angle. The ring-side contact angle does not seem to be significantly affected by gauge orientation, although at each orientation the same drift from YS to UTS demonstrated in Fig. 3 can be seen. Conversely, the gauge-side angle is more heavily affected by gauge orientation, and as the gauge comes closer to the top of the mandrel (orientation angle approaches 90°), a greater drift from YS to UTS is seen. This is to be expected, as $\theta_r = \theta_g$ when the orientation is 90°.

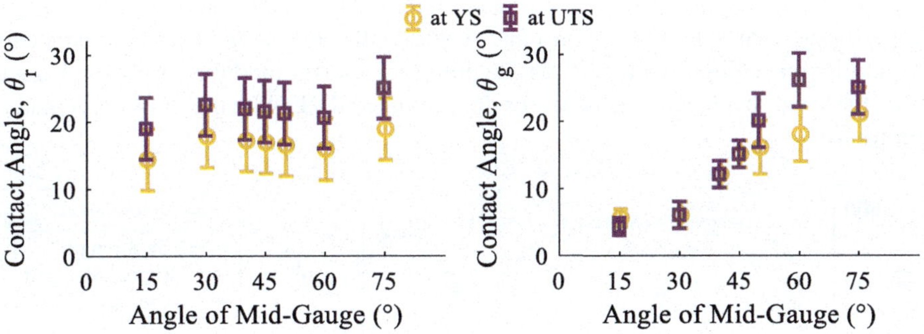

Fig. 4. Contact angles θ_r (left), and θ_g (right) as a function of gauge orientation measured by mid-gauge angle. Both YS increments (yellow circles) and UTS increments (purple squares) are shown. Error bars show the resolution of contact angle for the mesh at the given location.

The correction parameters were similarly investigated for changes to the clearance gap between the ring and mandrel, shown in Fig. 5. The x-axis in each plot shows the gap as a ratio of the original gap, or difference between mandrel and ring diameters (e.g., 0.5x the original clearance). These plots show that the gap affects the contact angle at yield, but not at peak load. As expected, a smaller gap correlates to a lower initial contact angle, as the ring requires less deformation to conform to the mandrel. The α ratio is similarly stable across gap sizes, with a drift from YS to UTS like that seen in Fig. 3. While the effects of gauge orientation on θ_g and of smaller gaps on θ_r at YS are noticeable, Table 1 shows the effect on strength is minimal.

4.4 Sensitivity of Parameters to Material Properties

To demonstrate broader applicability of the correction scheme, the effect of varying material properties must also be considered. To this end, the material definition in the

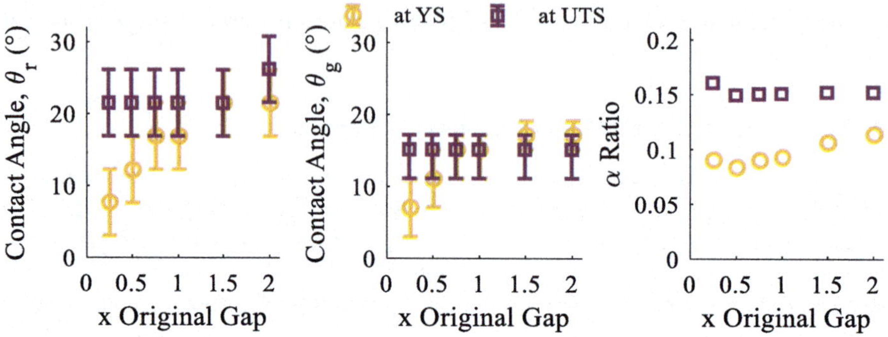

Fig. 5. Parameters θ_r (left), θ_g (center), and α (right) as a function of changing gap clearance, measured as a ratio of the original gap. Both YS increments (yellow circles) and UTS increments (purple squares) are shown. Error bars for θ_r and θ_g show the resolution of contact angle for the mesh at the given location.

FE model was changed and the effect on the correction parameters was examined. The elastic modulus was reduced by a third and increased by a factor of three while keeping the plasticity values K and n constant. Additionally, the strengths were reduced by half and increased by a factor of two, keeping the elastic modulus constant. Figure 6 shows the force-displacement curves for the simulated RTTs with these different material definitions.

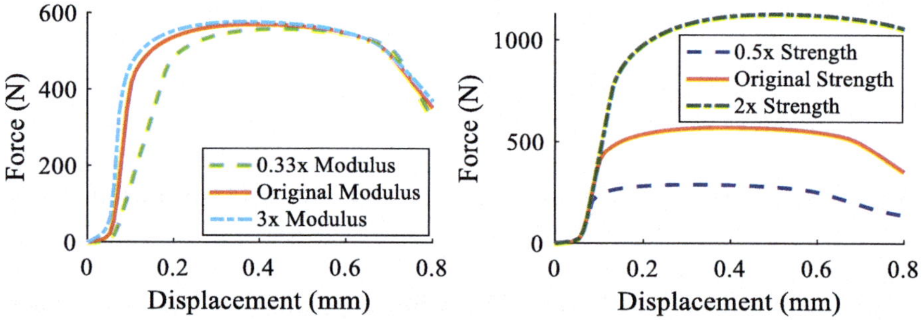

Fig. 6. Simulated RTT force-displacement curves for material models with different elastic moduli (left) and different strengths (right), used to investigate the effect of material properties on parameter sensitivity.

As before, the contact angles and radial-to-hoop force ratio were monitored for the simulated RTTs and are reported at both YS and UTS increments in Fig. 7. The results for the varied moduli are in the top row, with results for the varied strengths in the bottom row. The ring-side contact angle increases slightly with an increase in modulus. Because the ring-side stays in the elastic regime and doesn't experience plastic deformation, a stiffer modulus means the ring is more difficult to conform to the mandrel, and so less of the ring is in contact with the mandrel on the ring-side. However, the trend is reversed for strength variations: θ_r tends to decrease with an increase in strength. Effectively, a higher strength delays the onset of yield in the gauge, allowing the ring-side to elastically deform more before YS is reached.

Unlike ring-side behavior, the contact on the gauge-side is relatively stable for changes to both modulus and strength. This contact is primarily determined by deformation in the gauge; thus at the YS increment, the gauge has already deformed enough to conformed to the mandrel and θ_g is less susceptible to changes in material properties. The α ratio behavior tends to mimic that of θ_r, increasing as modulus increases and decreasing as strength increases. This may indicate that with more of the ring-side in contact with the mandrel, the contact on the ring-side and gauge-side are more evenly balanced and the resultant radial force at $\theta = 0$ drops, effectively decreasing α as well. This correlation between balanced contact angles and lower α ratios is also supported by the plots in Fig. 3: near YS the contact angles are nearly identical, but at greater displacements, the difference between the angles increases, as does α.

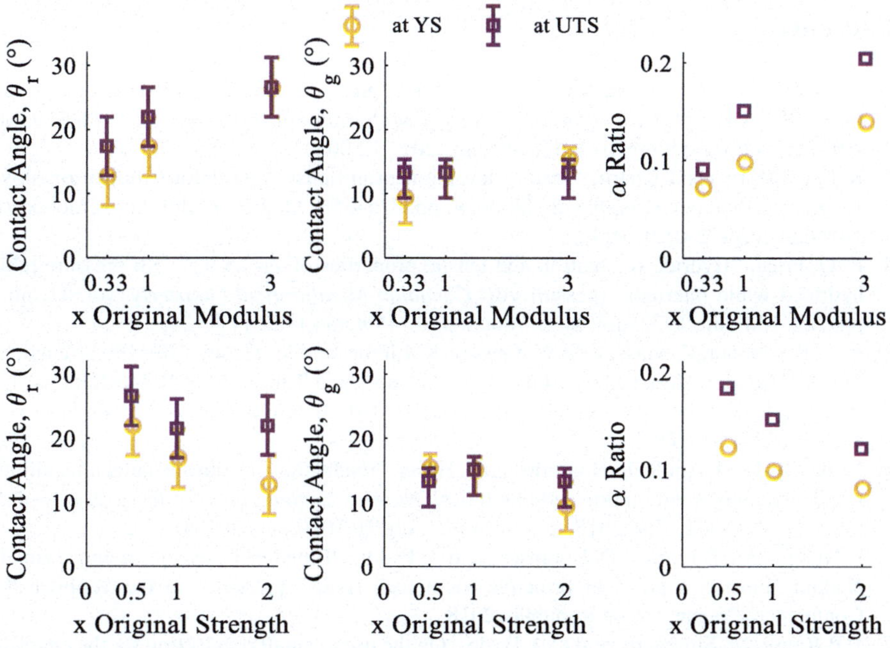

Fig. 7. Parameters θ_r (left column), θ_g (center column), and α (right column) as a function of elastic modulus (top row) and changing strength (bottom row), measured as a ratio of the original. Both YS increments (yellow circles) and UTS increments (purple squares) are shown. Error bars for θ_r and θ_g show the resolution of contact angle for the mesh at the given location.

5 Discussion and Conclusions

The RTT is a highly useful method for determining material strengths in the hoop direction. However, without correcting for friction, measured strengths are prone to significant errors, and so an accurate correction factor is needed, which is calculated in this work. While this correction offers greatly improved accuracy for a typical Zircaloy-4 LWR cladding sample, it must be demonstrated that the correction method is stable in a variety of test conditions and with different materials before it can be used more widely.

Generally, the changes investigated here have minimal effect on the corrected strength measurements; when paired with the sensitivities in Table 1, most parameter variations amount to a strength change of a few percentage points, less would be expected from random experimental variation. In cases where an experiment will deviate from the original RTT model, the results provided here can be used to accurately determine the correction factor. For example, an unintentional change in gauge orientation of 5° will cause an error on the order of 1%. Greater changes to orientation would be intentional, and Fig. 4 could be used to update the correction factor accordingly. Overall, the correction scheme is robust to minor or moderate changes and its use can be expanded to other claddings and more broadly to other tubular materials.

References

1. S. J. Zinkle, K. A. Terrani, J. C. Gehin, L. J. Ott, and L. L. Snead, "Accident tolerant fuels for LWRs: A perspective," *Journal of Nuclear Materials*, vol. 448, no. 1, pp. 374–379, May 2014, doi: https://doi.org/10.1016/j.jnucmat.2013.12.005.
2. K. Linga Murty and I. Charit, "Texture development and anisotropic deformation of zircaloys," *Progress in Nuclear Energy*, vol. 48, no. 4, pp. 325–359, May 2006, doi: https://doi.org/10.1016/j.pnucene.2005.09.011.
3. E. G. Price, "Hydride orientation and tensile properties of Zr-2.5 wt% Nb pressure tubing hydrided while internally pressurized," *Canadian Metallurgical Quarterly*, vol. 11, no. 1, pp. 129–138, Jan. 1972, doi: https://doi.org/10.1179/cmq.1972.11.1.129.
4. S. N. Dryepondt, C. Massey, M. N. Gussev, K. Linton, and K. Terrani, "Tensile Strength and Steam Oxidation Resistance of ODS FeCrAl Sheet and Tubes," Oak Ridge National Lab. (ORNL), Oak Ridge, TN (United States), ORNL/TM-2018/870, May 2018. doi: https://doi.org/10.2172/1507875.
5. M. Király, D. M. Antók, L. Horváth, and Z. Hózer, "Evaluation of axial and tangential ultimate tensile strength of zirconium cladding tubes," *Nuclear Engineering and Technology*, vol. 50, no. 3, pp. 425–431, Apr. 2018, doi: https://doi.org/10.1016/j.net.2018.01.002.
6. F. Nindiyasari, P. Pierick, D. Boomstra, and A. Pandit, "Ring tensile test of reference zircaloy cladding tube as a proof of principle for hotcell setup," presented at the TopFuel 2018 Conference, Prague, Czech Republic, 2018.
7. J. P. Rouse, M. Simonelli, and C. J. Hyde, "On the use of small ring testing for the characterisation of elastic and yield material property variation in additively manufactured materials," *Additive Manufacturing*, vol. 36, p. 101589, Dec. 2020, doi: https://doi.org/10.1016/j.addma.2020.101589.
8. S. M. S. Aghamiri, T. Sowa, S. Ukai, N. Oono, K. Sakamoto, and S. Yamashita, "Microstructure and texture evolution and ring-tensile properties of recrystallized FeCrAl ODS cladding tubes," *Materials Science and Engineering: A*, vol. 771, p. 138636, Jan. 2020, doi: https://doi.org/10.1016/j.msea.2019.138636.
9. F. Nagase, T. Sugiyama, and T. Fuketa, "Optimized Ring Tensile Test Method and Hydrogen Effect on Mechanical Properties of Zircaloy Cladding in Hoop Direction," *Journal of Nuclear Science and Technology*, vol. 46, no. 6, pp. 545–552, Jun. 2009, doi: https://doi.org/10.1080/18811248.2007.9711560.

10. R. S. Hansen, D. W. Kamerman, P. G. Petersen, and F. Cappia, "Evaluation of the ring tension test (RTT) for robust determination of material strengths," *International Journal of Solids and Structures*, vol. 282, Oct. 2023, doi: https://doi.org/10.1016/j.ijsolstr.2023.112471.
11. R. Hansen, "High-Magnification Digital Image Correlation Techniques for Aged Nuclear Fuel Cladding Testing," All Graduate Theses and Dissertations, Utah State University, Logan, Utah, 2021. [Online]. Available: https://digitalcommons.usu.edu/etd/8234
12. A. Khalfallah, Z. Ktari, C. Leitão, and J. V. Fernandes, "New Mandrel Design for Ring Hoop Tensile Testing," *Exp Tech*, vol. 45, no. 6, pp. 769–787, Dec. 2021, doi: https://doi.org/10.1007/s40799-021-00462-4.
13. C. P. Dick and Y. P. Korkolis, "Mechanics and full-field deformation study of the Ring Hoop Tension Test," *International Journal of Solids and Structures*, vol. 51, no. 18, pp. 3042–3057, Sep. 2014, doi: https://doi.org/10.1016/j.ijsolstr.2014.04.023.
14. M. N. Gussev, B. Garrison, C. Massey, A. L. Coq, K. Linton, and K. A. Terrani, "A correlation-based approach for evaluating mechanical properties of nuclear fuel cladding tubes," *Journal of Nuclear Materials*, vol. 574, p. 154192, Feb. 2023, doi: https://doi.org/10.1016/j.jnucmat.2022.154192.
15. Z. He, S. Yuan, Y. Lin, X. Wang, and W. Hu, "Analytical model for tube hydro-bulging tests, part II: Linear model for pole thickness and its application," *International Journal of Mechanical Sciences*, vol. 87, pp. 307–315, Oct. 2014, doi: https://doi.org/10.1016/j.ijmecsci.2014.05.010.
16. S.-K. Kim *et al.*, "Hoop strength and ductility evaluation of irradiated fuel cladding," *Nuclear Engineering and Design*, vol. 239, no. 2, pp. 254–260, Feb. 2009, doi: https://doi.org/10.1016/j.nucengdes.2008.10.024.
17. L. Jiang, J. J. Jonas, K. Boyle, and P. Martin, "Deformation behavior of two Mg alloys during ring hoop tension testing," *Materials Science and Engineering: A*, vol. 492, no. 1, pp. 68–73, Sep. 2008, doi: https://doi.org/10.1016/j.msea.2008.04.028.
18. Y. Yano *et al.*, "Ultra-high temperature tensile properties of ODS steel claddings under severe accident conditions," *Journal of Nuclear Materials*, vol. 487, pp. 229–237, Apr. 2017, doi: https://doi.org/10.1016/j.jnucmat.2017.02.021.
19. K. J. Geelhood, C. E. Beyer, and W. G. Luscher, "PNNL Stress/Strain Correlation for Zircaloy," Pacific Northwest National Laboratory (PNNL), Richland, WA (United States), PNNL-17700, Jul. 2018. [Online]. Available: https://doi.org/10.2172/969740.
20. M. A. Martín-Rengel, F. J. Gómez Sánchez, J. Ruiz-Hervías, L. Caballero, and A. Valiente, "Revisiting the method to obtain the mechanical properties of hydrided fuel cladding in the hoop direction," *Journal of Nuclear Materials*, vol. 429, no. 1, pp. 276–283, Oct. 2012, doi: https://doi.org/10.1016/j.jnucmat.2012.06.003.
21. T. M. Link, D. A. Koss, and A. T. Motta, "Failure of Zircaloy cladding under transverse plane-strain deformation," *Nuclear Engineering and Design*, vol. 186, no. 3, pp. 379–394, Dec. 1998, doi: https://doi.org/10.1016/S0029-5493(98)00284-2.
22. D. Morin, *Introduction to Classical Mechanics*. NY: Cambridge University Press.

Progress in the Modeling of High Burn up Structure: Application of the TRANSURANUS//MFPR-F Coupling to the NRC-192 Studsvik LOCA TEST

F. Kremer[1], A. Tidikas[2], and A. Slavickas[2](✉)

[1] Institut de Radioprotection Et de Sûreté Nucléaire, Severe Accident Department, St Paul-Lez-Durance, France
Francois.Kremer@irsn.fr

[2] Laboratory of Nuclear Safety, Lithuanian Energy Institute, Kaunas, Lithuania
Andrius.Tidikas@lei.lt

Abstract. The MFPR-F (Module for Fission Product Release – France) code coupled with the TRANSURANUS fuel performance code was applied to the Studsvik NRC-192 Loss of Coolant Accident experiment on a high burnup UO2 fuel rod. The presented study, conducted in the framework of the R2CA project, focuses on the formation of the High-Burnup Structure during irradiation and on the fission gas release during the LOCA transient. MFPR-F code supplements TRANSURANUS code with a more in-depth description of the local microstructure in the UO2 matrix, allowing to predict the onset of fuel restructuring [1]. This modelling is used in the present study to calculate the width of the HBS zone in the fuel rod at the end of irradiation, which is found to be 400 μm. Regarding the LOCA transient, the MFPR-F model for fission gas release from the HBS pores, based on an empirical parameter related to the pore over-pressurization, predicts that a large part the gas is released for over-pressure ranging from 150 to 300 MPa.

Keywords: High burnup structure · Fission gas release · LOCA

1 Introduction

The increase of the discharge burn up of nuclear fuel requires to improve the understanding of the behavior of high burnup fuel, in particular under accident conditions. To this end, numerical tools able to address the complexity of the phenomena involved in the formation of the High Burnup Structure (HBS) zone during irradiation, and in its role during LOCA, are necessary. The coupling of the fuel performance code TRANSURANUS and of the mechanistic meso-scale code MFPR-F offers that possibility. Indeed, the coupled codes have recently shown their ability to simulate LOCA transient [2]. Furthermore, the MFPR-F code models the formation and growth of HBS zone during irradiation, and the FP release from this zone during thermal transients [1]. Unlike most of the existing models for HBS formation which are based on the empirical concept of effective

J. Liu and Y. Jiao (Eds.): WRFPM 2023, SPPHY 299, pp. 330–339, 2024.
https://doi.org/10.1007/978-981-99-7157-2_34

burn-up [3, 4], the model of MFPR-F, derived from the work of Veshchunov & Shestak [5], relies on the calculation of radiation damage in the UO2 matrix, considered as the driving phenomenon for fuel restructuration [6].

In the present work, the TRANSURANUS//MFPR-F coupling has been applied to a high-burnup LOCA case, the NRC Studsvik-192 test, in order to test the predicting capabilities of MFPR-F on the fission gas release from the HBS zone. This study has been conducted in the framework of the R2CA project. The paper is organized as follows: the modeling of HBS in MPFR-F is first briefly described, then the simulation of irradiation of the reference rod is presented, focusing on the formation of the HBS zone predicted by MFPR-F. Finally, the results obtained for the NRC Studsvik-192 test itself are analyzed.

2 Modeling of HBS in MFPR-F

The mechanistic approach adopted in MFPR-F for modeling HBS formation has been described earlier [1, 5], and is here briefly recalled. The main hypothesis is that fuel matrix restructuration is induced by accumulation of dislocations generated by irradiation damage. The level of dislocation density at which the fuel matrix is fully restructured has been estimated at about ~10^{15} m^{-2} by experimental [6] and numerical means [7]. In MFPR-F, the model for formation of defects in UO2 matrix provides the evolution, under irradiation, of the dislocation density at each radial location in the pellet. When that quantity reaches the threshold of 2×10^{15} m^{-2} at a given radius, the fuel matrix is considered as restructured, resulting in a change of the local parameters for microstructure, porosity and gas. Regarding microstructure, the grain diameter is reduced to 0.9 μm, and the concentration of defects (including dislocations) is set to zero. The intergranular porosity is set to 15%, and further evolution of pore radius is fitted on experimental data [8]. Finally, the FG depletion from grain and its accumulation in HBS porosities observed in EMPA measurements performed on irradiated fuel rods [9] or highly enriched fuel discs [3] is reproduced in the MFPR-F HBS model.

MFPR-F also comprises a model for FG release from the HBS pores under thermal transient, based on the assumption that the release is driven by the over-pressurization of the pores. This assumption is grounded by thermal transient experiments carried out on very high burn-up fuel in a vacuum chamber [10]. In the current state of the model, the release occurs when the over-pressurization of HBS pores reaches a critical threshold ΔP_{crit}. This leads to the following criterion:

$$P_g = 2\gamma / R_{pore} + P_h + \Delta P_{crit}$$

with P_h the hydrostatic pressure, γ and R_{pore} the surface tension and radius of a pore, respectively, and $P_g = EoS(N, V, T)$ the pore gas pressure, determined by temperature T, pore volume V and the number N of gaseous atoms and molecules in the pore, via the equation of state EoS. . In the present work, several hard-sphere equations of state have been tested, such as Carnaham-Starling and Perkus-Yevik, adapted to the extreme conditions (very low molar volumes) encountered in the HBS pores [11].

3 Formation of HBS Zone Under Irradiation

3.1 Irradiation of the Reference Rod

The characteristics of the NRC Studsvik-192 test element (rodlet) are summarized in Table 1. The sample comes from a UO2 fuel rod irradiated during 4 cycles in a PWR. A simulation of this irradiation period is performed with the TRANSURANUS//MFPR-F coupling, providing a proper initial state for the subsequent LOCA simulation.

Table 1. Irradiated test element characteristics.

Parameter	Value
Height	300 mm
Free volume	10.4 cm3
Filling gas pressure	8.2 MPa
Initial ^{235}U enrichment	3.99% wt
Mean burn up	78 MWd/kgU

The irradiation simulation is based on a TRANSURANUS dataset prepared by SSTC NRS in the framework of the FUMAC project [12]. The irradiation history obtained in the simulation is shown in Fig. 1.

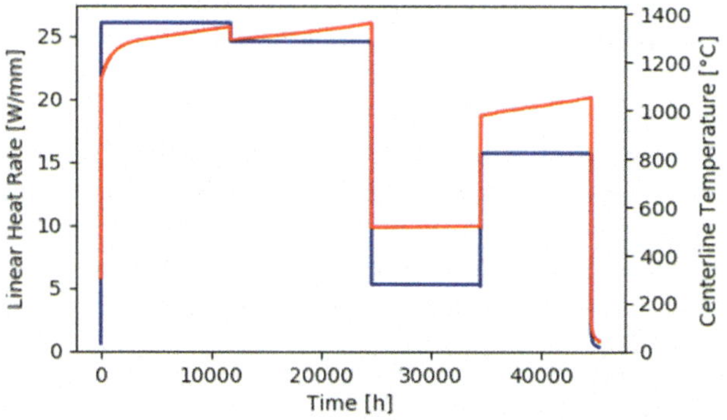

Fig. 1. Power history (W/mm, in blue) and pellet center temperature (°C, in red) at rod mid-section.

The final burnup of 72 MWd/kgU (part of test element) is sufficient for a HBS zone to develop. The dislocation density profile obtained at different instants during the last cycle of irradiation at rod mid-section is represented in Fig. 2. The profile is relatively flat except for radii higher than 3 mm, where the dislocation density increases with radius, as expected from the evolution of conditions toward the pellet rim (fission rate increase and temperature decrease). The dislocation density eventually reaches the restructuration threshold of 2×10^{15} m^{-2}, at a given radius delineating the interior limit of the HBS zone. As the burnup increases, the levels of dislocation density become more important, and the limit of the HBS zone moves toward the interior of the pellet, materializing the growth the HBS zone. Note that, the dislocation density inside the HBS zone being null, it is not represented on the figure.

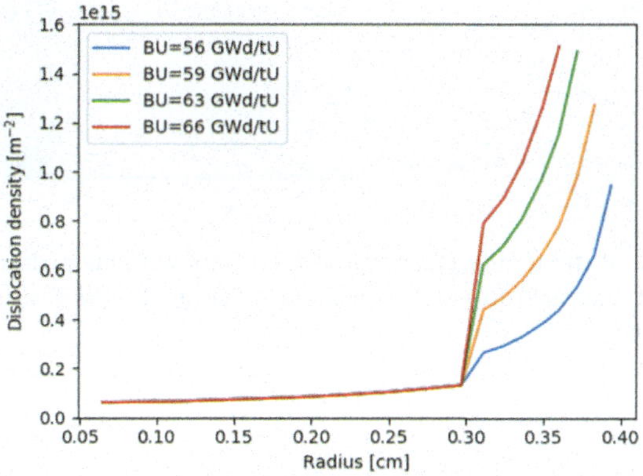

Fig. 2. Radial profiles of dislocation density [m^{-2}] calculated by MFPR-F at different instants (labelled by the pellet burnup) during the last irradiation cycle.

3.2 Fission Gas Repartition

The fission gas repartition over the pellet radius obtained at the end of irradiation is shown on Fig. 3. In the non-restructured zone of the pellet (white domain on the figure), most of the gas remained inside the grain. It can be noticed that in the central zone of the pellet, due to high temperature levels encountered during irradiation, part of the gas diffused out of the grain and have been released. In restructured zone (blue domain), a strong depletion of the intragranular gas occurred and resulted in a high quantity of gas trapped in the (intergranular) HBS pores, in line with experimental observations (see Sect. 2).

4 FG Behavior in HBS Zone During LOCA Transient

4.1 The NRC Studsvik 192 Test

The NRC Studsvik-192 test was performed in 2011 in the LOCA facility of the Studsvik nuclear laboratory in Sweden. The test element consists of a 0.3m section cut from the reference rod. The element is externally heated by infrared radiation in a clamshell furnace, where a flow of steam is applied. The test element is initially filled with Helium gas at 8MPa and 573 K, corresponding to the internal PWR rod pressure at the end-of-life. A temperature ramp is then applied at a rate of 5 K/s up to a maximum of 1446 K (peak cladding temperature) held during 5 s before cooling and quenching. The LOCA transient is displayed on Fig. 4.

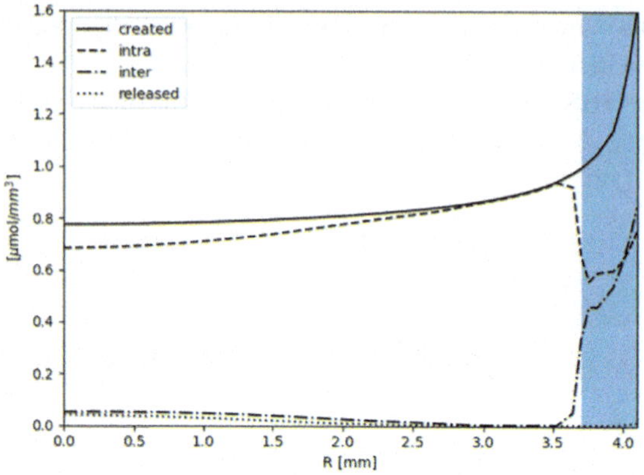

Fig. 3. Radial profile of the fission gas created (plain), intragranular (dash), intergranular/HBS pores (dash-dot), and released (dot) at the end of irradiation. The blue stripe corresponds to the extent of the HBS zone.

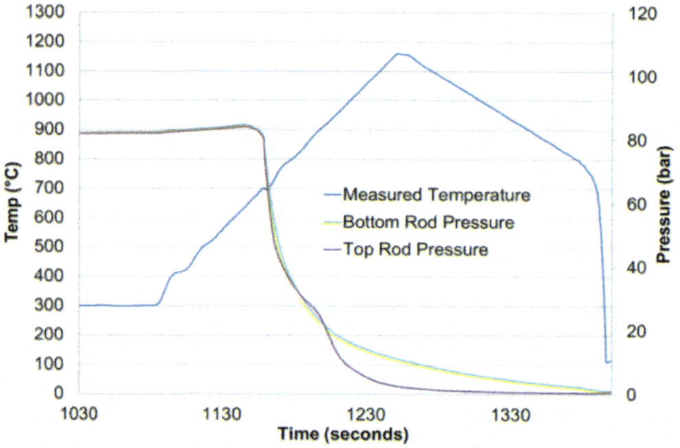

Fig. 4. Pressure and temperature evolution measured during the LOCA transient [3].

The internal pressure drop indicates a clad rupture 81 s after the start of heating, at a temperature of 981 K. The clad ballooning and rupture occurred slightly above mid-plane. The slit dimension is 9 × 22.7 mm. About half of the fuel mass was ejected from the rod as fine fragment, 32% of which are less than 0.125 mm in size.

4.2 TRANSURANUS//MFPR-F Simulations

The LOCA transient were simulated with the TRANSURANUS//MFPR-F coupling in restart mode, using the results provided by the irradiation calculation as initial state. As described in Sect. 2, the model for FG behavior in the HBS zone allows the usage of different equations of state. Simulations have thus been performed with either Perkus-Yevik (PY) and Carnaham-Starling (CS) equations of state to assess their impact on the results. Several values of the FGR threshold over-pressure ΔP_{crit} have also been tested.

Finally, a standalone TRANSURANUS calculation has been performed for reference. The simulations are summarized in Table 2.

Table 2. LOCA simulations and associated parameters of model for FG behavior in HBS.

Name	Eq. of State – FGR threshold over-pressure ΔP_{crit}
CS15	Carnaham-Starling – 150 MPa
PY15	Perkus-Yevik – 150 MPa
PY30	Perkus-Yevik – 300 MPa
TU	[Standalone transuranus calculation]

Thermo-mechanics. The thermo-mechanical behavior of the test element is very similar between the simulations, as seen on Fig. 5 representing internal pressure and external clad temperature evolution for the different simulations. Temperatures are indeed identical, whereas small differences are noticed, for coupled simulations, on the pressure rise before burst. The standalone TRANSURANUS simulation exhibits slightly higher pressure levels all along the transient, which can be explained by a different state at the end of the irradiation, in terms of pellet/clad deformation as well as fission gas release, resulting in a higher pressurization of the rod free volumes.

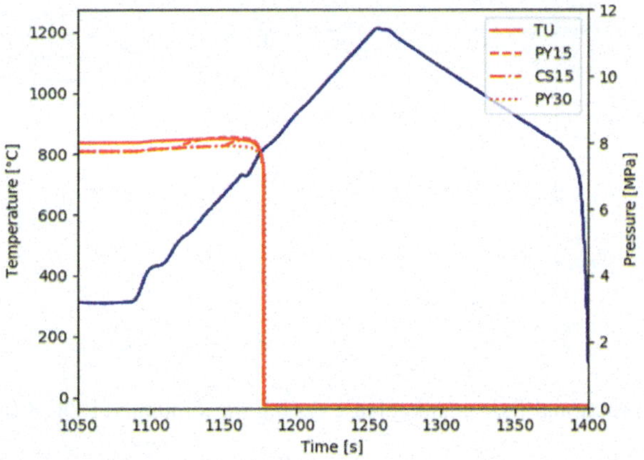

Fig. 5. Pressure [MPa, in red] and external clad temperature [°C, in blue] evolution during the simulated LOCA transients.

Table 3 presents the calculated burst times for various equations of state. The burst event is primarily influenced by the external/internal pressure difference, which is a constituent of the tangential stress, and the temperature of the cladding, which is a constituent of the burst stress of the cladding material. Under LOCA transient conditions, the rising temperature of the cladding is the main cause of burst propagation in the examined cases. Since the calculations are performed with the same cladding temperature, the variance in burst times is minimal. However, differences in the results can be attributed to the release of FG.

Table 3. Calculated burst time of the cladding.

Name	Burst time (s)
CS15	1173.4827
PY15	1173.6035
PY30	1174.5000
TU	1173.7399

Fission gas release. The internal pressure rise before burst is due to FG released from fuel and over pressurizing the gas in the rod free volumes. The differences observed in rod pressure evolution can thus been attributed to differences in the onset of FGR, represented on Fig. 6. It appears indeed that switching from CS to PY equation of state advanced the FGR onset (CS15 vs PY15 curves), whereas raising the over-pressure threshold from 150 to 300 MPa delayed it (PY15 vs PY30 curves).

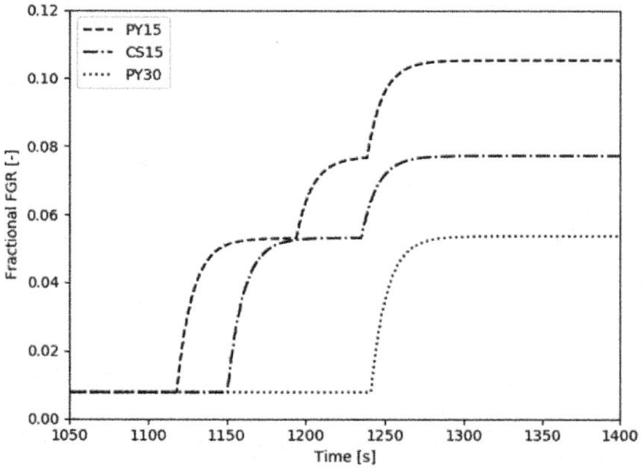

Fig. 6. Fractional fission gas release during the simulated LOCA transients.

HBS zone. A better insight in the FG behavior in the HBS zone can be obtained by comparing radial profiles of the FG repartition before and after the LOCA transient. Such repartition is represented on Fig. 7 for the different simulations, focusing on the external part of the pellet. The HBS zone extends from R = 3.7 to 4.1mm (blue domain). As mentioned in Sect. 3.2, before the LOCA transient (Fig. 7a) about half of the gas in the HBS zone is localized in the intergranular domain (in the HBS pores) and may be subsequently released after the transient. It appears from Fig. 7b–d that the release occurred in the exterior part of the HBS zone only, while in the interior part the gas remains in the pores. This behavior could be expected considering that, in the current state of the model, the pore gas content is higher in the exterior part of the HBS, hence a higher build-up of pore pressure during the transient. Accordingly, the extent of the HBS region where the release occurs depend on the parameters of the release model. For example, raising the over-pressure threshold from 150 to 300 MPa (PY15 vs PY30) decreases the extent of this region, as seen from Fig. 7b and c. Similarly, using the CS equation of state, which predicts lower pressure levels than the PY one, reduces the extent of the region, as seen from Fig. 7b and d (PY15 vs CS15).

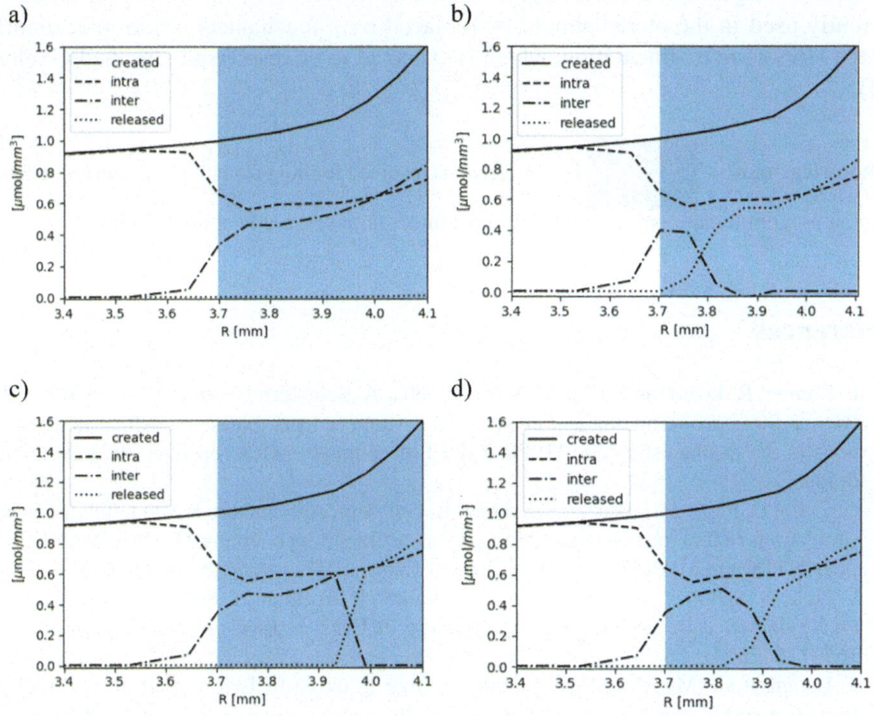

Fig. 7. Radial profiles of the fission gas repartition (see caption of Fig. 3); a) before LOCA, b) PY15, c) PY30, d) CS15

5 Summary and Conclusion

In this work, simulations of the NRC-192 Studsvik LOCA test were performed with the coupling of the fuel performance code TRANSURANUS and of the mechanistic meso-scale code MFPR-F. The coupled code was first used to simulate the irradiation of the reference rod and predicted the formation of a HBS zone 400 μm wide. Restart calculations were then performed to simulate the LOCA transient, using different set of parameters of the model for FG release from the HBS zone. All the calculations provided coherent results in terms of thermo-mechanical behavior of the rod, the evolution of central temperature and pressure in the free volume being almost identical to that obtained by a standalone TRANSURANUS calculation. Noticeable differences were observed solely in the pressure rise before burst, attributed to the fission gas release from the HBS zone, across the simulations. The FGRs were found to strongly depend on the over-pressure threshold ΔP_{crit}, and to a lesser extent on the equation of state applied in the HBS pores. This last point highlighted the importance of a precise evaluation of the pore pressure, which depends – in addition to the equation of state – on the initial gas content and volume of the HBS pores, for which analytical models such as [13] should be applied. Regarding the over-pressure threshold for FGR, the empirical parameter currently used in the model should be replaced by a mechanical criterion accounting for the HBS zone fragmentation, which is expected to be responsible for the FG release [14].

Acknowlegements. This project has received funding from the Euratom research and training programme through the R2CA Project under Grant Agreement no 847656.

References

1. F. Kremer, R. Dubourg, F. Cappia, V. Rondinella, A. Schubert, P. Van Uffelen, T. Wiss: High Burnup Structure formation and growth and fission products release modeling: new simulations in the mechanistic code MFPR-F. In: Proceedings EMRS Top Fuel 2018 conference, Prague
2. G. Zullo, D. Pizzocri, L. Luzzi, F. Kremer, R. Dubourg, A. Schubert, P. Van Uffelen: Towards simulations of fuel rod behaviour during severe accidents by coupling TRANSURANUS with SCIANTIX and MFPR-F. Annals of Nuclear Energy 190, *article accepted* (2023).
3. L. Holt, A. Schubert, P. Van Uffelen, C.T. Walker, E. Fridman, T. Sonoda: Sensitivity study on Xe depletion in the high burn-up structure of UO2. Journal of Nuclear Materials 452, 166–172 (2014).
4. T. Barani et al.: Modeling high burnup structure in oxide fuels for application to fuel performance codes. part I: High burnup structure formation" Journal of Nuclear Materials 539 (2020)
5. M.S. Veshchunov, V.E. Shestak: Model for evolution of crystal defects in UO2 under irradiation up to high burn-ups. Journal of Nuclear Materials 384, 12–18 (2009).
6. K. Nogita, K. Une., Radiation-induced microstructural change in high burnup UO, fuel pellets. Nuclear Instruments and Methods in Physics Research B91, 301–306 (1994).
7. V.G. Baranov, A.V. Lunev, A.V. Tenishev, A.V. Khlunov: Interaction of dislocations in UO2 during high burn-up structure formation. Journal of Nuclear Materials 444, 129–137 (2014).
8. F. Cappia, D. Pizzocri, A. Schubert, P. Van Uffelen, G. Paperini, D. Pellottiero, R. Macian-Juan, V.V. Rondinella: Critical assessment of the pore size distribution in the rim region of high burnup UO2 fuels. Journal of Nuclear Materials 480, 138–149 (2016).
9. R. Manzel, C.T. Walker: EPMA and SEM of fuel samples from PWR rods with an average burn-up of around 100 MWd/kgHM. Journal of Nuclear Materials 301, 170–182 (2002).
10. J.-P. Hiernaut et al.: Fission product release and microstructure changes during laboratory annealing of a very high burn-up fuel specimen » Journal of Nuclear Materials 377 313–324 (2008)

11. J. Noirot, L. Desgranges, J. Lamontagne: Detailed characterisations of high burn-up structures in oxide fuels. Journal of Nuclear Materials 372, 318–339 (2008).
12. Fuel modelling in accident conditions (FUMAC), Final report of a coordinated research project. IAEA CRP T12028 (2014–2018), IAEA-TECDOC-1889, IAEA, Vienna, Austria, 2018.
13. M.S. Veshchunov, V.I. Tarasov: Modelling of pore coarsening in the high burn-up structure of UO2 fuel. Journal of Nuclear Materials 488, 191–195 (2017).
14. J.Noirot, Y. Pontillon, S. Yagnik, J.A. Turnbull, T. Tverberg, Fission gas release behaviour of a 103 GWd/tHM fuel disc during a 1200 °C annealing test. Journal of Nuclear Materials 446, 163–171 (2014).

Modeling of the Gadolinium Fuel Tests with the Jasmine Fuel Performance Code

Xiaoyan Wei(✉), Yanan Zhu, Shengzhi Yang, Duoting Xu, and Xin Jin

China Nuclear Power Technology Research Institute, Shenzhen, China
xiaoyanwei126@163.com

Abstract. JASMINE is an advanced fuel rod performance code under development by China General Nuclear Power Cooperation (CGNPC). Basing on JASMINE1.0, JASMINE2.0 incorporates some state-of-the-art models or model features developed for UO_2 and (U, Gd)O_2 fuel pellets and CZ2 cladding, which is the zirconium alloy used in fuel rod researched by CGNPC. And it is used to accurately predict UO_2 and (U, Gd)O_2 fuel behavior with special emphasis for PWR reactors. Behaviors of (U, Gd)O_2 fuel during irradiation has been modeled in JASMINE2.0. These models consider the gadolinium effects on the power and burnup profiles, fission gas release (FGR), and other thermal-mechanical performances. Two Halden programme fuel rods and four GAIN programme fuel rods were used for the validation of JASMINE2.0. The comparison of online fuel central temperatures indicates that JASMINE2.0 has a conservative tendency in (U, Gd)O_2 fuel with the measured data. The validation results show that JASMINE2.0 has a good and a little conservative agreement with the measured FGR, rod internal pressure and free void volume.

Keywords: JASMINE2.0 · Gadolinium fuel rods · Fuel models · Experimental data

1 Introduction

The improvement of the fuel utilization's economy and the minimization of the nuclear waste demand the extension of the fuel cycle length for the commercial reactors. In order to raise the discharge burnup of the nuclear fuel, the uranium enrichment has to be increased [1]. To compensate the resultant larger reactivity disproportion in the core, adding gadolinium is a mature burnable poison design with good performance, which has been proven in nuclear power plants worldwide. A typical CPR fuel assembly contains up to about 24 Gd-bearing fuel rods. Gadolinium is mixed homogeneously in UO_2 with typical Gd_2O_3 concentrations in the range of 2–8 wt% [2]. These rods are essential for the optimization of core fuel management schemes. Therefore for fuel rod thermal-mechanical of fuel loads, the assessment of (U,Gd)O_2 fuel is an essential part of design analysis [3].

On the one hand, introducing gadolinium in UO_2 affects the material properties of the fuel. When Gd_2O_3 is homogeneously mixed with UO_2, the Gd atoms replace the

J. Liu and Y. Jiao (Eds.): WRFPM 2023, SPPHY 299, pp. 340–353, 2024.
https://doi.org/10.1007/978-981-99-7157-2_35

position of U atoms in the fluorite crystal structure of UO_2. Due to the scattering of photons with Gd atoms, the introduction of Gd in UO_2 decreases the effective diffusion coefficient [4], and lowers the thermal conductivity of fuel which leads to a higher fuel temperature than UO_2 for the same level of linear heat generate rate [5].

On the other hand, because of the higher absorption cross sections of Gd, the (U, Gd)O_2 burns gadolinium practically on fuel surface while fuel interior is completely shielded from thermal neutrons [3]. This effect leads to a gradual shift of the interface between burned and unburned gadolinium from pellet surface to its center, which is exhibited by radial power and burnup profiles.

Besides that, Gd^{+3} ions can also increase the concentration of oxygen vacancies and lead to a lower effective diffusion of fission product gases in the pellets [6]. This effect leads to a lower fission gas release and a larger gas swelling for (U, Gd)O_2 fuel than UO_2 fuel.

JASMINE is one of the most sophisticated fuel rod codes prepared for the modeling of different fuel types, and it considers the gadolinium effects, especially [7, 8]. The (U, Gd)O_2 fuel related models consider the gadolinium effects to fuel performance, such as thermal conductivity, radial power profiles, fission gas release and so on.

However, the experimental data could be used in the assessment of (U, Gd)O_2 fuel are so limited. Hereafter, six gadolinia irradiated fuel rods from Halden and GAIN programmes are used in JASMINE's gadolinium models validation.

2 Models

Thermal energy in the fuel rod converted from fission transfers as heat to coolant, during this process fuel rod undergoes thermal and mechanical transformation. JASMINE is built round a quasi-two-dimensional analysis of fuel and cladding. Physical models include fuel rod thermal and mechanical analysis, fuel rod fission gas release and internal pressure analysis, cladding corrosion and hydrogen analysis. Besides that, there is a material data bank for the fuel, cladding and coolant.

Hereafter, it gives the most important physical models in JASMINE, especially the gadolinium related models about thermal analysis, mechanical analysis, and FGR.

2.1 Thermal Model

Temperature is a key parameter that effects deformation, FGR, internal pressure, and other fuel behavior models. Therefore, accurate prediction of temperature is a prerequisite for good fuel rod modeling. Several factors contribute to accurately predict fuel temperature, such as corrosion, radial power distribution, fuel-cladding gap, contact pressure, etc.

Hereafter, the gadolinium fuel thermal conductivity model is therefore described, along with a brief description of the radial power profiles in JASMINE code.

Fuel thermal conductivity. The Halden [9] and the Lucuta et al. [10] relationships for (U, Gd)O_2fuel thermal conductivity have been widely used. As shown on Fig. 1, both relations have same tendency with temperature increasing, but they predict somewhat different fuel conductivity degradation with temperature and burnup. The thermal conductivity decreased by irradiation while it partly recovered after the thermal diffusivity measurement at temperatures up to about 1800 K [11].

The fuel thermal conductivity is a function of fuel temperature and burnup, and the (U, Gd)O_2 fuel thermal conductivity in JASMINE has the form:

$$\mathrm{k}_{100} = \frac{1}{A_1 + A_2 T + a_0 \mathrm{z}_{\mathrm{gd2o3}} + A_3 \mathrm{Bu} + A_4 f(T)} + g(T) \tag{1}$$

where k_{100} is the 100% dense fuel conductivity (W/m/K), T is temperature (K), $\mathrm{z}_{\mathrm{gd2o3}}$ is Gd_2O_3 weight fraction (−), Bu is burnup (MWd/kgU). The terms with parameters A_1 and A_2 represent the phonon contribution, where A_1 and A_2 are constants that relate to phonon-impurity and phonon-phonon scattering processes, respectively. The term with parameter a_0 represents the effect of low gadolinia contents [12]. The term with parameter A_3 represents the degradation of thermal conductivity with burnup. The function f(T) represents the radiation damage term at low temperatures and g(T) is the electronic conductivity at high temperatures.

For a fuel with given porosity, the following correction [12] is used

$$\mathrm{k}_p = (1 - \beta p)\mathrm{k}_{100} \tag{2}$$

where p is fuel porosity (−), with $\beta = 2.58–0.58T$.

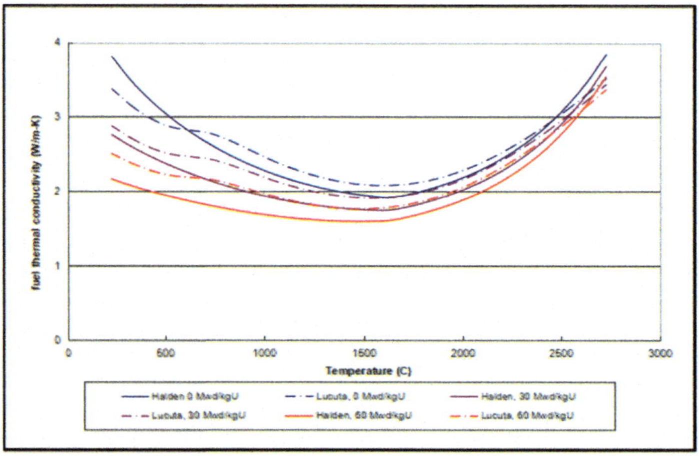

Fig. 1. Comparison between (U, Gd)O_2 fuel thermal conductivity of Halden and Lucuta et al.

Fuel Pellet radial power profiles. Accurate radial profiles of power and burnup are a prerequisite for reliable calculation of the radial dependency of the fuel thermal conductivity and porosity, in order to get realistic radial temperature distributions. The burnup profile also directly determines the radial profile of fission gas generated in the fuel pellet.

Since the differences of power and burnup profiles between UO_2 and (U, Gd)O_2 fuels are so immense, JASMINE uses different radial power distribution tables for UO_2

and (U, Gd)O_2 fuels. Both of them come from the 3-D core neutronic analysis code PCM, which is also designed by CGNPC [3].

The radial power profiles table contains the 3-D power coefficient about both the radial rings and burnup steps. The radial mesh must be sufficiently fine toward the pellet outer edge to correctly represent the peripheral rim effect. Better accuracy at high burnups is obtained with 20 rings. Figure 2 shows the radial power profiles given by PCM code for (U, Gd)O_2 fuel at different burnups.

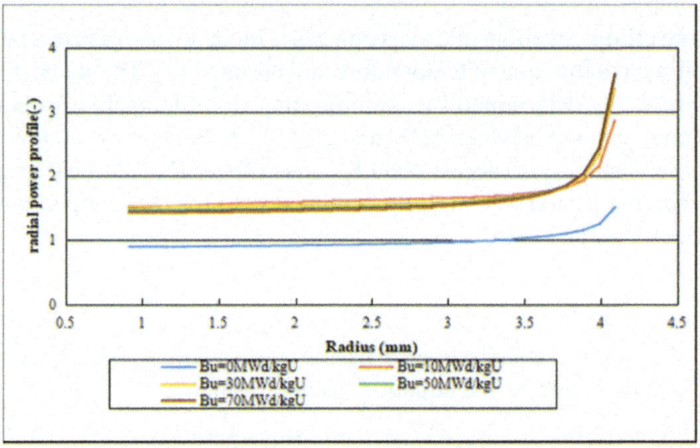

Fig. 2. Radial power profiles for (U, Gd)O_2 fuel at different burnups.

2.2 Mechanical Model

The mechanical analysis consists of the calculation of stresses and strains of the fuel and cladding and the corresponding deformations. JASMINE considers several components of strain: strains due to elastic, plastic, creep, irradiation growth and thermal expansion for cladding, as well as strains due to elastic, densification, solid swelling, gas swelling, relocation and thermal expansion for fuel pellets. In general, dynamic forces are not considered and the solution is therefore obtained by applying the principle conditions of equilibrium and compatibility together with constitutive equations.

The following assumptions are made on fuel rod geometry and material properties in order to simplify the mechanical analysis. (a) The plane-strain condition applies for axial direction, i.e. the axial elastic deformation is constant across the radius. (b) The fuel rod is assumed to be axially symmetric. (c) The complex material structure of fuel pellet and cladding can be described piece-wise by spatially variant elastic constants.

Fuel radial geometry. From the definition of the tangential strains, the fuel radial geometry for the solution of JASMINE mechanical analysis has the form

$$R_{fuel}^{inel} = R_{fuel}^{fab}(1 + \varepsilon_t^{dens} + \varepsilon_t^{solswl} + \varepsilon_t^{gasswl} + \varepsilon_t^{reloc})(1 + \varepsilon_t^{th}) \quad (3)$$

where R_{fuel}^{fab} is the fabricated radius, R_{fuel}^{inel} is the radius considered all the in-elastic strains, which includes the radial effects of densification ε_t^{dens}, solid swelling ε_t^{solswl}, gas swelling ε_t^{gasswl}, fuel relocation ε_t^{reloc}, and the fuel thermal expansion ε_t^{th}.

The fuel tangential elastic strains ε_t^{el} are then calculated relative to the fuel radial geometry defined by elastic stress-strain relations, so that the fuel radial geometry, accounting for all mechanical deformations, is given by the equation

$$R_{fuel} = R_{fuel}^{inel}(1 + \varepsilon_t^{el}) \tag{4}$$

Fuel gas swelling. Swelling due to fission gas is modeled using a correlation for unrestrained swelling as a function of temperature and burnup [13]. The shape of unrestrained isothermal curve was determined by assuming that (a) at temperatures below 1000 K, little swelling occurs; (b) between 1000 and 2000 K, bubbles grow at grain boundaries, creating volume changes; (c) above 2000 K dense (98%TD) columnar grains form and gas is removed, and fission gas swelling is reduced. So for fuel temperature larger than 2800 K, the equation is

$$S_g = 0 \tag{5}$$

where S_g is the volume change due to fission gas swelling (−). Otherwise, for fuel temperature less than 2800K, the equation has the form

$$S_g = \frac{2.261088 \times 10^{-34}\rho_0(2800 - T)^{11.73}e^{[-0.0162(2800-T)]}}{2.0736 \times 10^{-5}\rho_0}\left(1 - e^{\left(-2.0736\times 10^{-5}\rho_0 B\right)}\right) \tag{6}$$

where T is the fuel temperature (K), B is the local burnup of the fuel (MWd/kgU), ρ_0 is the fuel density (kg/m^3).

Fuel relocation. Under normal operating conditions, the fuel pellets, which are made of sintered ceramic materials, become fragmented as a result of thermal stress effects. This initial fragment shift phenomenon effectively increases the pellet radius [14]. Before gap closed, the pellet relocation strain, ε_{reloc}, is a function of pellet-cladding gap G and pellet radius R_c:

$$\Delta\varepsilon_{reloc} = \alpha\frac{G}{R_c} \tag{7}$$

After contact with the cladding, rearrangement occurs due to the effect of the stresses by the interaction between the pellet and the cladding. The relocation accommodation is described by:

$$\frac{d\varepsilon_{reloc}}{dt} = -\varepsilon_{reloc}\frac{P_c}{\beta} \tag{8}$$

The quantities α and β are the parameters, P_c is the contact pressure after gap closed, t is the time.

2.3 Fission Gas Release Model

The fission of atoms within fuel pellets creates noble gas atoms: primarily Xenon and to a lesser extent Krypton. These gas atoms are released from the fuel pellets by several mechanisms which can be classified into two main groups: a thermal release and an athermal release. The mechanisms involved in the fission gas release are complex [15]. Only the phenomena contained in JASMINE fission gas release model are described, (a) thermal release: diffusion of the fission gas produced in the fuel grains to grain boundary bubbles, with resolution at the grain boundary surface, (b) athermal release: recoil/knockout released directly to the rod void volume, (c) RIM release: gas released at the high burnup structure (HBS).

Diffusion equation solution with resolution at the grain boundary. For a spherical grain the time dependent diffusion equation, with uniform gas production within the grain and with gas in grain boundary bubbles resolved back into the grain, the diffusion equation is

$$\frac{\partial c(r,t)}{\partial t} = D(t)\left[\frac{\partial^2}{\partial r^2} + \frac{2}{r}\frac{\partial}{\partial r}\right]c(r,t) - \lambda c(r,t) + P(t) \tag{9}$$

where c(r, t) is the gas concentration in the bubble (moles/m^3), D(t) is the diffusion constant (m^2/s), λ is the decay constant for the unstable isotopes (1/s, = ln(2)/t½, where t½ is the isotope's half-life), and P(t) is the gas production rate (moles/m^3-s). The boundary conditions for Eq. (9) are the initial condition c(r, 0) = 0 and Speight-Turnbull grain surface boundary condition [16, 17].

$$c(a,t) = \alpha(t) n_{gb}(t) \tag{10}$$

where α is a dimensionless constant and n_{gb}(t) is the gas concentration in the grain boundary bubbles (moles/m^3). α can be expressed in terms of the probability b (s^{-1}) [17], that an atom in the grain boundary bubble will experience resolution and be deposited a distance δr into the grain, as

$$\alpha(t) = b(t)\delta_r a/6D(t) \tag{11}$$

where a is the grain radius, D is the diffusion coefficient (m^2/s).

Diffusion coefficient. The diffusion coefficient D [18] used in JASMINE has the following relationship

$$\begin{aligned} D &= D_1 + D_2 \\ D_1 &= 7.6x10^{-10} e^{-35000/T_K} \\ D_2 &= 1.38x10^{-16} R^{1/2} e^{-13800/T_K} \end{aligned} \tag{12}$$

where T_k is the local temperature (K), R is the local mass rating (W/gU).

Gas release to grain boundary bubbles. With these solutions for the diffusion equation, the gas concentration in the grain boundary bubbles n_{gb} (moles/m^3) and the gas released from the grain boundary bubbles n_{rel} (moles/m^3) are determined from the conservation equation

$$n_{rel}(\tau) + [\frac{1}{2} + \alpha(\tau)]n_{gb}(\tau) + n_0(\tau) = \int_0^\tau e^{-\lambda(t-t')}p(t')dt' = n_{tot}(\tau) \quad (13)$$

i.e., the gas released n_{rel} plus the gas in the grain boundary bubbles n_{gb} plus the gas retained in the grain n_0 is equal to the total amount of gas that has been generated by fission n_{tot}, accounting for the decay of the unstable isotopes, where the factor of ½ multiplying n_{gb} in the conservation equation accounts for the two grains on either side of the grain boundary bubble making equal contributions to the total gas in the grain boundary bubble.

From Eq. (13) we can get the fission gas release rate

$$R = \frac{n_{rel}}{n_{tot}} \quad (14)$$

3 Experiment Data

Well characterized irradiation experiments with gadolinium fuel are scantily available for the validation of models. Table 1 shows the general information of six gadolinium fuel rods from Halden and GAIN programme [19], respectively.

Table 1. General information of six gadolinium fuel rods.

Fuel rod name	Rod burnup (MWd/kgU)	Fuel density (%TD)	Enrichment (%)
IFA-515A2	54.73	95.36	13 U^{235}, 8 Gd
IFA-515B2	55.31	95.20	13 U^{235}, 8 Gd
GD0301	38.8	93.52	3.48 U^{235}, 3Gd
GD0302	37.9	93.52	3.48 U^{235}, 3Gd
GD0701	38.9	91.88	3.48 U^{235}, 7Gd
GD0702	39.0	92	3.48 U^{235}, 7Gd

The Instrumented Fuel Assembly IFA-515 was irradiated in the Halden Boiling Water Reactor between July 1994 and October 2000 [20]. The test rig contained six short LWR fuel rods instrumented with expansion thermometers for the assessment of the fuel central temperatures. The experiment aimed at the comparative investigation of the thermal performance of UO_2 and (U, Gd)O_2 fuel rods under identical conditions to very high burnups. The simulated maximum burnup of (U, Gd)O_2 rods was 55.31 MWd/kgU. Figure 3 shows the power history of IFA-515A2 and IFA-515B2. As represented in Fig. 3, these two fuel rod average linear power histories are less than 25 kW/m.

The GAIN Programme, which was an international program lead by Belgonucleaire, irradiated four rods with two different doping concentrations [21–23]. Rods GD0301 and GD0302 were doped with 3 wt% Gd pellets while rods GD0701 and GD0702 were doped with 7 wt% Gd pellets. All four rods were irradiated in BR3 to a final burn-up of approaching 40MWd/kgU. Moreover, rod GD0701 underwent 2 overpower transients in BR2. Nondestructive PIE was performed at the end of each irradiation. After final discharge, destructive PIE was performed for fission gas, free void volume and internal pressure. Figure 4 represents the power histories of four GAIN gadolinium rods. As represented in Fig. 4, these four fuel rod average linear power histories are no more than 25 kW/m, too.

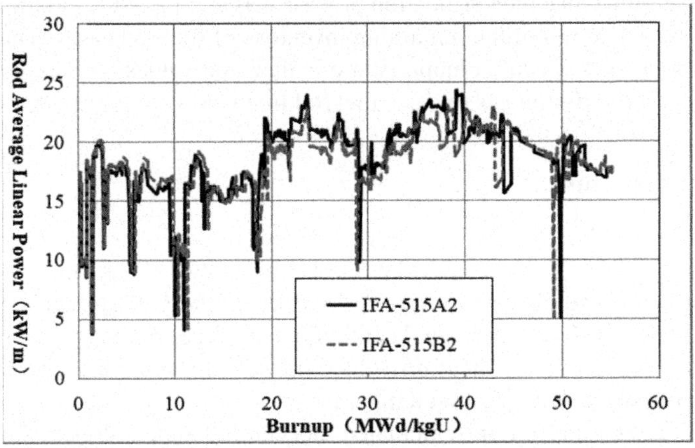

Fig. 3. Power histories for the IFA-515 gadolinium rods.

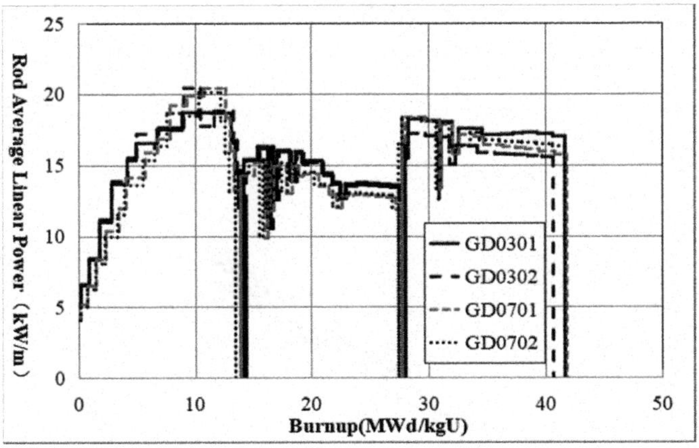

Fig. 4. Power histories for the GAIN gadolinium rods.

4 Validations

To demonstrate the predictive capability of JASMINE for PWR fuels under steady state and transient operating conditions, a fuel rod database was developed consisting of instrumented, test program and commercial fuel rods. The verification activity was conducted as an iterative benchmark testing and revision process. This process focused on the primary material properties and behavioral models. It was accomplished by using several selected subsets from the fuel rod database tailored to provide the relevant measured data for the following technical areas of evaluation: fuel temperature, fission gas release and cladding strain. Once the verification process was completed, validation testing was conducted as a single run of the remaining inventory of fuel rod cases. The final step in the validation process was the comparison of calculated values with experimental data for fuel temperature, fission gas release, fuel rod free void volume and internal pressure.

4.1 Fuel Temperature

The simulations of IFA-515A2 and IFA-515B2 were performed by JASMINE up to the rod burnups of 54.73 and 55.31 MWd/kgU, respectively. The performed computations indicated the gap closure in rods A2 and B2 at the burnup levels of 22 and 14 MWd/kgU, respectively and predicted negligible fission gas release ($<1\%$) in both rods. Consequently, the fission gas heat resistance between the fuel and cladding had only minor impact on the calculated fuel temperature.

Figures 5 and 6 represent the calculated fuel centre temperatures (axially averaged) versus the measured temperatures separately for the individual rods. The plotted points (341 data for rod A2 and 352 data for rod B2) correspond to the mean values of the data of each time intervals defined by CGNPC's history condensation code. These two figures indicate that the simulation of both gadolinium rods tend to overpredict the fuel centre temperature in a small compass, since almost all the points lie within a 10% accuracy band.

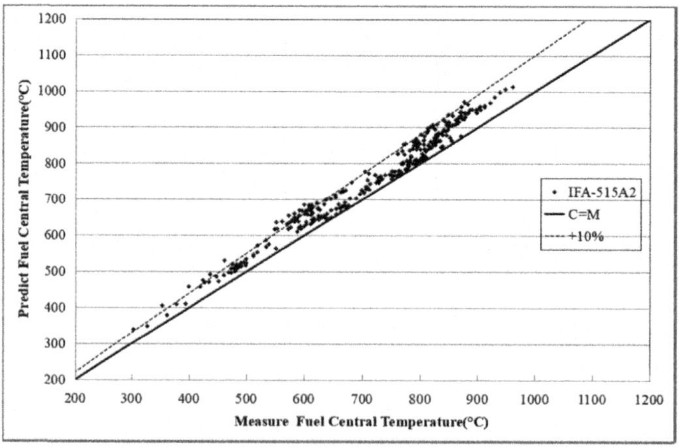

Fig. 5. Calculated versus measured gadolinium fuel central temperatures for rod IFA-515A2.

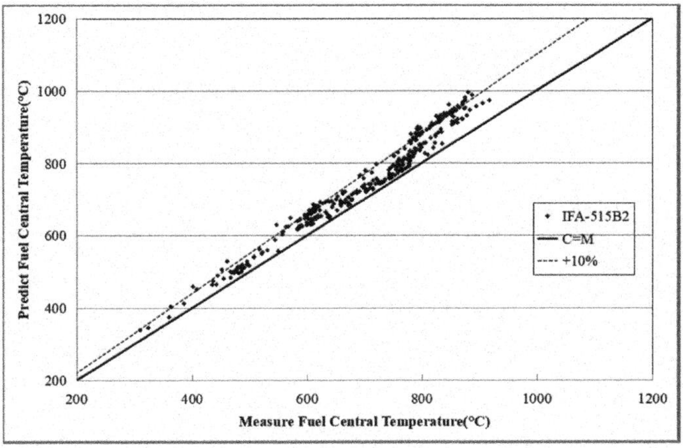

Fig. 6. Calculated versus measured gadolinium fuel central temperatures for rod IFA-515B2.

Figures 7 and 8 indicate the comparison between calculated and measured fuel temperatures as the function of rod burnup. Considering different burnup intervals, the tendency to excellently predict the fuel temperature comes into view during 10–40 MWd/kgU. The temperatures for these rods are approximately over predicted by 50 °C at BOL and over predicted by 80 °C at EOL. The over-prediction of fuel central temperatures at beginning may be due to the under predicted fuel thermal conductivity of gadolinium fuel at lower temperatures.

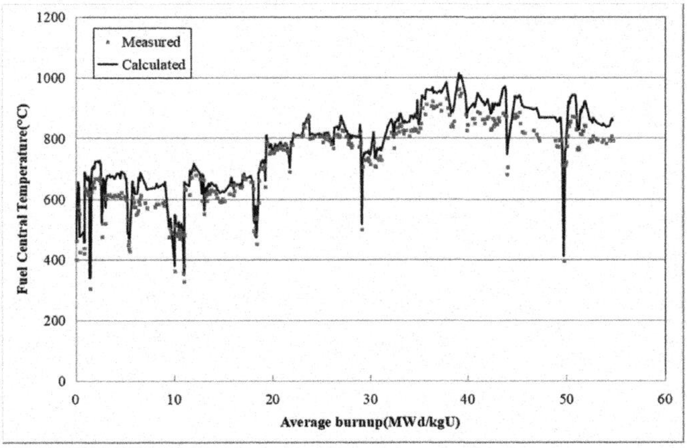

Fig. 7. Calculated versus measured gadolinium fuel central temperatures for IFA-515A2.

Table 2 represents the means and standard deviations (SD) for measured to predicted (M/P), and measured minus predicted (M-P), on fuel central temperatures during the whole irradiated life for rods IFA-515 A2 and B2. The M-P mean = −44.20 °C and SD = 24.22 °C represent the over-predictive capability of JASMINE models under a wide variety of thermal conditions. The uncertainty associated with the power histories

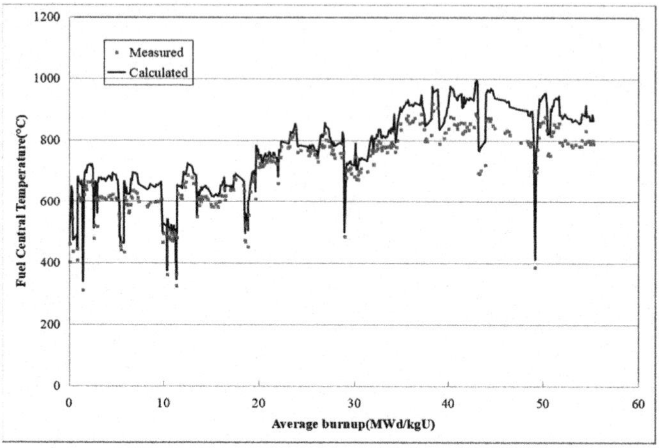

Fig. 8. Calculated versus measured gadolinium fuel central temperatures for IFA-515B2.

provided by Halden is reported to be approximately 0.05. And this level of uncertainty in power can cause fuel centerline temperatures to vary ±30–40 °C or more depending upon geometry, composition, and power level. Additional uncertainty inherent in the material properties and behavioral models contributes to the differences in prediction, as well as the temperature measurements uncertainty also contributes to the differences in measured data. Overall, the thermal performance of gadolinium fuel rods is judged to be a little over-predicted over a wide range of fuel rod conditions and provides an over-predicted foundation for the integral rods effects such as fission gas release, void volume and internal pressure.

Table 2. Statistic results of fuel central temperatures for two IFA-515 rods.

Fuel rod name	M/P		M-P(°C)	
	Avg.	Std. Dev.	Avg.	Std. Dev.
IFA-515A2	0.946	0.033	−40.31	22.34
IFA-515B2	0.937	0.028	−47.96	25.38
Total	0.941	0.033	−44.20	24.22

4.2 Fission Gas Release

Four gadolinium fuel rods were used for fission gas release model validation. As shown in Table 3, at EOL the measured FGR of these four GAIN rods were less than 1%, and the predicted FGR were less than 2%. That is because they were irradiated at lower linear power history, and the fuel temperatures are no more than the threshold for gas release, so it is difficult for these rods to enter the thermal release stage. Although the

FGR predictions by JASMINE have a good agreement with these experimental data, further gadolinium fuel FGR data is still needed for the validation of JASMINE model in the steady and transient FGR conditions.

Table 3. Statistical results of FGR for GAIN gadolinium rods.

Fuel rod name	Rod burnup (MWd/kgU)	FGR (%)		P-M (%)
		Measured	Predicted	
GD0301	38.8	0.23	0.87	0.64
GD0302	37.9	0.19	0.72	0.53
GD0702	38.9	0.66	1.63	0.97
GD0701	39.0	0.98	1.91	0.93

4.3 Fuel Rod Void Volume

An accurate prediction of the free void volume of a fuel rod is important in the calculation of the internal pressures along with the FGR prediction. Changes in fuel rod void volume with burnup are primarily due to the combined effects of fuel swelling, cladding creep, axial irradiated growth, and so on. Four well-characterized fuel rods from GAIN programme were used to assess the capability of JASMINE to accurately calculate fuel rod void volumes.

The free void volume results for these four $(U,Gd)O_2$ fuel rods are presented in Table 4. As they underwent different irradiated histories, there were different fuel temperatures, elastic and inelastic strains on fuel rods. Besides that, there was uncertainty in fabricated void volumes, therefore the void volumes at EOL are different. However, JASMINE has a little lower fuel thermal conductivity, which may result in higher fuel temperature, and larger fuel volume. That means JASMINE has a slight conservative result for simulating the integral fuel rod void volumes.

Table 4. Calculated and measured void volume for GAIN rods.

Fuel rod name	Void volume (cm^3)		M-P (cm^3)
	Measured	Predicted	
GD0301	5.95	4.25	1.70
GD0302	5.71	3.81	1.90
GD0702	6.55	6.08	0.47
GD0701	6.80	6.22	0.58

4.4 Internal Pressure

The gas pressure within a fuel rod changes continuously during irradiation. Since fuel rod internal pressure is a design criterion, it is necessary to ensure that the internal pressure predictions are accurate.

Table 5 shows the measured and JASMINE calculated internal pressure at EOL for four GAIN (U, Gd)O_2 fuel rods. The comparisons indicate that JASMINE can satisfactorily predict fuel rod internal pressure with a slight conservative tendency. As these four rods all have lower power histories, which mean that JASMINE predicted higher fuel temperature, larger fission gas release, lower void volume and higher internal pressure.

Table 5. Calculated and measured internal pressure for GAIN rods.

Fuel rod name	Internal pressure (MPa)		P-M (MPa)
	Measured	Predicted	
GD0301	2.28	3.58	1.30
GD0302	2.37	3.68	1.31
GD0702	2.65	3.28	0.63
GD0701	2.58	3.24	0.66

5 Conclusions

JASMINE2.0 has succeeded to simulate fuel rod performance with (U, Gd)O_2 fuel pellets. The thermal, mechanical and FGR models presented here and developed in JASMINE, will allow extending the validation range to burnup of 55.31 MWd/kgU for gadolinium fuel.

Even though well characterized irradiation experiments with gadolinium fuel are scantily available for the models. Six gadolinium doped fuel rods were used for the validation on the fuel central temperature, FGR, fuel void volume and internal pressure. The validation tests represent that JASMINE has a good agreement with the experimental data. However, the coverage of these data are so limited, especially in FGR, further gadolinium fuel data are still needed for the validation of JASMINE models.

In the near future, some hot-cell examinations will be performed on irradiated (U, Gd)O_2 fuel rods, and further researches will be conducted on mechanism models of fission gas release and gas swelling. Besides that validations on JASMINE will be continued and extended with other available experimental data, to verify the new (U, Gd)O_2 models.

References

1. Burnable Absorber Fuel, Characteristics and use of Urania-gadolinia fuels. IAEA-TECDOC-844, 1994.
2. M. Schlieck, H.-D. Berger, A. Neufert. Optimized Gadolinia Concepts for Advanced in-core Fuel Management in PWRs. Nuclear Engineering and Design 205 (2001) 191–198.
3. JingGang Li, JingHan Peng, Chao Wang, Jun Chen. Investigating core axial power distribution with multi-concentration gadolinium in PWR. Nucl SCI TECH (2022) 33:138.
4. M.Durazzo, et al. Phase studies in the UO_2-Gd_2O_3 system. Journal of Nuclear Materials 400(2010) 183-188.
5. Mutsumi HIRAI and Shinji ISHIMOTO. Thermal Diffusivities and Thermal Conductivities of UO2-Gd2O3. Journal of Nuclear Science and Technology, 28[11], pp995~1000 (November 1991).
6. M.Hirai, J.H. Davies, R. Williamson. Diffusivities of fission gas species in UO_2 and $(U,Gd)O_2$ nuclear fuels during irradiation. Journal of Nuclear Materials 226(1995) 238-251.
7. Xiaoyan Wei, et al. Modeling of Fuel Rod Behavior and Recent Advances of the JASMINE Code. Proceeding of the 25th Internal Conference on Nuclear Engineering. ICONE25–67144.
8. Xin Jin, et al. JASMINE: A Fuel Rod Thermal-Mechanical Performance's Code. TopFuel 2016.
9. W. Wiesenack, in: Proceedings of the International Topical Meeting on Light Water Reactor Fuel Performance, Port land, USA, 1997, p507.
10. P. G. Lucuta. A Progmatic Approach to Modelling Thermal Conductivity of Irradiated UO_2 Fuel: Review and Recommendations, J. Nucl. Mater, 232 (1996) 166-180.
11. Masaki Amaya, et al. Thermal conductivity of irradiated UO_2 and $(U,Gd)O_2$. Journal of Nuclear Materials 300 (2002) 57-64.
12. A. R. Massih, S. Persson and Z. Weiss, Modelling of (U, Gd)O2 fuel behaviour in boiling water reactors, J. Nucl. Mater. 188 (1992) 323-330.
13. C.M. Allison, G.A. Berna. A Library of Materials Properties for Light-Water-Reactor Accident Analysis. NUREG/CR-6150,1993.
14. Caillot I, Delette G., Julien B., Couty J.C. Impact of fuel pellet fragmentation on pellet-cladding interaction in a PWR fuel rod: results of the RECOR experimental programme. 14th International Conference on Structural Mechanics in Reactor Technology, SMiRT 14, Lyon, France, August 17–22, 1997
15. D. R. Olander, Fundamental Aspects of Nuclear Reactor Fuel Elements, TID-26711-P1 (1974).
16. J.A. Turnbull and E. Kolstad, "Investigations on Radioactive and Stable Fission Gas Behaviour at the Halden Reactor," Fission Gas Behaviour in Water Reactor Fuels, Seminar Proceedings, Cadarache, France, September 2000, 369 – 380.
17. Background and Derivation of ANS-5.4 Standard Fission Product Release Model, NUREG/CR 2507, 1982.
18. L. C. Bernard, J. L. Jacoud. An Efficient Model for the Analysis of Fission Gas Release. J. Mater. 302(2002) 125-134.
19. Improvement of Computer Codes Used for Fuel Behaviour Simulation (FUMEX-III), IAEA-TECDOC-1697, 2013
20. Terje Tverberg, Masaki Amaya. Study of thermal behaviour of UO_2 and $(U,Gd)O_2$ to high burnup (IFA-515). HWR-671, 2001.
21. Hoffmann H and N Kraus. 1984. Fabrication and Characterization Data of BN2 UO_2-Gd_2O_3 Fuel Rodof the GAIN Programme. GAIN Topical Report GN 84/10.
22. Manley AJ, PD Kennedy et.al. 1989. Post Irradiation Examination of Three Rods Comprising Task 3GAIN Topical Report GN 89/50.
23. Reindl J, G Bart. et al. 1991. Destructive Examinations on the BN2, GE5, and NF17 Fuel Rods 9, 10, 11, 23, 24, and 33 of the GAIN Programme. GAIN Topical Report GN 91/60.

A Survey of Worldwide Fuel Cycle Design Approaches and Their Implications on Plant Operations and Safety Analyses

J. Strumpell[1], R. Kliewer[1](✉), J. O'Brian[1](✉), N. Garner[1](✉), B. Holden[1](✉), L. Gerken[1](✉), N. Vollmer[2], M. Zilly[2](✉), and S. Zheng[3](✉)

[1] Framatome Inc, 3315 Old Forest Road, Lynchburg, VA 24501, USA
{john.strumpell,rod.kliewer,james.obrian,norman.garner, brandon.holdan,lisa.gerken}@framatome.com

[2] Framatome GmbH, Paul- Gossen Strasse 100, 91052 Erlangen, Germany
{nico.vollmer,matias.zilly}@framatome.com

[3] Framatome SaS, 2 Rue Professeur Jean Bernard, 69007 Lyon, France
songhui.zheng@framatome.com

Abstract. Fuel cycle design approaches vary significantly across the globe for LWR plants. This is driven by unique approaches to economic models, different operational modes, and variations in the regional safety criteria. Cycle length can vary from a few months up to 2 years. Other factors such as taxonomy, technical requirements and regulations have a strong influence as well. This paper assimilates available information to compare the various approaches worldwide and evaluate their impact on operational and safety analysis. In addition, there is a lot of technical work underway to evaluate increasing burnup and enrichment beyond currently approved limits. Environmental and political reasons such as lowering CO_2 production or reducing independence on other energy sources – both foreign and domestic can play an important role. Current fuel cycles as well as future evolution owing to ongoing improvements will be considered. This paper will be based on Framatome's global experience for nuclear core design and safety analysis for PWRs and BWRs. Clearly this paper focuses on the technical part and shows how flexible nuclear power plants are today and how they will open the door to address emerging special requirements. With this operating space of nuclear power, it contributes to low carbon dioxide production, security of supply of electricity, highest safety standards and strong economic value for Nuclear Power Plant (NPP) operators.

Keywords: Fuel cycle · CO_2 reduction · Cycle length

1 Operating Strategies

1.1 Operating Strategy – USA

In the US today, nuclear contributes approximately 19% of the total electricity generation and most nuclear reactors operate either 18-Month or 24-Month cycles. All but one BWR and only nine of the PWR reactors currently operate 24-Month cycles. Several

J. Liu and Y. Jiao (Eds.): WRFPM 2023, SPPHY 299, pp. 354–366, 2024.
https://doi.org/10.1007/978-981-99-7157-2_36

additional PWR reactors are in the process of transitioning to 24-Month cycle operation. Also, significant effort is ongoing to enable the use of increased enrichment and high burnup that would support many more PWR reactors in transitioning to 24-Month cycle operation. The average duration of refueling outages in the US is approximately 26 days, with a typical planned range of between 15 days to 36 days. A steam generator replacement outage can extend to approximately 90 days. Strategically, transitioning a reactor from 18-Month to 24-Month cycle operation reduces the operating cost and increases electrical generation by eliminating one outage every six years.

Primarily, the US reactor fleet operates as base load, meaning the reactors operate at full power for the duration of the operating cycle. A few reactors have performed flexible power operations over the past few years to accommodate electricity demand changes and price fluctuations. However, going forward it appears that utilities are exhibiting a strong desire to return to base load operations by distributing the power generated by the reactor to various options, including electricity generation, hydrogen production, or other potential needs.

To meet the energy demands of these operating cycles, typical PWR core designs follow an in-in-out, low leakage loading pattern, where fresh fuel is "checker-boarded" with once-burned fuel in the center region of the core and loaded in a "ring-of-fire" pattern around this central region. Twice-burned fuel is "checker-boarded" with once-burned fuel in peripheral core locations to complete the low leakage design. After the interior locations are filled, the remaining once-burned fuel is loaded in peripheral locations along with twice-burned fuel to complete the low leakage design. To keep fuel costs low, each fuel cycle is optimized in terms of number of fresh fuel assemblies (FA), uranium enrichments and locations, and burnable absorber content and locations to meet the specific energy demands of the operating cycle.

BWRs also utilize low leakage loading patterns and "checker-boarded" fresh and once-burned fuel in the interior of the core and twice-burned fuel on the periphery of the core. Additionally, fresh fuel is kept out of the second ring of fuel from the periphery by placing up to three fresh fuel bundles in select control cells (groupings of four bundles around a cruciform cross-section control blade) toward the core periphery. Typical reload batch size for a high-power density BWR operating on 24-month cycles ranges between 36 and 42% of the core size with about 20% of each reload operating only two 24-month cycles and the remainder discharged after three cycles. To keep fuel costs low, BWR cycles utilize significant flexibility to axially and radially zone enrichment and gadolinia. Further they utilize split-batches to burn the core evenly while achieving the specified burnup limits, taking into account the fuel's exposure history and local peaking conditions across the core. BWR operating strategies also take advantage of the ability to manage the fast and thermal neutron spectrum by controlling core flow and the elevation of the boiling boundary within the core. This permits conversion of some of the U238 (greater that 95% of the U) to fissile Pu239 and then burning that Pu239 to generate most of the thermal power in the last 20% of the fuel cycle.

1.2 Operating Strategy – France

In France, nuclear electricity is a major part of the total electricity production. It contributes about 70% of the total electricity generation. The cycle length varies between

12 and 18 months. Due to the specific situation, the unique operator of all the nuclear power plants needs maintaining the stable operation plan to guarantee the electricity grid's need. In the other hand, the nuclear electricity production has the big impact on the electricity grid, according to the ask of the grid, operator should have a great flexibility to provide the different power level. This means that each reactor should have the quick load-follow operating capacity. In general, a lot of reactors will have full power operation during the day when grid's ask is high and have low power operation during the night when grid's ask is low.

To maintain the stable operation, the same kind of reactors adopt the same fuel management with the same or very similar loading pattern to simplify the reload safety justification with two advantages:

- Each cycle has very similar operating situation and the operating schedule is easily established,
- Reload study can be performed at the last moment, very often close to the end of cycle allowing consideration of the very realistic situation.

1.3 Operating Strategy - Germany and German Export Market

While electricity production from nuclear is no longer available in Germany, it did contribute about 30% of the electricity production in the 1990s. The typical cycle duration in both the BWR and PWR fleet was 12 months with loading strategies to achieve low neutron leakage with small reload batch fractions on the order of 25% or less. The largest part of the nuclear reactor fleet in Germany was built by KWU, and all KWU reactors in operation outside Germany form part of the German export market.

A uniqueness of these KWU built reactor types is the combination of the reference in-core measurement system (Aeroball) and the continuous-operation self-powered neutron detectors. This provides constantly available high fidelity information of the core status which has been integrated into the safety concept from the beginning. The concept allows for load-follow and long-term low power operation and made stretch-in operation for end-of-life cycles in Germany possible. Another specificity is the possibility for reshuffling in the core which makes very short outages possible in the range of 7–15 days.

Germany. The heterogeneous energy market with a large fraction of wind power in Northern Germany, different levels of electricity consumption with high demands in industry-rich Western and Southern Germany on the one hand and differences in nuclear regulation (e.g., MOX usage was allowed in only in some of the German NPPs) makes it difficult to generalize the operation strategy. Specially the years after the decision of the phaseout from nuclear electricity production show some peculiarities. The gross electricity production was limited by law, with the option to transfer from one plant to another while the dates of ultimate shut-down were fixed for each NPP. This led to short and very short cycles without fresh fuel on the one hand and unconventionally long cycles with more than 500 Extended Full Power Days (EFPD) on the other hand.

German Export Market. The German Export Market is diverse, ranging from legacy 2- and 3-loop PWRs, modernized BWRs and more recently, the EPR. For the older plants, both fuel economy as well as requirements for long-term operation (e.g., limitations to the allowed neutron fluence of the reactor pressure vessel) play a role. For new builds like the EPR, the large size of the core increases the potential design space for core optimization. Key to addressing the diverse customer requirements are highly qualified core design staff working with modern methods and codes which advance continuously through application and R&D.

2 Fuel Cycle Design Approaches

2.1 Fuel Cycle Design Approach – North America

In North America the cycle design is tailored to maximize the cycle energy output while minimizing both the fuel cost and the outage cost within allowable safety limits. This approach tends to focus on minimizing the batch size while maintaining a longer cycle length or uprating the base power output. These cycles range from 18 to 24 months, with recent trends striving to bring more units to 24-month operation. Improvements in integral burnable absorbers has allowed for the lowering of peaking limits and a corresponding increase in rated power. The following trends continue to drive the core design approach today:

- Utilize available margin to minimize batch size thereby minimize fuel cost
- Operate longer cycle lengths to reduce outage cost
  - Cycle length set by customer
- Design a custom low leakage loading pattern each cycle to maximize discharge fuel burnup
  - Most plants utilize axial low enriched blankets to further reduce leakage

These basic goals result in a unique core reload each cycle, for each plant. Though a basic set of checks tends to ensure a given loading pattern is viable, a larger set of safety checks vs. the analysis of record and detailed fuel performance analyses are required each reload. These goals, and the fuel related improvements, have resulted in the 4.95 w/% enrichment limit becoming increasingly challenged over time (Figs. 1 and 2).

As recent trends have driven more utilities to move to or evaluate 24-month cycles, this has necessitated an increase in batch size and/or U-235 enrichment. Efforts to bring higher enrichment and burnup limits to the market can offset this (Fig. 3).

Both PWRs and BWRs strive to minimize the radial peaking factor, but the strategy for axial power control differs significantly. While the PWR design spreads power around evenly for the entire cycle, the BWR design purposefully creates a significant shift in the axial power for the purposes of Pu-239 generation. This Pu-239 generation offers significant benefits in cycle economics. BWRs design in this behavior with axially varied enrichment and neutron poison whereas PWRs are typically loaded axially symmetric with low enriched blankets (Figs. 4 and 5).

The engineering effort is somewhat higher when custom loading patterns are created each reload as compared to using standardized designs. Therefore, a larger set of checks are performed to ensure the safety analysis remains valid. The payoff from a custom design is a finer degree of optimization which results in increased fuel utilization.

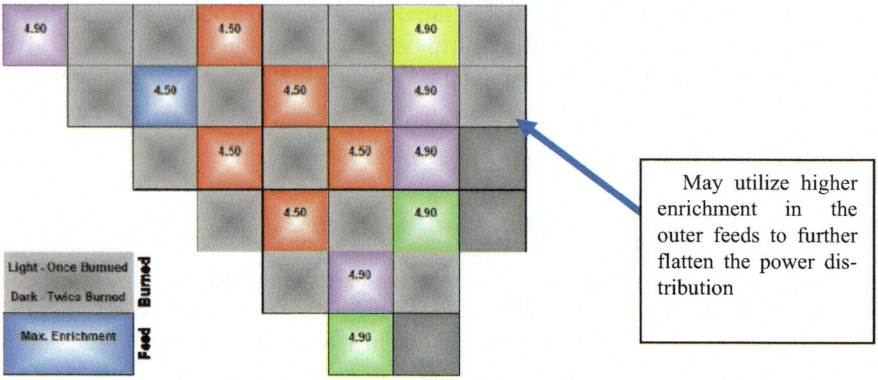

Fig. 1. Typical PWR 18-month 85 feed cycle design for W-4Loop PWR

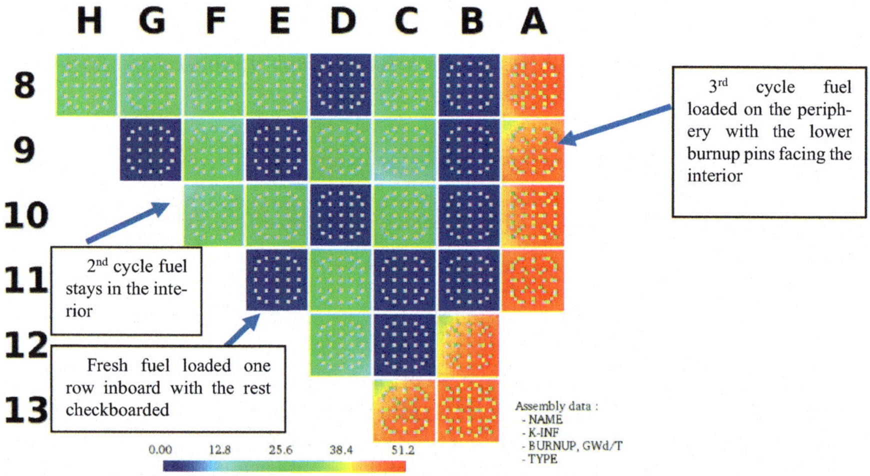

Fig. 2. Typical PWR loading scheme – burnup map

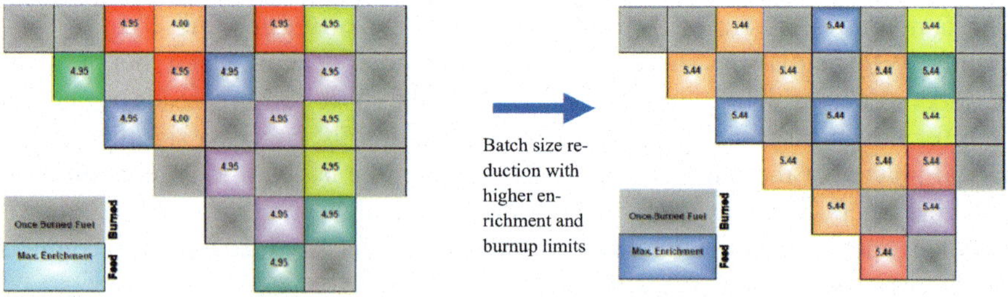

Fig. 3. 24-month designs for W-4Loop PWR

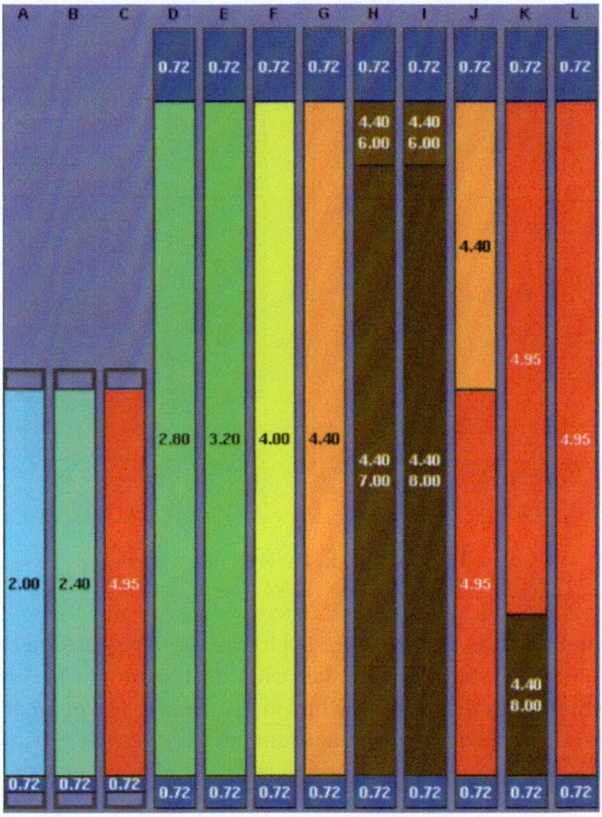

Fig. 4. Typical BWR axial loading

2.2 Fuel Cycle Design Approach – France and Its Export

As reference fuel supplier, Framatome, in close cooperation with customers, provides a large choice of fuel cycle design according to the electricity production plan. The key parameters for the fuel cycle design are safety and economy. For safety, core loading patterns are optimized to minimize the maximal core power peak and increase the operating margin. For economics, fuel assemblies are designed to reach high burnup to improve the fissile material utilization. One of the specifications of fuel cycle design and safety studies is the standardization. This standardization aims to perform the whole safety study on a series of fuel cycles to define a set of key safety parameters. After that, for the reload cycles, safety justification is performed by checking the key safety parameters. This method speeds up a lot the reload safety studies allowing to realize the reload design close to the end of cycle with the most representative conditions.

In practice, when the customer decides to change the fuel management, Framatome defines the representative fuel cycles for future operation consistent with the customer's requirements: cycle length, operating strategy, targets of the enrichment, the maximal burnup of fuel assemblies and fuel rods. In general, the fuel management is defined with the following steps:

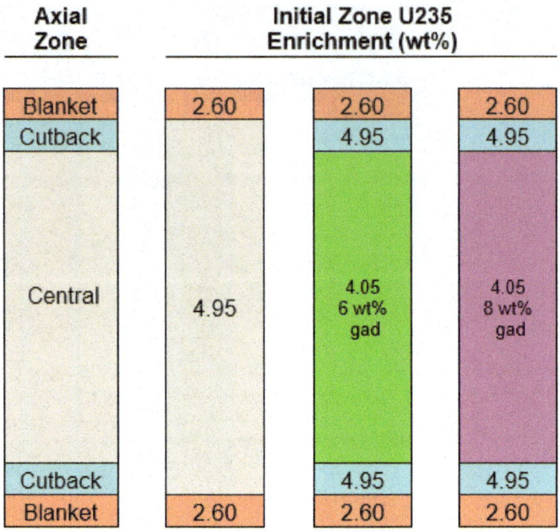

Fig. 5. Typical PWR axial loading

An equilibrium cycle as the reference cycle based on the customer's requirement.

The transition cycles start from the last cycle of the current fuel management and pursues the equilibrium cycle of the new fuel management. The number of the transition cycles varies according to the new fuel management characteristics. Usually, it is between 3 and 5 cycles. The criterion is that the last transition cycle converges to the equilibrium cycle regarding the reload characteristics and the results. Some flexibility is built in for cycles with reload characteristics different from those of the equilibrium cycles to cover the non-standard situations.

As an example, Fig. 6 illustrates a theoretical fuel cycle scheme defined for a typical fuel management. In this example, two operating conditions are considered: one with a stretch-out and another without stretch-out. For each operating condition a series of transition cycles, equilibrium and flexibility cycles are defined.

- cy0: representative cycle of the current fuel management (initial cycle)
- Eq and Eq': equilibrium cycles of the new fuel management
- Ti and Ti': transition cycles
- F1 and F2: flexibility cycles (shorter cycle F1 with less fresh fuel assemblies after the stretch-out operation and longer cycle F2 with more fresh fuel assemblies after an anticipated shutdown of Eq cycle.

The safety studies are divided in two steps:

- Generic safety study
- Reload safety study.

Generic Safety Study. The neutronic data are calculated for all cycles of the fuel management. For each parameter, the envelope value including uncertainty is retained for safety studies. A set of general neutronic data are built to be used for all the accident

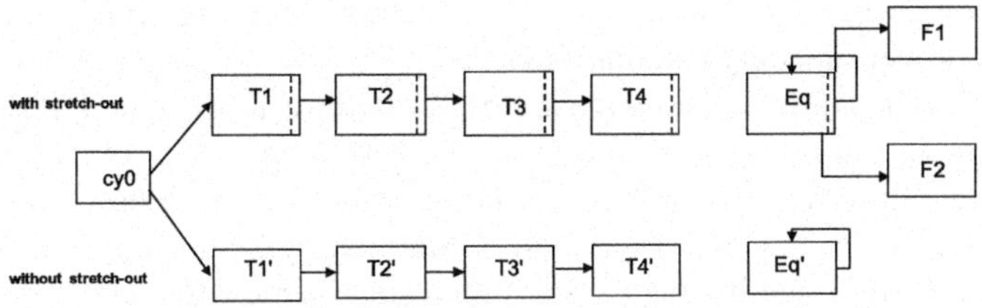

- cy0: representative cycle of the current fuel management (initial cycle)
- Eq and Eq': equilibrium cycles of the new fuel management
- Ti and Ti': transition cycles
- F1 and F2: flexibility cycles (shorter cycle F1 with less fresh fuel assemblies after the stretch-out operation and longer cycle F2 with more fresh fuel assemblies after an anticipated shutdown of Eq cycle

Fig. 6. Typical fuel cycle scheme for a fuel management

transient studies. In addition, for each accident, a set of specific neutronic data are also defined and used for each accident transient studies. The safety studies are performed for each cycle of the fuel management.

After the safety studies, the General Key Safety Parameters (GKSP) and Specific Key Safety Parameters (SKSP) are defined. These KSPs will be verified during the reload study of each effective cycle.

Reload Safety Study. For each effective cycle, the specific loading pattern is defined based on the onsite operating situation (previous cycle length, operating condition and assembly burnup). The loading pattern should respect the main characteristics of the theoretical cycles to have the similar core power distribution: number of fresh fuel assemblies and positions, enrichment, burnable absorbers. All the neutronic data specified in the GKSP and SKSP are then calculated for this cycle:

- If all the KSP limits are respected, the safety of this reload cycle is justified,
- If some KSP limits are not respected, additional safety justifications are performed by considering the effective cycle neutronic data.

With this two-step methodology, the reload study can be realized in a very short time.

2.3 Fuel Cycle Design Approach – German Export Market

While dependent on the individual customer and regulatory requirements, the most frequently realized reload safety case in fuel cycle design offered in the German export market is based on a reference In Core Fuel Management (ICFM) and the derived Reload Safety Analysis Checklist (RSAC) with associated limits of Key Safety Parameters (KSP).

For typical reloads, according to customer requirements, a core design is proposed, sufficiently similar to the reference ICFM. The reload safety justification is fulfilled, if all KSP are within the predefined limits. If individual KSP violate the limits, the reload safety can be justified by additional calculations. The RSAC approach offers sufficient design space to generate a reload according to customer and regulatory requirements with a limited number of safety calculations.

Particularly quick safety assessments are performed for KWU type PWR plants for which a preliminary reload safety evaluation is performed during cycle n − 1, and the reload safety evaluation for cycle n is performed during the outage based on the actual EOC burnup of cycle n − 1 and, potentially, minor design changes. A high level of automation ensures high-quality analyses within a small-time window.

Figures 7 sketches 1/8th sectors of typical reload patterns for 12-month cycles in PWRs with 177 fuel assemblies in near-equilibrium loading, where the number in the assemblies denotes the irradiation period.

Fig. 7. Typical 177 FA PWR loading

Figures 8 displays an equilibrium loading of a PWRs with 157 fuel assemblies, 8 of which are shielding assemblies designed to reduce the neutron fluence on core barrel and reactor vessel.

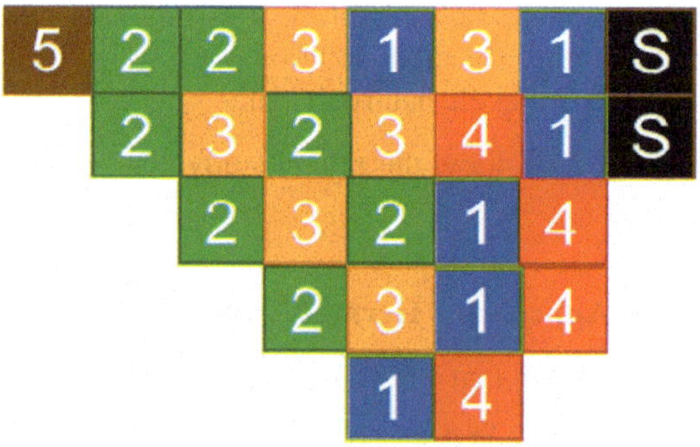

Fig. 8. Layout of an equilibrium cycle PWR with 157 FA

The general approach for BWR core designs is the same as described in the Section on North America.

3 Safety Analysis Implications

Safety analysis is performed to calculate the predicted performance of a plant during accidents with its respective systems and fuel cycle designs. The results of the safety analysis are used to demonstrate that the safety requirements established by the regulator are met. The safety analysis methods utilized in a specific region depend largely on the regulatory structure, rather than the type of fuel cycle design. There are various regulatory agencies across the world and the specifics of the regulations differ, but because of the similarity in plant design and operation, the phenomena for which they protect are universal. Similarly, while the methods developed by various licensees or vendors may differ, they must all address the same phenomena. The impact of the fuel design is more on the phenomena or challenges. For example, a specific 18-month loading core design may challenge the LOCA limits. Another 18-month fuel cycle design maybe have adequate margin to the LOCA limits but challenge the clad strain criterion. Thus, there are rarely large changes in future cycle designs once a fuel cycle design that maximizes the economic benefits and maintains desirable margin to safety criteria has been established. In certain instances, the fuel design could also require more explicit consideration of phenomena previously considered to be negligible. A good example of the effects of fuel cycle design is using higher enrichments to achieve more economical 24-month core designs, with higher EOC burnups. This could be achieved via higher power throughout the cycle in the once-burnt fuel. In this type of design, the rod internal pressure criterion may be more challenged. Additionally, the potential for dispersal of fine fuel fragments into the RCS during the LOCA may no longer be negligible. Because safety analysis is only a prediction of true plant performance, safety analysis changes could refine the internal pressure prediction but would not alone improve the true margin. Changes to fuel cycle design or fuel rod design could be required. In the case of fuel dispersal, methodology changes or supplemental analyses could be envisioned to support evaluation of the potential impact.

4 Economic Implications

The main driver for changing fuel cycle length could be the economic advantage for the LWR – of course keeping safety as a priority as well as legal and regulatory limits. In this part of the present paper, we assume that all safety, legal parameters etc. are kept and focus exclusively on the economic aspect of different fuel cycle length.

For illustration, we will discuss in detail a 12-month cycle (12mo) and a 24-month cycle (24mo) for a typical PWR with 1000 MWe and 193 FA. For this study we will use data available on the internet or based on experience.

In this example, the 12mo PWR replaces less than 20% of the fuel assemblies per reload – this leads to an average use of more than 5 years per fuel assembly. In a 24mo fuel cycle, typically 50% of the fuel assemblies are replaced per reload – this leads to an average burnup of 4 years per fuel assembly.

In this example, the 12mo cycle uses the fuel 25% more efficient than the 24mo cycle – leading to a significantly lower fuel cost, including disposal costs etc.

However, the 24mo cycle has one less outage, which saves costs and allows electricity production in this time. Furthermore, less enrichment of the uranium would be required; we assume 4.5wt% instead of 4.9wt%.

Uranium costs (price is ~100€ / kg from [1]):

Assuming 5000 kg uranium (before enrichment) are required per fuel assembly, this means 0.5M€ per fuel assembly in a PWR.

- 12mo cycle, reload of 40 FA per year = 20 M€
- 24mo cycle, reload of 100 FA per 2 years = 50 M€

Bottom line: the short cycle would save 5 M€ per year in uranium cost

Remark: considering less enrichment, uranium consumption for 24mo cycle would be 10% less

Enrichment costs:

Typical values for 12mo cycles are 4.9 wt% enrichment, while 24mo cycle might require 4.5 wt% enrichment. Enrichment costs are taken from [2].

With feed assay of 0.711 wt% Uranium 235 and SWU price of 90 € per Specific Work Unit (SWU), tail assay of 0.23 wt% and the example above:

- 12mo cycle: 8 SWU/kg EUP with a feed quantity of 9.7 kgU as UF6 are needed
- 24mo cycle: 7.2 SWU/kg EUP with a feed quantity of 8.9 kgU as UF6 are needed.

That means with the assumption of 500 kgU per FA and the example above:

- For 12mo cycle: 8 SWU/kg × 90 €/ SWU × 500 kg × 40 FA = 14.4M€ (per year)
- For 24mo cycle: 7.2 SWU/kg × 90 €/SWU × 500kg × 100 FA = 32.4M€ (per 2 years).

Bottom line: the 24 m cycles would save 1.8M€ per year in enrichment costs and around 10% less feed uranium

Fuel assembly costs:

Based on [3], FA prices are around 350 € per kgU.

- 12mo cycle: 40 FA each 500 kgU means 7M€
- 24mo cycle: 100 FA each 500 kgU means 17.5M€

Bottom line: the 12mo cycle would save around: 1.75 M€ per year for FA fabrication

Final storage costs:

Final storage is around 600 €/kgU, with the above-mentioned assumption.

[https://www.reutersevents.com/nuclear/high-costs-proliferation-concerns-feed-doubts-over-waste-recycling]

- 12mo cycle: 40 FA × 500 kgU × 600 €/kgU = 12 M€
- 24mo cycle: 100 FA × 500 kgU × 600 €/kgU = 30 M€

Bottom line: 12mo cycle would save around 3M€/ year spent fuel costs

Loss of production:

With current electricity pricing of around 120 €/MWh [4], this results in 120 €/MWh × 24h × 1000 MW = 2.88 M€ per day.

Bottom line: Assuming 5 days outage per year are saved for a longer cycle, this would result in 14.4 M€ per year higher income –or loss in 12mo cycle.

In summary: in this example, it seems to be favorable to go for longer cycles (Fig. 9). However, considering the huge variation in all parameters starting from uranium cost, including disposal cost and finally electricity pricing, individual case studies need to be done – not taken into account are outage costs, leaking fuel assemblies, tax, human error and many more parameters which might influence the fuel cycle costs.

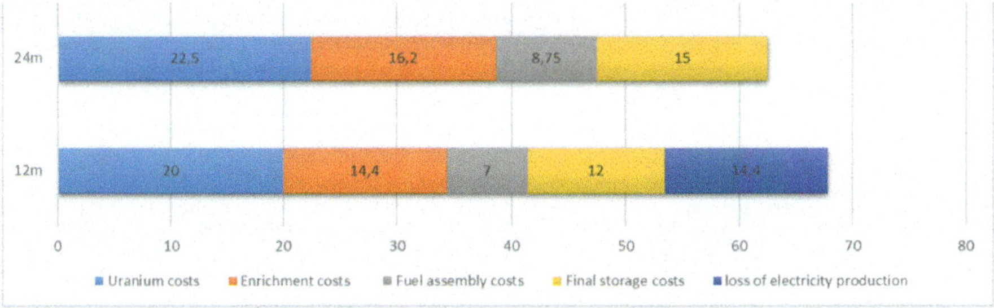

Fig. 9. Yearly fuel cycle costs comparison

Summary of economic implications. Fuel cycle length impacts profitability of an NPP. Uranium and enrichment service are commodities and need to be ordered wisely. Electricity prices and prices for final storage cannot be influenced easily. Therefore, fast outages and a robust fuel assembly optimized for your reactor seem to be important for an optimized economic fuel cycle. This is the basis for the most economic cycle length for your reactor.

5 Conclusion

This compilation and analysis of fuel cycle designs from around the world demonstrates two key principles for supporting the continued utilization of nuclear power to help the world to achieve its clean energy goals. First, the wide range of cycle lengths successfully analyzed and operated safely throughout the world proves the flexibility of nuclear power plants to adapt to their local situation. Moreover, the economic analysis gives concrete evidence that optimization of "system" economics (i.e., assessing the full economic picture of the machine and/or its role in the utility business model) provides a better tool for informing executive leadership on preferred operating strategies rather than "local" optimization of system subsets such as over weighting the importance of fuel cycle costs.

References

1. Cameco web site, https://www.cameco.com/invest/markets/uranium-price
2. Urenco web site, [https://www.urenco.com/swu-calculator]
3. Jim O'Brian, Framatome, TopFuel paper. 2021 in Santander
4. Eex website, https://www.eex.com/de/

Author Index

J. Liu and Y. Jiao (Eds.): WRFPM 2023, SPPHY 299, pp. 367–369, 2024.
https://doi.org/10.1007/978-981-99-7157-2

The manufacturer's authorised representative in the EU is Springer Nature Customer Service Centre GmbH, Europaplatz 3, 69115 Heidelberg, Germany. If you have any concerns regarding our products, please contact ProductSafety@springernature.com

Printed and bound by CPI Group (UK) Ltd, Croydon, CR0 4YY
01/10/2026
02225868-0011